Wilfried Roetzel · Peter J. Heggs
David Butterworth (Eds.)

Design and Operation of Heat Exchangers

Proceedings of the EUROTHERM Seminar No. 18,
February 27 – March 1 1991, Hamburg, Germany

With 236 Figures

Springer-Verlag
Berlin Heidelberg NewYork
London Paris Tokyo
Hong Kong Barcelona Budapest

Prof. Dr.-Ing. Wilfried Roetzel
Institut für Thermodynamik
Universität der Bundeswehr Hamburg
Hostenhofweg 85
2000 Hamburg 70
Germany

Prof. Dr. Peter John Heggs
Dept. of Chemical Engineering
University of Bradford
Bradford BD7 1DP
U. K.

Eur. Ing. David Butterworth
Heat Transfer and Fluid Flow Service
Building 392.7
Harwell
Oxfordshire OX 11 0RA
U. K.

ISBN-13: 978-3-642-84452-2   e-ISBN-13: 978-3-642-84450-8
DOI: 10.1007/978-3-642-84450-8

Library of Congress Cataloging-in-Publication Data
Eurotherm Seminar (18th : 1991 : Hamburg, Germany)
Design and operation of heat exchangers
proceedings of the Eurotherm Seminar no. 18, Hamburg, February 27 – March 1, 1991
W. Roetzel, P. J. Heggs, D. Butterworth (eds.).
(Eurotherm seminars ; v. 18)
Includes bibliographical references.

1. Heat exchangers--Congresses.
I. Roetzel, W. (Wilfried). II. Heggs, P. J. III. Butterworth, D. (David). IV. Title. V. Series.
TJ263.E93   1991
621.402'5--dc20   91-38271

Typesetting: Camera ready by authors
61/3020-5 4 3 2 1 0 – Printed on acid-free paper.

Preface

The Eurotherm Committee was created in 1986 from member countries of the European Community. It has the purpose of organising and coordinating scientific events such as seminars and conferences in the thermal sciences. The series of Eurotherm Seminars established by the Committee has become a popular forum for high-level scientific and technical interchange of ideas in a wide range of specialist topics. While the presentation and publication of papers at the Seminars are encouraged, the primary aim is to stimulate discussion and liaison between specialist groups. The present Chairman of Eurotherm is Professor C.J. Hoogendoorn of the Technical University, Delft (Fax [NL] 15, 783251). Information on future Seminars is available from the Secretary, Keith Cornwell, Heriot-Watt University, Edinburgh (Fax [UK] 31, 451, 3129).

This particular Seminar No. 18 on the Design and Operation of Heat Exchangers was the first one on this topic and was held at the Universität der Bundeswehr Hamburg (University of the Federal Armed Forces Hamburg) from February 27 to March 1 in 1991. The seminar was an international event and was attended by more than 60 scientists not only from countries of the European Community such as Belgium, France, Germany, Great Britain, and the Netherlands but also from other countries such as Canada, China, India, Israel, Romania, Soviet Union, Sweden and the United States of America.

In this proceedings volume thirty seven conference papers are published dealing with various aspects of the design and operation of heat exchangers. The first four papers are of general interest and not restricted to special flow arrangements or geometries. The next six contributions refer to shell and tube heat exchangers. In the following chapter ten papers are presented which deal with cross-flow heat exchangers. Another group of six refers to plate heat exchangers. The next two papers deal with heat storage in regenerators and in the soil. The last chapter contains nine papers on multiphase heat transfer in various applications such as power plant condensers, heat pipe and direct contact heat exchangers. Thus, a great variety of theory and practice is offered in this volume and thanks are due to all contributors.

The conference could not have been so successful without the efforts of many people in Hamburg, Bradford and Harwell. Most of the work involved in organizing the conference had been done by Bernhard Spang, who really did a good job. Further, the conference chairmen were assisted during the seminar by the sessional chair persons J. Buxmann,

M. Fiebig, L.E. Haseler, S. Kakac, F. Lauro, E. Marschall, J.K. Nieuwenhuizen, J.W. Rose and J. Taborek, who stimulated and directed fruitful discussions. For this we are most grateful.

Finally, thanks are due to the president of the Universität der Bundeswehr Hamburg, Prof. Dr. H. Homuth, and to the dean of the Fachbereich Maschinenbau (Department Mechanical Engineering), Prof. Dr.-Ing. L. Gaul, who supported the conference with good will and help at many occasions.

The response to our invitation shows the continued broad interest in the field of the design and operation of heat exchangers and it is hoped that other seminars on the same or similar topics will follow.

W. Roetzel
P. J. Heggs
D. Butterworth

Contents

Plate Heat Exchangers

Regenerators

Multiphase Systems

General Problems

Selected Problems in Heat Exchanger Design

Jerry Taborek
Consultant[1], Virginia Beach, VA 23451

SUMMARY
First, the various meanings of the heat exchanger design concept are briefly analyzed. Criteria for selection of heat exchanger types and their components are discussed. Next, a group of problems which are not easily identified in the course of the usual design process, is analyzed. These include items like pressure drop to heat transfer conversion effectiveness, pressure drop utilization, problems connected with surface over-design due to exaggerated safety factors or summer/winter operations, analysis of under-designed exchangers, differential vs. integral condensation, and similar.

HEAT EXCHANGER DESIGN CONCEPTS
The classical expression for determining the area of an heat exchanger - the thermal design problem - is defined by the fundamental equation :

$$A = \int_0^Q \frac{dQ}{U_x \, (T_h - T_c)_x} \approx \sum \frac{\Delta Q}{(U \, \Delta T)_{\Delta Q\text{-av}}} \tag{1}$$

where :   A is the heat exchanger area required to perform heat duty Q (W),
U_x and $(T_h - T_c)_x$ are the "local" values of the overall heat transfer coefficient U and the effective temperature difference between the hot and cold fluid.

The practical integration of this equation is performed by a stepwise procedure in suitable increments ΔQ, and is required for correct solution whenever U and the term $(T_h - T_c)_x$ change appreciably with the flowpath (area A). Such situation occurs typically under the following conditions :

a.   condensation of vapor mixtures and in presence of noncondensible gases, where U_x can vary by an order of magnitude;
b.   boiling of fluid mixtures with progressive depletion of the light component;
c.   flow boiling, which is function of vapor fraction and composition.

---

[1]) This paper is based on projects supported by B-JAC International Inc., Midlothian, VA 23112. Basic research was performed at the University Karlsruhe under the Alexander v. Humboldt Award.

4

d. in no-phase-change processes, when the flow regime transits from laminar to turbulent, or a large change of U_x exists from inlet to outlet for viscous fluids (minor changes are corrected by procedures of Roetzel [1]).

Graphical integration of Eq.1 was developed by Colburn and Hougen [2] in 1934 for solution of partially condensing vapors, but in the pre-computer era it was employed only for important designs, because of the considerable computational effort. Presently, computer programs use stepwise calculations on a routine basis. The calculation is shown in Fig.1A,1B, representing solution of the classical example from [2].

> A mixture of steam (1.14 kg/s) and nitrogen (0.35 kg/s) entering at 95 C and exiting at 40 C, is condensed in a counterflow exchanger. The coolant is water at 25 C entry and 60 C exit temperature. Under these conditions, a small part of the steam is still uncondensed at the exit (0.019 kg/s).

Figure 1A shows the temperature profile and the steep decline of the local overall coefficient U, as the vapor becomes more saturated with the gas. Figure 1B shows the solution of the Σ term in Eq.1, in steps which decrease (for better accuracy) in the region of rapid change of U. Notice that in this case about 80 % of the heat duty is performed in only 60 % of the exchanger area. Severe error would result if the calculation would be based on average conditions only.

It is of fundamental importance (but rarely mentioned in heat transfer texts) to realize that Eq.1 can be formally integrated *only if a number of restrictions are satisfied*, as discussed by Gardner and Taborek [3], the main ones being :

a. The coefficient U is *constant over the entire area* and the flow is purely counter-current or co-current (no multiple tube-passes);
b. the thermal history of any particle of either stream is identical, i.e. no bypassing or stratification;
c. there is no change of phase in only a part of the exchanger, i.e. no subcooling, desuperheating or partial boiling;
d. in shell and tube exchangers with cross-flow baffles, the heat transferred in one baffle compartment is small compared to the overall, i.e. the number of baffles is large (> 5 from analysis in [3]).

Under these conditions, formal integration of Eq.1 yields two possible solutions as shown in Eq.2 :

Fig. 1A

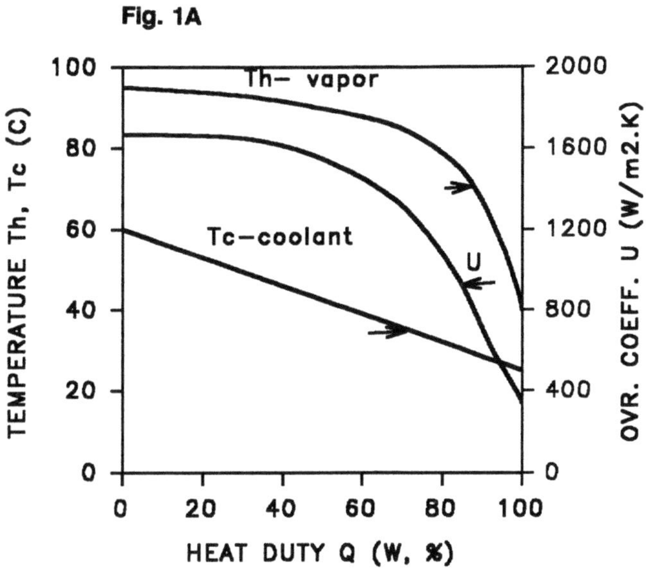

Fig. 1B

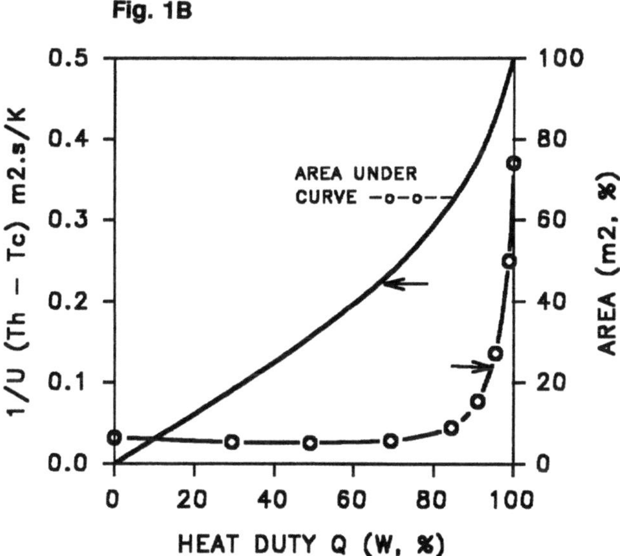

Fig.1. Solution of Eq.1 for steam/N_2 mixture [2] by incremental calculations.

6

a. The log. mean temperature difference ΔT_{LM} formulation. A correction factor F must be developed for multi-pass (mixed flow) systems which deviate from counter-flow;
b. The effectiveness ϵ-NTU method, based on ΔT_{max}, the inlet temperature difference. The mixed flow effect is absorbed in the derivation of the ϵ term for each system.

$$Q = A \ U \ \Delta T_{LM} \ (F) = \epsilon \ R \ \Delta T_{max} \tag{2}$$

where $\epsilon = f(NTU, R, \text{flow arrangement})$, $NTU = (U\ A)/m\ c_p$, and R = heat capacity ratio of the two fluids, as per conventional definitions. In most heat transfer texts, equation 2 is used as the "thermal design equation", as the heat exchanger area A can be determined directly if U is estimated from tables or experience, making the design problem appear trivial. The difficulty in "real" industrial designs comes from the fact that U is a complex function of permissible pressure drop and design geometry.

Thus the "real design problem" is to find such most suitable arrangement of the constructional elements (e.g., shell and baffle type, tube passes, tube length etc.) which would result in the least expensive unit, while satisfying all operational requirements and restrictions, as discussed below. For no phase change fluids, this will include full utilization of the available pressure drop, resulting in maximum heat transfer coefficient. As heat exchangers are basically pressure vessels, the design must also comply with demands of mechanical design, as dictated by various Standards.
A satisfactory solution of these problems requires trial-and-error calculations and most often compromises, which must be resolved by engineering judgement. The main concepts used in such design process are commented on in the text below.

HEAT EXCHANGER TYPES AND COMPONENT SELECTION METHODOLOGY
To design an heat exchanger which would transfer a given amount of heat Q (kW) can be accomplished in numerous, virtually infinite ways. In practice, the field of candidate choices of heat exchanger types is quickly narrowed down by the various restrictive demands on constructional, operational and process criteria. The most important ones are listed below :

a. pressure, temperature and corrosion/erosion resistance, explosion or toxic safety, environmental considerations (possibility of fluid leaks), and hence need for special designs or materials;
b. probable size, rugged or light duty, probable design accuracy;

c. available pressure drop and effectiveness of pressure drop utilization (vacuum);
d. fouling tendency, ease and frequency of cleaning, repairs;
e. is compactness of design (cars, aircraft, ship) required ?
f. process type : no-phase-change, condensation, boiling. Special considerations are imposed by these processes (like venting of non-condensible gases etc.), for which only some design types are suitable.

Evaluation of the above criteria and their, often contradictory interactions, will determine the selection of acceptable exchanger type(s) and their design elements. Preempting on the following text, it becomes apparent that tubular designs, especially shell and tube exchangers (ST), are by far the most versatile. For this reason more attention is paid here to ST exchangers, which also permit a custom-made composition of standard constructional elements. On the contrary, design types like plate and plate-fin compact exchangers are built from a limited selection of pre-designed constructional elements, their characteristics being proprietary to individual manufacturers.

However, the above statement should not be interpreted as an universal endorsement of shell and tube exchangers, which have their own share of disadvantages, e.g. poor compactness, with one of the highest volume-to-area ratios of all designs (surpassed only by double-pipe exchangers). Continued improvements of alternate types like gasketed and welded plate, spiral and compact plate-fin exchangers, makes such types a viable or superior alternative in specific duties, which were traditionally the domain of shell and tube design.

Only the most important items which determine the selection of the HEX design types and their components are briefly discussed below. For detailed treatment of the frequently complex problems encountered in the industrial practice, refer to the exhaustive text by Yokell [6].

1. Fluid Pressure : If the pressure of either fluid exceeds about 15 bars, the design is restricted (with exception of some special designs) to tubular exchangers, mainly shell and tube. If the overall surface is relatively small, banks of double-pipe exchangers can be considered, especially if the use of longitudinal fins is indicated. For moderate pressure ranges a number of other designs is potentially applicable, if other considerations discussed below, permit. The most frequently used types are spiral, gasketed plate, plate-fin and tube-fin, each having its own limits and area of best application [4,5].

2. Temperature, Corrosion and Safety : High temperature or corrosivity of fluids require special materials, not suitable for some construction types. Flammable or toxic fluids require special designs, usually offered only in tubular construction [4,5,6].

3. Fouling and Cleanability : Most fluids develop fouling of various forms (crystalline, sedimentation, bacterial, polymerization etc.), which presents a heat transfer resistance and increases pressure drop. The selection of design fouling resistances is still an "art", the designer having often available only the very limited TEMA Tables, which do not represent fouling as function of the main parameters (flow velocity, wall temperature). To be on the "safe side", designers often exaggerate fouling without properly realizing the consequences of the resulting over-design (see below).

Various design and operational provisions such as high flow velocity, limited wall temperature, corrosion resistant materials, water quality control, etc. decrease fouling tendencies, but ultimately most exchangers must be cleaned. The ability to perform cleaning efficiently is often a decisive factor in exchanger type selection, as the cost of cleaning, usually associated with production interruption, can be considerable.

Recent developments in the cleaning technology should be noticed. Large power plant condensers use continuous cleaning by recirculating rubber ball system. Chemical cleaning techniques consist of periodically recirculating a solvent in a closed circuit over the fouled surfaces. This method is used successfully for some deposit types and permits cleaning of surfaces without external access (shell-side of fixed tubesheet exchangers, passages in plate-fin exchangers), and is becoming more popular.

However, majority of fouling deposits, in particular the frequently used cooling water, must be cleaned mechanically by rotating brushes, water or steam jets etc., and thus require appropriate external access. Very effective tube-side cleaning is performed by scrapers propelled by high water pressure [7]. Water jets permit now cleaning of U-tube constructions, thus eliminating the need for more expensive types with removable bundles. Provisions for ease of access and assembly are essential in all cases.

4. Maximum permissible pressure drop : Regardless of heat exchanger type, the specification of the dp-max value determines to a large extent the design of no phase-change heat exchangers, as the heat transfer coefficient increases with pressure drop. The limits of dp-max are primarily determined as a fraction of the absolute pressure of the fluid, usually 10 - 20 %, or 1.5 bar, whichever is smaller.

Within such limits, dp-max is related to economic considerations : higher flow velocity results in higher heat transfer coefficient and hence smaller (lower cost) exchanger, often with decreased fouling rates (lower maintenance cost); these advantages are balanced by higher cost of pumping power. The interactions are usually quite complex and dp-max values derived from experience are often used, e.g. cooling water systems are usually designed to 0.7 to 1.2 bar pressure drop.

For no phase change fluids it is imperative that dp-max is utilized to full possible extent. This is particularly true for cases like low pressure gases, where the pressure drop must be kept to a minimum value. Selection of a design with best pressure drop to heat transfer conversion characteristic is then essential, as discussed later. However, the dp-max specified does *not have to be fully used*, if not contributing sufficiently to higher U-value or fouling control. For example, decreasing baffle spacing below about 25 % of shell diameter diverts the flow into inefficient leakage streams and contributes little to heat transfer.

5. Maximum Flow Velocity : The flow velocity derived from dp-max considerations may be too high for possible erosion or vibration damage. For erosion control, the use of harder, more costly materials is the design alternative. In some cases this can be inter-connected with the economics of corrosion control. For example, a power plant condenser with sea water is limited to 2 m/s with Cu-Ni tubes, but could operate at up to 4 m/s with titanium tubes. Higher cost of tubing is here offset by higher U - value (smaller unit), and virtual absence of corrosion and hence no need for re-tubing. High shell-side velocity can cause erosion damage at the inlet (use of impingement plates or flow distributors), or tube vibration, which must be prevented at all cost.

6. Minimum Flow Velocity : Because of the strong effect of flow velocity on fouling, the need for a minimum acceptable velocity must be respected even if higher cost exchanger results. This is illustrated in the following example. An organic stream is cooled with tower water on the tube-side. A small overlap of the outlet temperatures permits two design alternatives : a multi tube-pass unit with LMTD correction of 0.8; or a single tube-pass, in counterflow. As the water resistance $1/\alpha$ is minor compared to the organic stream, the least expensive design is the counterflow. However, in single tube-pass, even with the maximum tube length, the water velocity falls below 1 m/s, thus causing potentially a severe fouling problem. Proper design would use two (or more) tube passes within dp-max limits, the higher unit cost being offset by decreased fouling and lower cost of cleaning.

7. Condensation : Shell and tube exchangers are without doubt the most versatile equipment for condensation duties, permitting operation in horizontal or vertical position, outside or inside of tubes. Sizes can range from small to extremely large and designs for minimum pressure drop, often required in vacuum operation, are possible. If non-condensible gases are present, their *proper "venting" is a major problem* requiring special design provisions, thus eliminating many design types regardless of cost. Spiral exchangers are also very suitable for some condensation duties, as are some other types, within restrictions of pressure drop or other limitations [4,5].

8. Boiling : Tubular exchanger construction of various types predominates in process boiling service because of pressure and fouling. Gasketed plate design is sometimes used as evaporators (food industry), and plate-fin is used in special applications such as cryogenics [5].

9. Construction Materials : The materials needed for process requirements (temperature, corrosion etc.) may not be compatible with the manufacturing process of many HEX types (e.g. plate fin), thus limiting HEX type selection [4,5,6].

10. Mechanical Design, Codes : A heat exchanger of whatever design type is basically a pressure vessel which must exhibit mechanical integrity of construction, welded connections, tube-to-tubesheet joints and numerous other demands required by safety, many of them affecting thermal design [4,5,6]. Every design is subject to various rules and restrictions defined by a number of National Codes (ASME, TEMA, AD-Merkblätter-DIN, CODAP, etc.) with a bewildering array of minute variations, which cause nuisance in design. Hopefully the time is near for Codes unification.

ANALYSIS OF SPECIAL PROBLEMS

The rest of this paper is devoted to analysis of selected design problems which are not easily identified from the design specifications. This is especially important if the design is performed with a computer program, which often produces a single, presumably optimum solution. However, such solution is based on fixed input of shell type, baffle type, tube diameter and layout, the computer program varying only shell diameter, tube passes, baffle spacing and (possibly) tube length. Careful study of the intermediate results may indicate that other combinations of the input items or other HEX type would lead to a superior solution. Some of the problems encountered in such analysis are discussed below.

1. **Conversion of pressure drop to heat transfer :** The best utilization of pressure drop is in axial tube flow, either inside tubes or in longitudinal flow in tube bundles. Under such conditions, *the frictional forces are expended directly at the contact with the heat transferring wall.* Other flow arrangements, such as flow across tube banks, result in higher heat transfer, but at the cost of much higher pressure drop. This is caused by eddy currents, where part of the friction is consumed away from the contact wall, and hence does not contribute to heat transfer. If cross-flow baffles are used, various degree of secondary inefficiencies occur, such as flow turn-around, contraction and expansion, flow stratification, bypasses etc., further decreasing the pressure drop conversion effectiveness. The various baffle types which the designer can select are shown in Fig. 2.

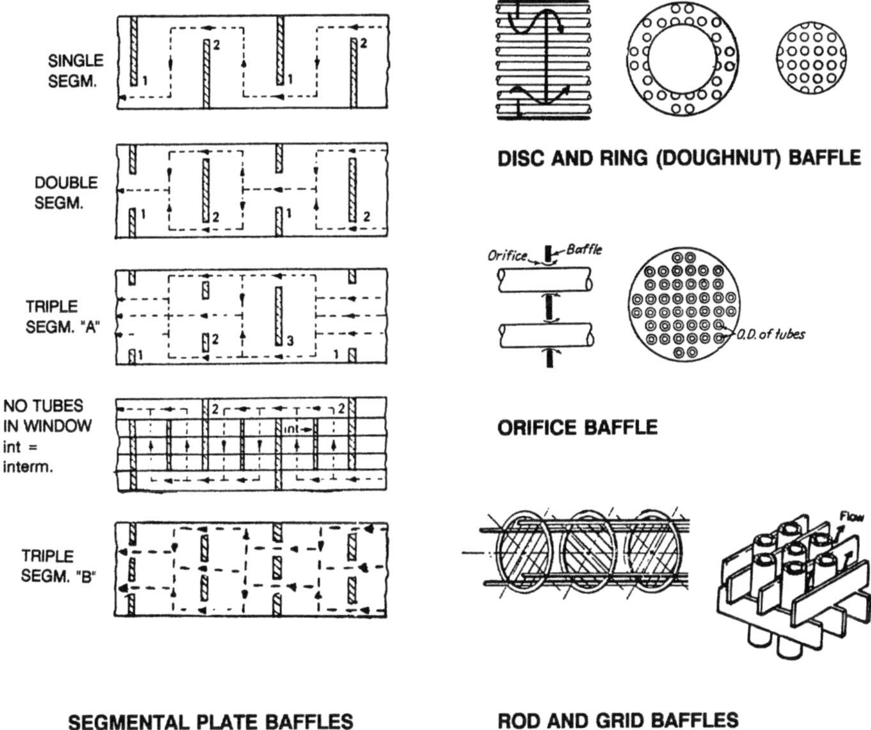

Fig. 2. Schematic representation of various baffle types.

Briefly, the segmental, double-segmental and triple-segmental baffles have pressure drop ratios of about 1 : 1/2 : 1/3. The disc-and-doughnut baffles are similar to the double segmental baffles. The rod and grid baffles are only tube supporting structures in a longitudinal flow field, their pressure drop being about that of triple-segm. baffles. The orifice baffles were used in laminar flow in the 1940's, but later abandoned due to tube vibration effects. For detailed characteristics see [4,5 and 9].

A very instructive comparison on pressure drop to heat transfer conversion was extracted from data obtained with a test heat exchanger by Short [8]. The tested configurations include (symbols in Fig.3) pure longitudinal flow (LONG), segmental baffles (SEG), disk and ring (donut) baffles (DSDO) and orifice baffles (ORIF). To this data set was added calculated performance from correlations (under identical flow conditions), for an ideal tube bank (IDTB), and for longitudinal flow with grid supports (GRIDB). While the pressure drop difference between the various flow arrangements exhibited a spread by a factor of up to 200, heat transfer varied only by a factor of about five (5). The performance comparison as ratio of the heat transfer coefficient to pressure drop versus flow rate of water in the reference exchanger is shown in Fig. 3.

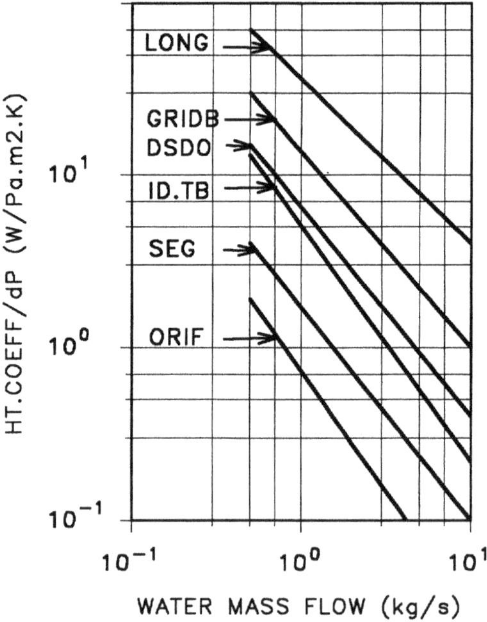

Fig. 3.   Effectiveness of pressure drop to heat transfer conversion for various flow types.

The results show, as expected, that the pure longitudinal flow is by far the most effective system, followed by grid or rod baffles (longitudinal flow with grids), the disk and donut baffles, and the ideal tube bank. Segmental baffle performance is substantially lower due to ineffective and pressure drop-parasitic flow components, with the orifice baffle being the lowest, because of high contraction/expansion losses. The high performance of the disk and ring baffles is due the radial flow between the bundle center and periphery, which eliminates bundle bypass, and uses much lower cross-flow mass velocity than segmental baffles. While the disc and donut (ring) baffle type was used in US in the 1940's, it is rarely used now, mainly because of manufacturing problems (compared to dbl-seg baffles), and absence of dependable methods of calculation. The following conclusions can be made from these evaluations :

a. If low pressure drop on the shell-side is essential (and X-shell is excluded), consider using rod or grid baffles, with correlations as suggested in [9,10];
b. For higher dp-max, consider using disk and ring baffles or double- and triple-segmental baffles;
c. Use segmental baffles only if ample pressure drop is available, the resulting higher heat transfer coefficient would result in less expensive exchanger or higher flow velocity is desired for fouling control.

2. Utilization of available pressure drop. In all no-phase-change processes (and in some phase-change), heat transfer is proportional to flow velocity and hence pressure drop expended. As designs are commonly governed by specified maximum pressure drop, it is imperative that as much as possible of the pressure drop is utilized, but only if it results in a design with overall better characteristics. Computer program results must be carefully watched in this respect and human intervention is indicated.

The control of fine-tuning pressure drop utilization is much more effective on the shell-side, where the designer has to his disposition a variety of shell types, baffle types and spacing and, ultimately, tube layout pitch. On the tube-side only number of tube passes and usually limited variation of tube count/tube length is available. If tube passes are increased from NTP1 to NTP2, pressure drop changes as $(NTP1)^3/(NTP2)^3$, or eight times for a change between 1 and 2. Thus low pressure drop stream must be sometime placed on the shell-side, other considerations permitting. It often requires complex manipulation of the constructional elements to obtain proper design. Plate exchangers, being high pressure drop devices, are very sensitive to pressure drop utilization. Surprisingly many cases are poorly designed in this respect.

3. Pressure drop limited design. This characterizes a design where the exchanger size had to be increased (or unit in parallel added), only to accommodate available pressure drop. Once such an increase of size occurs during the computer design process and it is identified as being due to pressure drop alone, it is a "*pressure drop limited design*" and corrective measures must be considered. However, not in all programs is such identification easy or even possible, and in all cases the corrective measures include changes in constructional component specifications, i.e. basic data input and hence new trial design. However, a review of the pressure drop specification should be made first, to determine if it needs to be revised, sometime even by small amount. If shell-side is limiting, a trial selection of alternate construction elements is indicated, usually shell and/or baffle type. Tube-side pressure drop is much more difficult to control, mainly through tube passes and tube length.

4. Operation of over-surfaced units. Specification of unrealistically high fouling resistances or safety factors leads to heavy over-surface of exchangers which must, however, operate also at initial clean conditions. As a rule, increasing the size of the exchanger *more than about 30 percent due to fouling calls for a review of the fouling resistance* estimation. Similarly, exchangers using cooling water or air are designed to 95 percent of the highest summer temperature, and *will over-perform the rest of the year.* Only in some operations (heat recovery units, steam condensers in vacuum, etc.) over-performance in general may be acceptable (or even useful), while in most cases the over-performance must be absorbed by downstream units, leading to possible problems or requiring elaborate control provisions to avoid process upsets.

To pre-empt unpleasant surprises and prepare for necessary controls, each exchanger *should be evaluated also at clean conditions and at winter prevailing temperatures.* The problem is illustrated by an example of an actual case, sea water cooling demineralized water from 78 C, in a large 1-1 exchanger. The sea water temperature changes from 5 C in winter to 32 C in peak summer, when also maximum fouling is expected. In Figure 4 are shown the temperatures of both fluids, clean and fouled (CL, FO), at summer and winter conditions (S, W). Notice the wide range of outlet temperatures between summer/winter and clean/fouled operation. In some cases over-cooling of the hot stream may result in freezing, and the coolant may reach boiling temperatures, if proper controls are not provided. The simplest way to correct the winter over-performance is to decrease the coolant flow rate, but the consequence would be a rapid increase of fouling at the resulting low flow velocity. Two other solutions are possible in serious cases :

a. part of the hot stream is piped to bypass the exchanger, holding the outlet temperature virtually constant; the decrease of velocity must not affect fouling.

b. if the above arrangement is not practical (condensers), then part of the coolant can be recirculated (keeping the flow velocity constant for fouling control) and thus heated, so that the hot stream outlet temperature stays at a desired level. This is rather expensive but effective, and used sometime for cooling water control under winter conditions when the process stream must not be over-cooled.

Other typical example group are tube-side vertical thermosiphon reboilers, which are often designed to unrealistically large fouling resistances, which may never materialize. Under clean conditions the overdesign results in erratic operation. If the heating medium is steam, the pressure can be gradually increased as fouling progresses. However, if the unit is heated by process vapor with fixed pressure (heat recovery), a simple liquid seal loop can be installed, which will flood bottom part of the shell-side with condensate and thus effectively decrease the heated surfaces [6]. The condensate level can be decreased as fouling progresses.

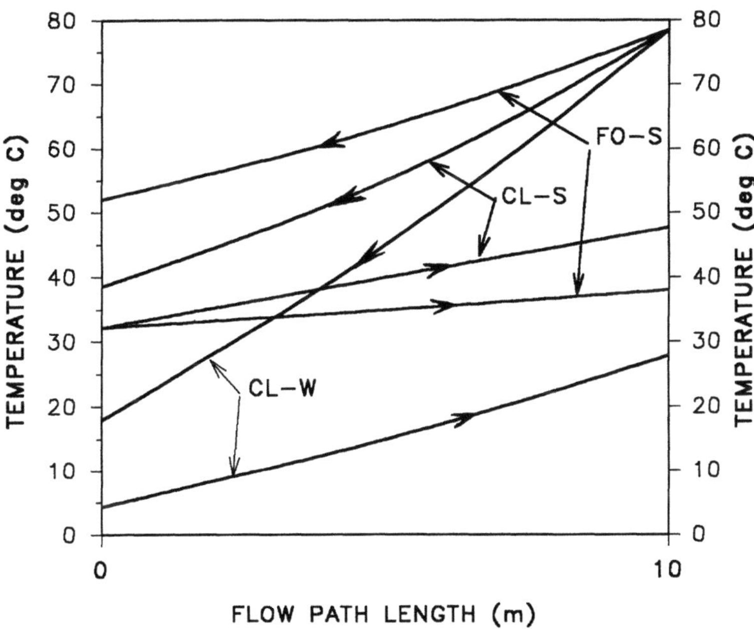

Fig. 4. Temperature profile in an exchanger, clean and fouled (CL,FO) and under summer and winter conditions (S,W) of cooling water.

5. Treating Under-performing Exchangers.   While most exchangers are over-designed due to exaggerated fouling, combination of circumstances may conspire to produce an exchanger which does not perform to specified conditions. Simple errors in design and in manufacturing must be excluded first by a careful check. However, an amazing variety of other reasons can be the cause, the few examples cited below being typical of what can be expected :

a. incorrect construction or assembly of the tube bundle (baffle positioning etc.);
b. in no-phase change processes or boiling, unexpected development of severe fouling in early stages of operation, due to a variety of reasons;
c. in boiling processes, vapor accumulation (blanketing) in parts of the surface, because of improper flow arrangement;
d. in no phase change operations, partial boiling (freezing) occurs, which was not expected, and is due to higher (lower) wall temperature as a result of overdesign and under no fouling conditions;
e. in condensation, many problems are encountered, see Section 6 below;
f. stream composition is different from that considered in design;

However, if higher duty is required regardless of reasons, and before resorting to replace the exchanger, consider the following remedies :

g. if the shell-side fluid is the controlling resistance, retubing with low finned tubes may increase the performance substantially;
h. if the tube-side fluid is controlling, increasing the number of tube-passes is sometimes possible. The use of twisted tape or wire loop inserts [13] will increase the coefficient substantially, if the flow is in proper Reynolds number range.
j. in some cases both modifications may be helpful.

6. Poor Performance of Condensers   While large number of reasons can be responsible for poor performance of condensers, three items stand out :

a. Accumulation of non-condensible gases : due to inadequate venting provisions, gas accumulation will gradually decrease the condenser effective area and hence performance, which is sometime incorrectly explained as fouling. In some cases, presence of gases may occur which was not expected from the process analysis, usually products of reactions etc.

b. Differential condensation : If vapor mixture of two or more components remains in close contact with the condensate throughout the condensing path, equilibrium exists and we term it as "integral" condensation. On the other hand, if the condensate is separated from the condensing vapor as it is formed, equilibrium exist only between the remaining vapor and the condensate formed at that point. The vapor composition approaches that of the low boiling component at the outlet, *resulting in much lower mean temperature difference* than would be the case for integral condensation. In extreme cases the dew point at outlet may be below the coolant temperature [11], making the condenser outright inoperative.

This is termed "differential" condensation and occurs in horizontal shell-side condensation with E, J or G shells. This design should be avoided for moderate and wide range boiling mixtures at all cost. Use vertical condensers, X-shells or other designs, e.g. plate exhangers. A case study for a 50/50 mixture of ammonia-water was presented by Mariott [11], and for a hydrocarbon mixture by Bell in [12], shown here in principle as Figure 5.

c. Inadequate drainage of the condensate : too small nozzle or adverse pressure conditions for drainage can cause raising level of the condensate in the shell, rendering part of the surface inoperative. When the condensate level reaches certain height, drainage will be activated and the process repeat itself, resulting in erratic performance.

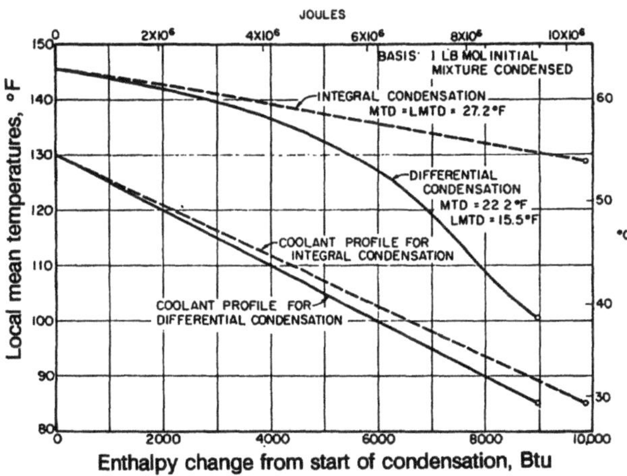

Fig. 5. Integral and Differential condensation T - H profile for C_4- C_5 mixture [12]

REFERENCES

1. Roetzel, W., <u>VDI Wärmeatlas</u>, Sec. Cb, 6th ed., VDI Verlag, Düsseldorf, 1991.

2. Colburn, A.P. and Hougen, O.A., "Design of Cooler Condensers for Mixture of Vapors with Noncondensing Gases", Ind.Eng.Chem., vol.26, no.11, pg.1178, 1934.

3. Gardner, K., and Taborek, J., "Mean Temperature Difference - A Reappraisal", AIChE Journal, vol.23, pp. 777-786, 1977.

4. <u>Heat Exchanger Design Handbook</u>, E.U. Schlünder ed., vol. 3 and 4. Hemisphere Publ., N.Y., 1983 (updates 1989) [2].

5. Sauders, E.A.D., <u>Heat Exchangers : selection, design, construction</u>, John Wiley, New York, 1988.

6. Yokell, S., <u>A Working Guide to Shell-and-tube Heat Exchangers</u>, McGraw-Hill, New York, 1990.

7. CONCO Systems Inc., Verona, PA 15147.

8. Short, B., University of Texas Publ. No. 4324, 1943.

9. Taborek, J., "Longitudinal Flow in Tube Bundles with Grid Baffles" in <u>Heat Transfer - Philadelphia 1989</u>, AIChE Symp.Ser. vol.85, 1989.

10. Taborek, J., <u>Heat Exchanger Design Handbook</u>, E.U. Schlünder ed., vol. 3, Sec. 3.3.12. Hemisphere Publ., N.Y., 1983 (to be published 1991).

11. Mariott, J., Heat Transfer Eng., vol.10, no.4, 1989.

12. Bell, K.J., Lectures on Heat Exchangers, Oklahoma State Univ., 1980. also in <u>Heat Exchanger Design Handbook</u>, E.U. Schlünder ed., vol.3, Sec. 3.4.4, Hemisphere Publ., N.Y., 1983 (updates 1989)[2].

13. Heatex Inserts, CAL GAVIN Corp., Birmingham, UK.

---

[2]) Same text was re-published as <u>Hemisphere Handbook of Heat Exchanger Design</u>, Hemisphere Publ. Corp., NY, 1990.

Compact New Formulae for Mean Temperature Difference and Efficiency of Heat Exchangers

Holger MARTIN

Institut für Thermische Verfahrenstechnik der Universität Karlsruhe

Summary

Recently a system of very compact formulae for the well known efficiency – NTU relationships for heat exchangers of various flow configurations has been presented by the author [1, 2]. The advantages of these new formulae in the calculations required in thermal design of heat exchangers will be demonstrated for a number of flow configurations. Shell-and-tube exchangers with multiple passes, with split flow and two tube passes, as well as a number of cross-counterflow configurations will be shown to be treated very much easier using the new formulae. In combination with the cell-method, as developed by Gaddis and Schlünder, the new formulae may be used as efficient engineering tools in computer programs. The application of now widely available software, such as a spreadsheet calculation program for easy execution of heat exchanger design calculations with the cell method will also be demonstrated in this context.

Thermal design of heat exchangers

The basic design equation for heat exchangers is usually written as:

$$\dot{Q} = (U\,A)\,\Delta T_m \tag{1}$$

In nondimensional form this same relationship becomes:

$$\boxed{\epsilon_i = N_i\,\Theta} \tag{2}$$

The efficiency

$$\epsilon_i = \frac{\dot{Q}_{ij}}{(\dot{M}c_p)_i\,(T_{i,in}-T_{j,in})} = \frac{T_{i,in}-T_{i,out}}{T_{i,in}-T_{j,in}} \qquad (i,j = X,Y), \tag{3}$$

or, the normalized change of temperature of fluid i is equal to the product of the number of transfer units N_i and the normalized mean temperature difference Θ:

$$N_i = (UA)/(\dot{M}c_p)_i \tag{4}$$

$$\Theta = \Delta T_m/(T_{i,in}-T_{j,in}). \tag{5}$$

While the determination of overall heat transfer coefficients U is treated in great detail in many texts on heat transfer, the calculation of the correct mean temperature difference $\Delta T_m = (T_x - T_y)_m$ between the to fluids X and Y is very often given much less attention. People in practise often use the well-known logarithmic mean temperature difference $\Delta T_m = \Delta T_{LM}$ even in those cases, where the logarithmic mean is not the correct integral average.

Simple flow configurations

An investigation of the influence of flow configuration on heat exchanger performance, i.e., mean temperature difference θ, or efficiency ϵ has shown, through the comparison of simple configurations, like stirred tank, parallel flow and counterflow and the various crossflow configurations, that the type of flow configuration is crucial in the range close to thermal equilibrium (N >> 1) (see Figure 1).

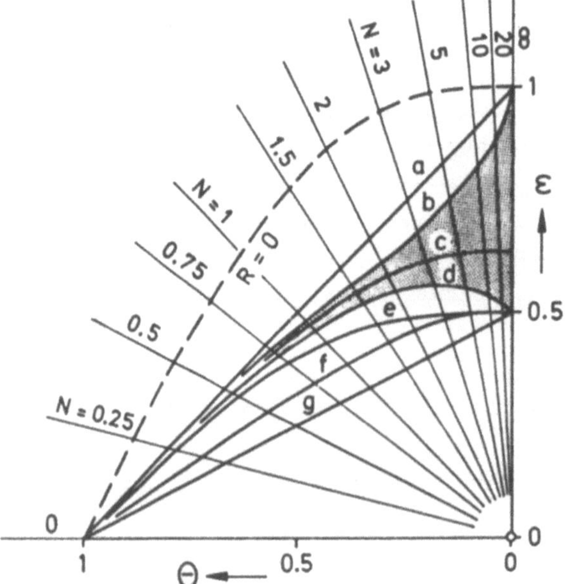

Fig. 1. Efficiency ϵ as a function of mean temperature difference θ for simple flow configurations with NTU (N) as a parameter.
Full lines: R=1 **a** counterflow; **b** crossflow, ideal; **c** crossflow, one side laterally mixed; **d** crossflow, both sides laterally mixed; **e** parallel flow; **f** stirred tank, one side; **g** stirred tank, both sides
Broken line: R=0 (not for **g**, in this case the curve for R=0 coincides with **a**)

This is particularly so for equal or nearly equal flow capacities on both sides.
For small number of transfer units (N < 1) or short relative residence times of
the fluids in the apparatus (... should NTU not be better interpreted as "nondimensional time unit"?) the transfer performance is affected far less by the flow
configuration than by N itself. Often the flow configurations occuring in real
heat exchangers can be represented by cascades of interconnected "cells" or sub-
exchangers, in each of which a simple configuration is realized. In such cases,
the formulae that have been derived for the normalized mean temperature difference Θ or the normalized change in temperature ϵ of the simple configurations
can be applied with good approximation, as shown in greater detail in [1]. It
seems to be convenient for a quick analysis of such equipment, to keep these formulae ready for application in a simple and compact form.

For the design problem the form $\Theta(\epsilon_x, \epsilon_y)$ would be best suited. This function can
be represented by the logarithmic mean of the temperature differences at both
ends of the apparatus for the simplest configurations only: stirred tank, parallel
and counterflow.

The formulae for simple flow configurations are conventionally given in many
textbooks in the $\epsilon(N, R)$ form. For asymmetric cases, such as crossflow, one side
laterally mixed, two formulae are needed depending on whether the reference
NTU for the mixed stream is meant to be N $(=N_1=X)$ or RN $(=N_2=Y)$.

Table 1: Mean temperature difference Θ for some simple flow configurations

Configuration	$\Theta(X, Y)$	$1/\Theta(X, Y)$
Stirred tank, both sides	$1/(1+X+Y)$	$1+X+Y$
one side (X)	$1/[X+Y/(1-\exp(-Y))]$	$X+\varphi(Y)$
Parallel flow	$\dfrac{1-\exp[-(X+Y)]}{X+Y}$	$\varphi(X+Y)$
Counterflow	$\dfrac{1-\exp[-(X-Y)]}{X-Y\cdot\exp[-(X-Y)]} \quad (X\neq Y)$ $\qquad 1/(1+N) \qquad (X=Y=N)$	$\varphi(X-Y)+Y$
Crossflow, both sides laterally mixed	$\dfrac{1}{\dfrac{X}{1-\exp(-X)}+\dfrac{Y}{1-\exp(-Y)}-1}$	$\varphi(X)+\varphi(Y)-1$
one side (X) laterally mixed	$\dfrac{1-\exp(-X[1-\exp(-Y)]/Y)}{X}$	$\varphi(Y)\cdot\varphi\left(\dfrac{X}{\varphi(Y)}\right)$

Writing the formulae in the form $\Theta(X, Y)$ avoids the need for an a priori definition of the reference NTU. Table 1 is a compilation of the formulae for the most important simple flow configurations. By multiplying Θ with X or Y, the required non-dimensional change in temperature ϵ_x or ϵ_y of the corresponding stream is easily obtained.

In Table 1 one often comes across a term such as $x/(1 - e^{-x})$. In order to write the formulae in a more compact form, this term may be denoted as a function

$$\boxed{\varphi(x) = \frac{x}{1 - e^{-x}}} \tag{6}$$

As may be seen from the right hand column of Table 1, substituting X, Y, the sum (X+Y) or the difference (X-Y) respectively for the variable x, extremely compact expressions can be obtained for the reciprocal of the normalized mean temperature difference.

The characteristics of the function $\varphi(x)$ are the following: At x=0, it has a limiting value of unity,

$$\boxed{\varphi(0) = 1} \tag{7}$$

which may be seen from a series expansion of the exponential function. Its slope at $x = 0$ is 1/2 and it tends to its argument for large values of x

$$\boxed{\varphi(x) \rightarrow x \qquad (x \gg 1)} \tag{8}$$

Thus the compact formulae in the right hand column of Table 1 and their limiting cases are very easily handled in practice.

For example, one can find the normalized temperature change ϵ_i of both streams in **parallel flow** from the very simple formulae

$$\epsilon_x = \frac{X}{\varphi(X+Y)} \tag{9}$$

and

$$\epsilon_y = \frac{Y}{\varphi(X+Y)} \tag{10}$$

and their limiting values may be immediately arrived at, if the behaviour of $\varphi(x)$ is kept in mind. Here, the argument x is the sum of the NTUs and the physical meaning of φ is the ratio of the maximum temperature difference $(T_{x,in} - T_{y,in})$ to the mean temperature difference $(T_x - T_y)_m$ for parallel flow.

For **counterflow**, the difference of the NTUs replaces the sum. While two formulae, for $X \neq Y$ and $X = Y$, are needed in the conventional way of writing as in the central column, a single and much simpler one suffices in the new way (right hand column of Table 1).

The efficiencies are now obtained from:

$$\epsilon_x = \frac{X}{\varphi(X-Y) + Y} \tag{11}$$

and

$$\epsilon_y = \frac{Y}{\varphi(Y-X) + X} \cdot \tag{12}$$

For equal flow capacities in counterflow, C=−1, the NTUs (X = Y = N) and effi-
ciencies ($\epsilon_x = \epsilon_y = \epsilon$) are equal, and ϵ for this case, with Eq. 7, becomes:

$$\epsilon = \frac{N}{1+N} \qquad (C = -1), \tag{13}$$

while for one medium, say stream Y, with infinite flow capacity (condensation,
evaporation) C = 0, one easily obtains:

$$\epsilon_x = \frac{X}{\varphi(X)} \qquad (C = 0) \tag{14}$$

As the flow direction is of no importance when C = 0, the same result can be got
for parallel flow from Eq. 9. Note that due to the property of the function $\varphi(x)$
according to

$$\varphi(x) = \varphi(-x) + x \tag{15}$$

the subscripts in the θ-formula for counterflow can be freely interchanged.
Therefore, the denominators in Eqs. 11 and 12 are always equal

$$\varphi(X-Y) + Y = \varphi(Y-X) + X \tag{16}$$

The formula for **crossflow, one side** (X-side) **laterally mixed**, results in the
efficiency expression

$$\epsilon_x = \frac{Z}{\varphi(Z)}, \qquad \text{with:} \quad Z = \frac{X}{\varphi(Y)} \cdot \tag{17}$$

For equal flow capacities $X/\varphi(Y)$ can reach a maximum value of unity for X = Y =
N → ∞ and its maximum efficiency becomes:

$$\epsilon_{max} = 1/\varphi(1) = 0.632 \tag{18}.$$

The function φ also figures in the case of crossflow over n rows of tubes [1].

<u>Multipass heat exchangers</u>

For shell-and-tube heat exchangers with even numbers (2m) of tube-side passes
and many baffles, typified by a shell-side stream laterally mixed, the compact
$1/\theta$ form can be written using the function φ :

$$\boxed{\frac{1}{\theta} = \varphi(Z_m) + \varphi(Y) - \varphi(Y/m) + \frac{X+(Y/m)-Z_m}{2}} \qquad Z_m = \sqrt{X^2+(Y/m)^2} \tag{19}$$

This general equation has been found by empirically generalizing the known results for two, four, and an infinite number of passes [1]. In fact this equation is
mathematically correct also for 2m = 6, 8, 10, etc. passes, as may be seen from

a comparison with a generalized solution obtained by Kraus and Kern in 1965 [3]: This reference has been brought to my attention only recently by B. Spang, Hamburg:

$$\epsilon = \frac{2}{1 + R + \dfrac{2}{n}\sqrt{1+(nR/2)^2}\ \coth \dfrac{NTU\sqrt{1+(nR/2)^2}}{n} + \dfrac{2}{n}\,f(z)} \tag{20}$$

$$f(z) = \frac{jz^j+(j-2)z^{j-1}+(j-4)z^{j-2}+\ \cdots\ -(j-4)z^2-(j-2)z-j}{1+z+z^2+z^3+\ \cdots\ +z^{j-1}+z^j}$$

$$z = \exp(2NTU/n) \qquad\qquad j = (n-2)/2 \quad (\,= m-1\,)$$

The equality of ϵ as calculated (with $\epsilon = N\,\Theta$) from Eq. 19 and from the much more complicated Eq. 20 due to Kraus and Kern can in fact be shown mathematically (see Appendix C in the forthcoming translation of [1]).

Without rewriting the known solutions for two and four passes in the new way such a simple closed form of the generalized solution for (2m, m, m) would probably not have been found.

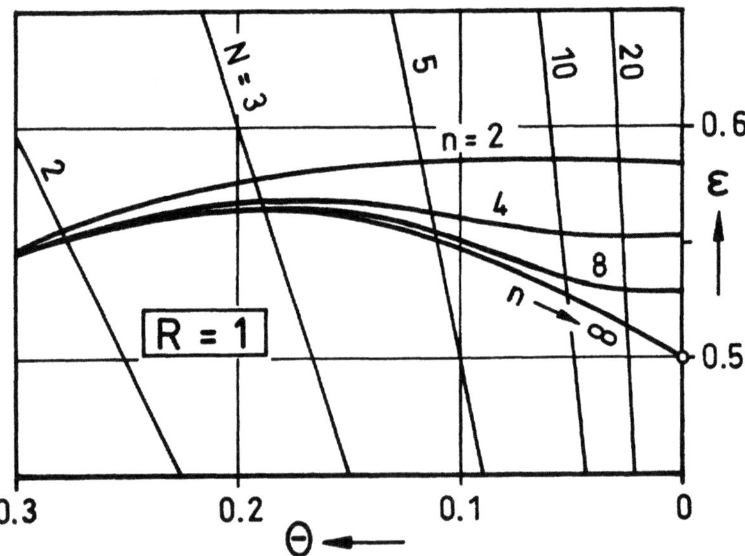

Fig. 2. Efficiency ϵ as a function of mean temperature difference Θ for heat exchangers with even numbers of tube-side passes and one shell side pass, laterally mixed, at equal flow capacities R=1, with N as a parameter.

Figure 2 shows the efficiencies of heat exchangers with even numbers of tube-
side passes and one shell side pass, laterally mixed, at equal flow capacities
plotted versus the mean temperature difference Θ with N as a parameter. From
this it can be found that the multipass heat exchangers with even numbers of
internal passes lie in the lower crossflow region between curve **d** (in Fig. 1) for
crossflow, both sides mixed ($2m \rightarrow \infty$) and the curve for $2m = 2$ from Eq. 19.
For odd numbers of internal passes the efficiency is higher or lower than for the
next lower even number depending on flow direction (more counterflow passes
are of course better). Though the problem in principle has been solved analytically for any number of passes [4], a similar general formula for odd numbers of
passes could not yet been found.

<u>Split-flow, two tube passes</u>
As an example for the application of the new formulae, the following flow configuration with a shell side split flow (TEMA G shell) [5] with two tube passes is
treated as shown in Figure 3:

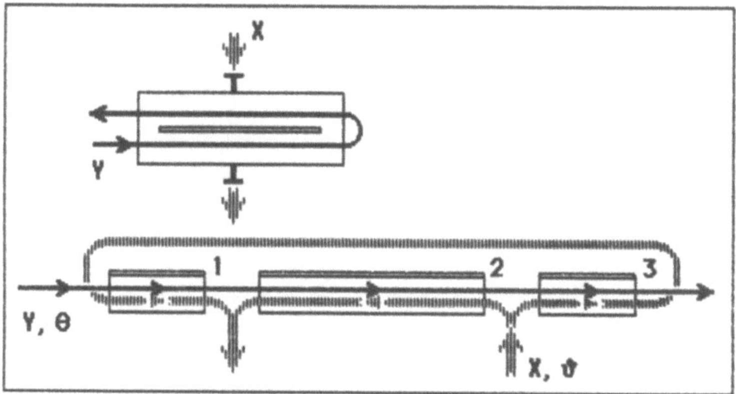

Fig. 3. Shell-side split flow (TEMA G shell) with two tube-side passes

For the configuration shown, with Θ and ϑ as the normalized temperatures of the
Y and X streams respectively, the efficiencies of the three cells, denoted by E_i, ϵ_i
in place of $\epsilon_{y,i}$, $\epsilon_{x,i}$ ($i=1, 2, 3$) for shortness, may be found easily from the formulae for parallel and counterflow heat exchangers:

$$\frac{\vartheta_1'' - \vartheta_3''}{\theta' - \vartheta_3''} = \frac{\vartheta_3'' - \vartheta'}{\theta_2'' - \vartheta'} = \epsilon_3 = \epsilon_1 = \frac{X/2}{\varphi[(X/2)+(Y/4)]} \qquad (21, 22)$$

$$\frac{\theta' - \theta_1''}{\theta' - \vartheta_3''} = \frac{\theta_2'' - \theta''}{\theta_2'' - \vartheta'} = E_3 = E_1 = \frac{Y/4}{\varphi[(X/2)+(Y/4)]} \qquad (23, 24)$$

$$\frac{\vartheta_2'' - \vartheta'}{\theta_1'' - \vartheta'} = \qquad \epsilon_2 = \frac{X}{\varphi[X-(Y/2)]+(Y/2)} \qquad (25)$$

$$\frac{\theta_1'' - \theta_2''}{\theta_1'' - \vartheta'} = \qquad E_2 = \frac{Y/2}{\varphi[X-(Y/2)]+(Y/2)} \qquad (26)$$

These are six equations for six unknown temperatures.

The subscripts i = 1, 2, 3 denote the cell number, the superscripts ′ and ″ are used for inlet and outlet respectively.

$\theta' = \theta_1' = 1$ and $\vartheta' = \vartheta_2' = \vartheta_3' = 0$ was used as a normalization. X and Y stand for the NTUs of the shell side and the tube side streams respectively.

The solution for $1 - \theta'' = \epsilon_y$ is easily found to be:

$$\boxed{\epsilon_y = 1 - \frac{(1-E)^2(1-P)}{1-2(X/Y)E^2(1-P)}} \qquad E = \frac{Y/4}{\varphi[(X/2)+(Y/4)]} \qquad (27)$$

$$P = \frac{Y/2}{\varphi[X-(Y/2)]+(Y/2)}$$

This result looks much simpler than the formulae given by Taborek on page 1.5.2-14 of the Heat Exchanger Design Handbook [5] for the same problem.

Its correctness may be checked by comparison with Figure 11 on 1.5.2-13 in [5]. For equal flow capacities, X = Y = N one can easily find from Eq. 27 with N→∞:

$$E_\infty = 1/3 \quad P_\infty = 1/2 \quad \text{and thus:} \quad \mathbf{E_{\infty, R=1} = 0.75}.$$

For Y = 2X (R = X/Y = 1/2) (tube side NTU greater, i.e., shell side stream with the greater capacity) the corresponding limiting efficiencies are:

$$E_\infty = 1/2 \quad P_\infty = 1 \quad \text{and therefore:} \quad \mathbf{E_{y\infty, R=1/2} = 1}.$$

Both limits are in perfect agreement with Figure 11 on 1.5.2-13 in [5]. This configuration, as pointed out by Taborek in [5] obviously enables higher efficiencies compared to the usual two tube pass-one shell pass exchanger, while the shell-side pressure drop is nearly the same.

<u>Cross-counterflow configurations</u>

For **counter-directional cross-counterflow** with n passes, with every second pass of the laterally mixed X-stream crossing the unmixed continuous Y-stream in the opposite direction, the formulae in the $1/\theta$ form may be written:

$$\boxed{\frac{1}{\theta} = \frac{X}{1 - A_n(z,b)}}\ , \qquad \text{with}\ \boxed{z = \frac{X/n}{\varphi(Y/n)}}\ , \qquad \boxed{b = 1 - (Y/2X)z} \qquad (28)$$

and $A_n(b,z)$ from Table 2:

Table 2: Function $A_n(z,b)$ for counter-directional cross-counterflow with n passes:

n	$1/A_n(z,b)$	$\epsilon_{\infty,\,X=Y}$	
1	e^z	0.6321	
2	$(1-b)+b\,e^{2z}$	0.7616	(0.7746)
3	$(1-b)[1+b(1-2z)]e^z+b^2e^{3z}$	0.8246	(0.8375)
4	$(1-b)b[2+b(1-(z/4)e^{2z})]+b^3e^{4z}$	0.8615	(0.8730)

The numbers in brackets in the right hand column of Table 2 are the corresponding maximum efficiencies for a countercurrent cascade of n crossflow elements, one side laterally mixed. In this case the Y-stream is thought to be laterally mixed between each element.

Co-directional cross-counterflow, with every pass of the laterally mixed X-stream crossing the unmixed Y-stream in the same direction leads to even higher efficiencies.

With $a = (Y/X)z$, the corresponding formula for $A_n(z,a)$ with n=2 passes is:

$$1/A_2(z,a) = e^z(e^z - az).$$

From this the maximum efficiency with $X=Y=N \rightarrow \infty$ can be found via $z_\infty = 1$, $a_\infty = 1$ to be

$$\epsilon_{\infty,\,X=Y} = 0.7859,$$

which is to be compared to the values given in the second line of Table 2.

Table 3: Spreadsheet calculation for counter-directional cross-counterflow
N =15, R =1 N_{cell} =5N/100=0.75, ϵ_{cell} =0.414 (ideal crossflow), ϵ=0.898

Row	Left	1	2	3	4	5	Right
		0	**0**	**0**	**0**	**0**	
1	349	205 / **144**	120 / **85**	70 / **50**	41 / **29**	24 / **17**	
2	349	264 / **229**	190 / **159**	132 / **108**	89 / **72**	59 / **47**	
3	349	299 / **278**	241 / **217**	186 / **163**	139 / **119**	101 / **85**	102
4	349	320 / **307**	277 / **259**	230 / **210**	184 / **165**	143 / **126**	
5	349	332 / **325**	302 / **289**	264 / **248**	223 / **206**	183 / **166**	
		325	**289**	**248**	**206**	**166**	

Row	Left	1	2	3	4	5	Right
1		304 / **310**	290 / **289**	290 / **277**	319 / **286**	399 / **331**	564
2		317 / **315**	322 / **312**	345 / **325**	392 / **361**	467 / **427**	564
3	349	342 / **334**	362 / **347**	396 / **375**	447 / **421**	507 / **484**	564
4		374 / **362**	402 / **386**	440 / **421**	486 / **467**	531 / **517**	564
5		406 / **393**	438 / **422**	474 / **459**	512 / **499**	544 / **536**	564
		393	**422**	**459**	**499**	**536**	

Row	Left	1	2	3	4	5	Right
1	782	621 / **554**	539 / **505**	506 / **492**	503 / **502**	517 / **523**	
2	782	688 / **648**	612 / **580**	562 / **542**	537 / **527**	531 / **529**	
3	782	727 / **704**	666 / **641**	615 / **593**	578 / **563**	558 / **549**	564
4	782	750 / **736**	705 / **686**	659 / **639**	619 / **603**	590 / **578**	
5	782	763 / **755**	731 / **718**	693 / **677**	656 / **640**	624 / **610**	
		755	**718**	**677**	**640**	**610**	

Row	Left	1	2	3	4	5	Right
1		735 / **741**	721 / **720**	724 / **710**	757 / **722**	839 / **771**	1000
2		750 / **747**	755 / **745**	780 / **760**	830 / **798**	905 / **866**	1000
3	782	776 / **767**	796 / **781**	833 / **811**	884 / **859**	945 / **921**	1000
4		808 / **796**	837 / **821**	876 / **857**	922 / **904**	967 / **954**	1000
5		842 / **828**	874 / **858**	911 / **895**	949 / **936**	981 / **973**	1000

				898			

The huge amount of possible cross-counterflow configurations makes it difficult to provide analytical solutions for all cases that may arise in practise. So it may be often easier and faster to use numerical methods, such as the cell-method developed by Gaddis and Schlünder (see e. g. in [1], or also in [5], vol.1). Together with the now widely available spreadsheet calculation programs and the new compact formulae this method provides a valuable engineering tool for heat exchanger design calculations. Table 3, as an example, shows such a spreadsheet calculation for a counter-directional cross-counterflow configuration with four passes, each with five rows of laterally mixed crossflow elements. The spreadsheet allows to show the local outlet temperatures of both streams exactly at the position, where it occurs in the real configuration. So the result is presented in a form, which is easy to understand and to compare with other configurations.

Conclusions

The well-known formulae for the relationship between efficiency of a heat exchanger and the number of transfer units can be written in a much more compact form by introducing a simple auxiliary function $\varphi(x) = x/(1 - e^{-x})$ and the notation $1/\Theta(X,Y)$. The resulting formulae are often very easy to keep in mind and the limiting values of maximum possible efficiencies can be found quite rapidly. Together with the cell-method developed by Gaddis and Schlünder and with the possibilities of spreadsheet programs, the new formulae may be used by practising engineers as efficient tools for thermal design of heat transfer equipment.

References

1. Martin, H.: Wärmeübertrager. Stuttgart, New York: Georg Thieme Verl.1988 (an english translation: Heat Exchangers will be published by Hemisphere Publ. Co. Washington, DC in 1991).
2. Martin, H.: Simple New Formulae for Efficiency and Mean Temperature Difference in Heat Exchangers. Chem. Eng. Technol. 13 (1990) 237-241.
3. Kraus, A. D.; D. Q. Kern: The Effectiveness of Heat Exchangers With One Shell Pass and Even Numbers of Tube Passes. ASME paper 65-HT-18 (1965).
4. Roetzel, W.; Spang, B.: Analytisches Verfahren zur thermischen Berechnung mehrgängiger Rohrbündelwärmeübertrager. VDI-Fortschrittsberichte Reihe 19, Nr.18. Düsseldorf: VDI-Verlag 1987.
5. Schlünder, E. U. (ed.) Heat Exchanger Design Handbook. 5 volumes. Washington, DC: Hemisphere Publ. Co. 1983 (updated by supplements).

The Multi-Dimensional Thermalhydraulics Code TRIO
Applications to Heat Exchangers

by P. MERCIER et M. VILLAND
(Groupement pour la Recherche sur les Echangeurs Thermiques)
CENG/GRETh - 85 X - 38041 GRENOBLE CEDEX - FRANCE

1. INTRODUCTION

The prediction of thermal performances of a heat exchanger is a major step in the design of heat exchangers which influences the subsequent steps of thermomechanical studies and cost evaluation.

Classical methods commonly used in industry provide a quick solution to many common problems. Some integral types are based on drastic assumptions like heat exchange coefficients which are constant throughout the process, or simplified flow configurations (co or counter current).

Other methods have been developped with the objective of more precision in the rating ; a more local approach can be used by dividing the heat exchanger into several zones, and applying local physical correlations in each of them. Such methods are widely used today because they allow for simple software developments of correct precision and wide versatility. CETUC and CEPAJ developped by GRETh are among these [3], [6].

Until the last decade, the complex geometries in which the fluid flows and heat transfer occures avoided any detailed calculations. With the development of multidimensional thermohydraulic computation tools it is now possible to predict the details of the flow and heat transfer and to obtain more realistic representations of their characters. These methods are based upon the solution of the 2D and 3D Navier-Stokes equations in flow domains partially blocked by internal structures (tube bundles, plates, grids,...).

These tools initially developped in the nuclear industry have been extended to tackle many problems in heat exchanger industries for example :

-    evaluation of internal flow distribution, on shell or tube side ;

-    evaluation of the thermal loading on the structures in rated conditions (steady states and transient).

2. PHYSICAL MODEL

2.1 Shell-side

The 3D statistically averaged conservation equations for mass, momentum, energy or any other passive scalar are used. Turbulent fluxes are modellized with turbulent diffusivities evolving form the two equation turbulence model (k,eps).

For heat exchanger applications, the physical model for the shell side deals with the multidimensional flow through a porous medium.

The local transport equations for mass, momentum and energy are integrated over control volumes which can be partially or totally obstructed by solid structures. Many possibilities can thus be described with the geometrical parameters which result from the volume integration : porosities of the volume and lateral surfaces, wetted area per unit of volume. In this way, tubes or baffles can be implemented in the spatial discretization.

The constitutive laws (friction factor and heat transfer at the tube wall) for the closure of the model have to be choosen appropriately, according to the flow configuration and the geometry of the tubes.

2.2   Tube side

The physical model for the tube side deals with a set of one dimensional flows in parallel tubes with common inlet and outlet pressures. The distribution of the flow between the tubes is a result of the solution of the dimensional momentum equations for each tube.

The fluid in the tube side may be :

- an incompressible, slightly dilatable fluid for single phase applications ;

- a two phase water steam mixture for applications to steam generators.

In each control volume, tubes are assumed to be under the same conditions and are represented by a single tube.

3.   NUMERICAL METHOD

The local conservation equations for the shell side and the tube side are spacially discretized according to a finite volume approach.

For incompressible flow, the main features of the SOLA techniques [1] are used to solve the discrete equations :

- staggered grid : the main variables are located on a staggered mesh : pressure and temperature are located at cell centers, surrounded by nodes carrying each component of the velocity. It follows that continuity and energy equations are integrated over the same control volume V, the components of the momentum equation are integrated over control volumes displaced from V by half a mesh size in each direction.

- transient solution : the time discretization is semi-implicit ; the convective and diffusive terms of momentum and energy balance are explicit, i.e. evaluated at the former time step. Thus a time step limitation (the Courant condition) must be respected. The continuity and the pressure field are implicit, evaluated at the new time step. Thus the pressure solution never violates the continuity condition ; the mass balance of the fluid is automatically verified in every cell, and therefore in the whole area of calculation.

- linear system : as a result of the two characteristics mentioned above, the time discretized balance equations for mass and momentum lead to a linear system between the pressures nodes. The matrix contains only geometric information and is time independant while friction and singular pressure drops are explicitly discretized. A direct inversion procedure is performed according to the properties of the matrix (symmetric positive definite).

4.   APPLICATIONS TO INDUSTRIAL PROBLEMS

4.1   Some specificities of industrial problems

In order to solve industrial problems in the area of heat exchangers, numerical methods have to take into account some more frequently-incountered particularities :

- a 3D representation in most applications ;

- a geometrical complexity due to several aspects : the presence of thermally active zones corresponding to tube bundles and the presence of solid walls like shell, baffles or deflectors ;

- a high Reynolds number to obtain optimal efficiency ;

- the need for the most appropriate constitutive laws to regulate numerical integration over volumes suite larger than the local phenomena scale of length.

4.2   A local application of TRIO

By "local application" we mean a refined modelisation of a specific geometry which is of particular interest for comprehension knowledge of flow behaviour. TRIO software is then used as a tool for thermalhydraulic analysis.

The example of plate and frame heat exchanger can be used as an illustration. Local studies of hydraulic phenomena have been performed under the following assumptions :

- a 2D modelisation with a area representing half a corrugation step ;

- a re-injection of boundary condition profiles between inlet and outlet sides of the area (velocity and pressure) ;

- the use of a (k, ε) turbulence model.

A comparison with experimental results is shown in figure 1 ; it concerns the velocity field in a corrugated channel showing the general behaviour of the main flow and the presence of recirculation flow under low Reynolds conditions.

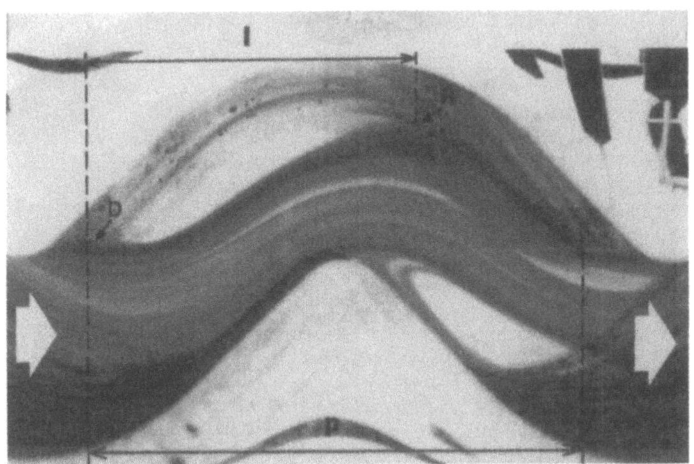

Experimental visualisation of fully developped flow

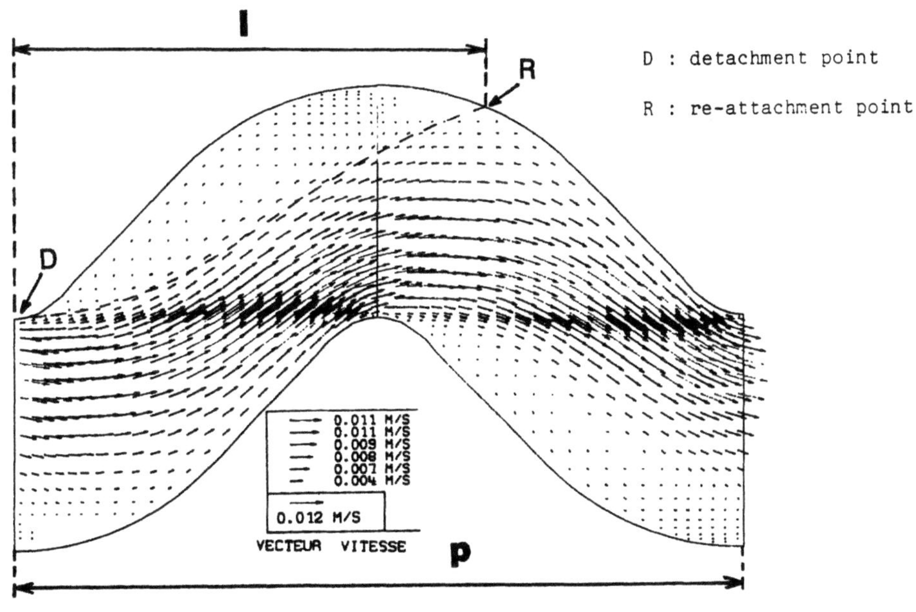

Velocity field by TRIO

Figure 1 : Comparison between TRIO and experimental results at Re = 150

4.3 Application to compact heat exchanger

a)   <u>Entrance effects</u>

Many problems concern entrance effects in tube or plate H.X., due to their effects on thermal performances. Fluid distribution depends mainly on two factors : the geometric shape of collectors related to the position of the inlet nozzle and the hydraulic resistance encountered by the flow. That resistance is directly connected to singular pressure drops and friction losses in the different channels (tubes or plates).

The illustration (figure 2) compares two different entrance effects, and two flow distributions in a crossflow heat exchanger :

- on the left side (1), the real geometry with true friction laws in each channel fitted with turbulators ;

- on the right side (2), in order to appreciate the divergence from reality, the same geometric data except for the friction law in the channel, where the classical law for smoth channel was used.

Nearly the same flow configuration in the inlet collector is observed, but a very different distribution of the fluid in the channels appears.

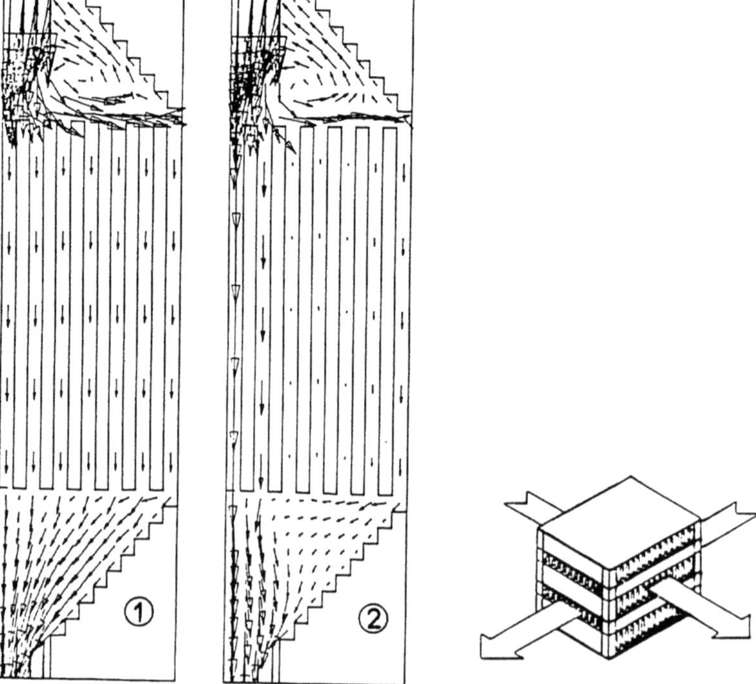

Figure 2 : Entrance-effect in a plate H.X.

b)   <u>Modelisation of compact heat exchangers</u>

The thermal design of such exchangers is usually done using classical methods, but the contribution of thermalhydraulic modelisation has several applications :

- to determine local conditions of temperature fluids, on each part of the plate fin surface ; this information is essential to proceed with a thermomechanical analysis ;

- to determine thermal performances in case of non-uniform inlet distribution, that is not possible under uniform distribution assumptions of classical design software.

Results shown in figure 3 have been obtained in a cross flow configuration in finned channels. The ten lines drawn between 320°C and 180°C for hot fluid 75°C and 300°C for cold fluid allow us to determine :

- diagonal orientation of isothermal fluid lines ;

- non-uniformity of outlet temperatures.

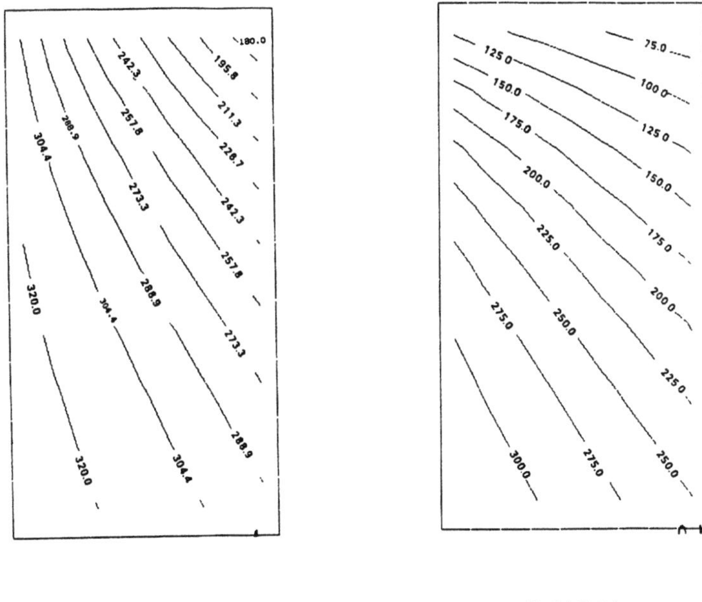

Hot fluid   Cold fluid

Figure 3 : Temperature field in a plate H.X., with crossflow configurations.

4.4   Applications to tubular H.X.

The qualification of the TRIO model for tubular H.X. has been supported by LMFBR studies [5] ; it is of great importance to be confident in the code's predictions of flow fields and temperature fields on both shell and tube sides.

Current applications of the software have led to studies of internal flows in shell and tube heat exchangers. Among them, applications to baffle spacing problems have given answers to the following questions :

- what is the fluid distribution between two adjacent baffles ?

- is that distribution modified by geometrical data ?

- do leakages perturb the flow ?

The figures 4 and 5 illustrate the influence on the velocity field of the parameters :

- baffle cut f

- baffle spacing e, for a same baffle cut ratio (f/d)

In figure 4, the best regular flow seems to be obtained when the baffle cut is approximately the same size as the baffle spacing. A larger spacing emphasizes a jet effect near the baffle tip and promotes the development of a large recirculation behind the baffle.

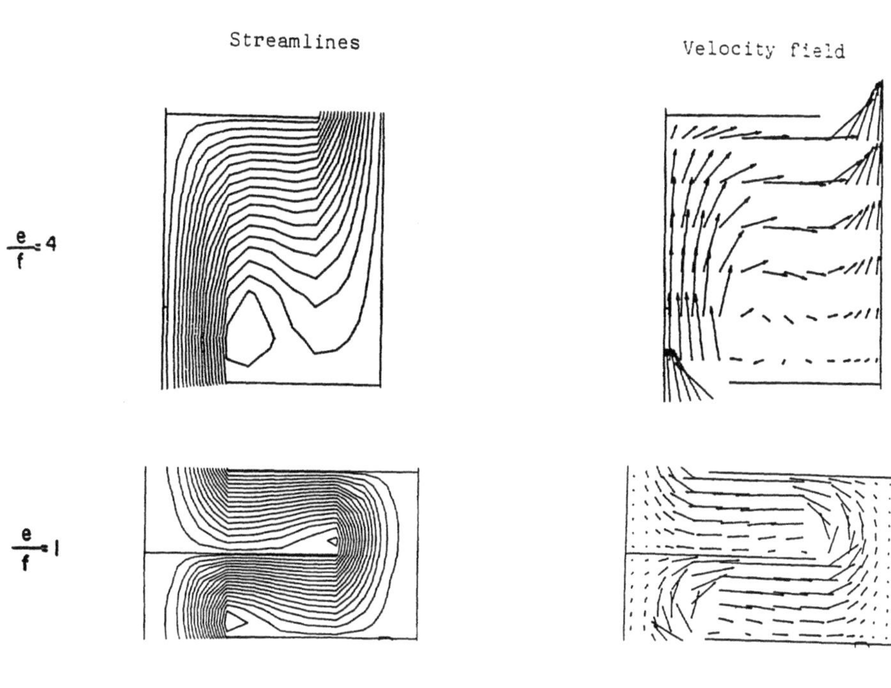

Figure 4 : Influence of baffle spacing

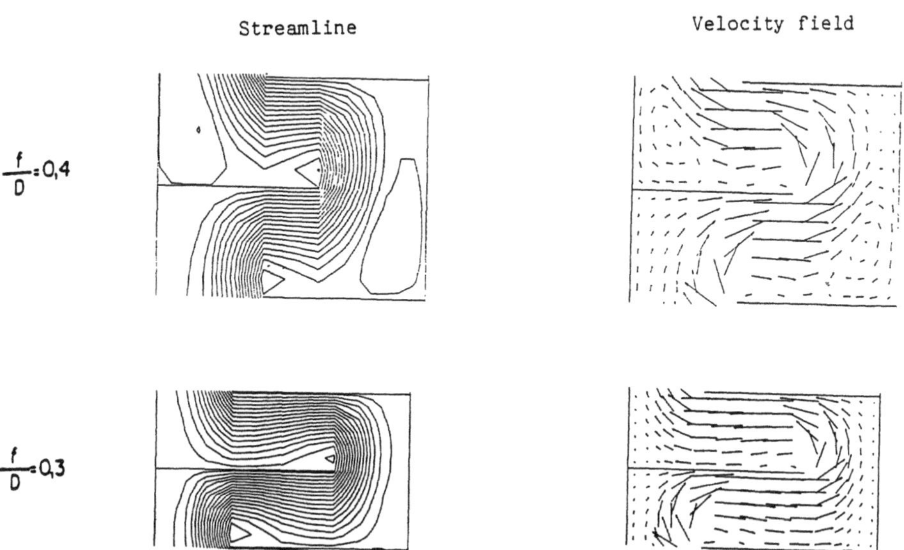

Figure 5 : Influence of baffle cut

5. CONCLUSION

TRIO has been applied to various situations in the H.X. area, and the quality of the results has been tested in specific experiments.

In the future, this code may be more widely used in heat exchangers, with applications in both the thermal design of H.X., and the improvement on the performance of specific parts through local applications

6. REFERENCES

[1]   HIRT, C.W., NICHOLS, B.D. and ROMERO, N.C. : SOLA, a numerical solution algorithm for transient fluid flows. LASL report, L.A. 5852 (1975).

[2]   HARLOW, F.W. and AMSDEN, A.A. : J. of Computational Phys., 8, 197 (1971).

[3]   R. VIDIL, G. RATEL, J.M. GRILLOT : Thermal performances of plate and frame heat exchangers. The CEPAJ software. EUROTHERM Seminar n° 18, Hamburg, (1991).

[4]   G. RATEL, P. MERCIER, G. ICART : Heat exchanger behaviour in transient conditions. EUROTHERM Seminar n°18, Hamburg (1991).

[5]   D. GRAND, B. MENANT, P. MERCIER, M. VILLAND : Numerical modelling of thermal hydraulics in H.X. of LMFBR. IARH Meeting, Lausanne (1987).

[6]   P. MERCIER, G. RATEL : Dimensionnement des échangeurs à tubes et calandre - Logiciel CETUC. Revue Générale de Thermique n° 313 (Janvier 1988).

[7]   B. MENANT, P. MERCIER, M. VILLAND : logiciel de thermohydraulique TRIO-VF - Application aux échangeurs de chaleur. Revue Générale de Thermique n° 340 (avril 1990).

Heat Exchanger Control by Stream(s) By-Pass

P.J. HEGGS and I.M. ABID
Department of Chemical Engineering
University of Bradford, BRADFORD, BD7 1DP, U.K.

Summary

The ϵ-Ntu methodology is used to present criteria for the successful
implementation of by-pass control of two stream heat exchangers. The
theory has been installed in an algorithm for investigation of particular
heat exchangers. An example of a 1:1 countercurrent exchanger is used to
investigate the use of by-pass for flow disturbances to both streams.
For negative disturbances with respect to the design flowrates, the
algorithm predicts the by-pass flows that achieve the desired outlet
temperatures. By-pass will be ineffective for positive disturbances.

Introduction

Heat exchangers are usually designed to perform a given heat duty, and it
is generally assumed that the inlet temperatures and the process conditions remain constant. Unfortunately, fluctuations in inlet temperatures
and stream flowrates frequently occur, and these disturbances affect the
outlet temperatures from the exchangers. Even for situations, where the
inlet temperatures and flowrates remain relatively constant, the outlet
temperatures will vary due to fouling of the heat exchanger surfaces.

Often it is necessary to control the outlet temperature of one stream, or
both streams, due to downstream process equipment constraints. One
method of control that has particular merit, is to by-pass one or both
streams around the exchanger and blend the by-pass flow with the outlet
flow from the exchanger to achieve the specified outlet temperature(s),
Shinskey [1]. However, it is not possible to predict a priori whether
by-pass will provide the control. The only exchanger parameter that will
remain fixed is the transfer area of the original design. Any variation
of the process parameters will require a performance calculation to
predict the resultant outlet temperature changes. The ϵ-Ntu design
methodology is an ideal tool for this prediction.

For hand calculations, it is possible to use the ESDU charts [2] to
obtain the values of the thermal effectiveness, ϵ, for the particular
exchanger configuration and flow path arrangement. Computer applications

will require predictive formulae for the effectiveness, and many do exist for the more popular exchangers found in industry.

This investigation looks into the effects of process parameter changes on the outlet temperatures from heat exchangers and the utilisation of the ϵ-Ntu methodology in the prediction of the by-pass flow as a means of control.

The ϵ-Ntu Representation For Any Heat Exchanger

Any two stream heat exchanger can be represented by the schematic diagram, Figure 1. The hot fluid with a heat capacity rate, C_h, enters the exchanger at an inlet temperature, $T_{h,in}$ and leaves at $T_{h,out}$. The cold fluid with a heat capacity rate C_c, enters the exchanger at $T_{c,in}$ and leaves at $T_{c,out}$.

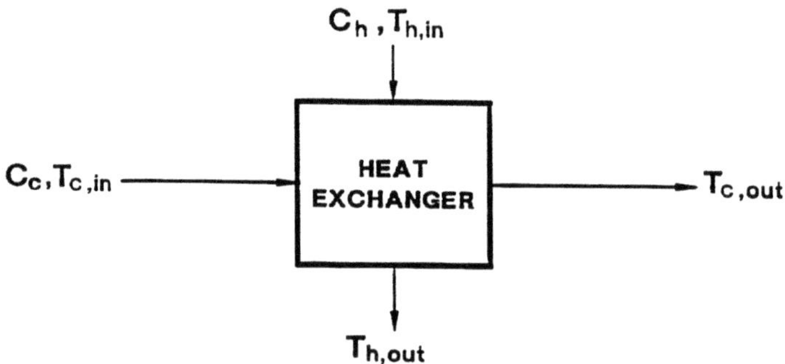

Fig. 1 Schematic diagram of a heat exchanger

The thermal effectiveness of an exchanger is defined as follows:

$$\epsilon = \frac{\text{Actual heat transferred}}{\text{Thermodynamic maximum transfer}} \tag{1}$$

and can be written in terms of temperatures:

$$\epsilon = \frac{C_h}{C_{min}} \frac{(T_{h,in} - T_{h,out})}{(T_{h,in} - T_{c,in})} = \frac{C_c}{C_{min}} \frac{(T_{c,out} - T_{c,in})}{(T_{h,in} - T_{c,in})} \tag{2}$$

In equation (2) it is assumed that there is no heat loss or gain from surroundings, and it is a simple rearrangement of the definition that will provide expressions for the outlet temperatures, i.e.

$$T_{h,out} = \left[1 - \epsilon \, \frac{C_{min}}{C_h} \right] T_{h,in} + \epsilon \, \frac{C_{min}}{C_h} \, T_{c,in} \qquad (3)$$

$$T_{c,out} = \epsilon \, \frac{C_{min}}{C_c} \, T_{h,in} + \left[1 - \epsilon \, \frac{C_{min}}{C_c} \right] T_{c,in} \qquad (4)$$

Equations (3) and (4) are general representations of the outlet temperatures in terms of the inlet temperatures, the heat capacity rates and the thermal effectiveness.

The thermal effectiveness of an exchanger is a function of the following variables

$$\epsilon = fn(Ntu, \ C^*, \ configuration, \ flow \ arrangement) \qquad (5)$$

where $\quad Ntu = UA/C_{min}$, $\qquad (6)$

$$C_{min} = min[C_h, C_c] \qquad (7)$$

and $\quad C^* = C_{min}/C_{max}$. $\qquad (8)$

Hence for any process change and/or stream by-pass, the values of the number of transfer units, Ntu, and the ratio of the heat capacity rates, C^*, will alter and correspondingly, the value of the effectiveness will be different. It is imperative, that the overall heat transfer coefficient is re-calculated for each change of flowrate of either, or both, streams.

Once the new value of the effectiveness has been obtained, the equations (2) and (3) can be used to predict the outlet temperatures. If $T_{h,out}$ is less than the desired value of the hot outlet temperature, by-pass of either stream could possibly provide the control, and likewise, if $T_{c,out}$ is greater than the desired value of the cold outlet temperature. If $T_{h,out}$ is greater than desired, or $T_{c,out}$ is lower than the desired; by-pass cannot be utilised for control purposes.

By-Pass Flow Scheme for a Heat Exchanger

Figure 2 represents a schematic diagram of a heat exchanger with by-pass on both streams. Assuming that the flowing heat capacity rates are independent of temperature, the final outlet temperatures after by-pass can be written as follows:

42

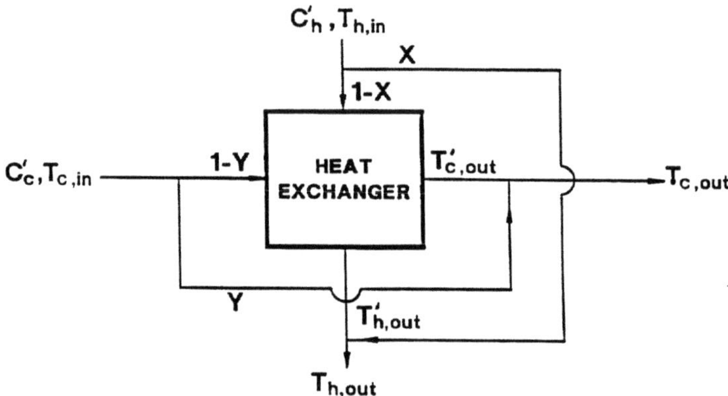

Fig. 2 Schematic diagram of a heat exchanger with by-pass
on both streams

$$T_{h,out} = X\, T_{h,in} + (1-X)T_{h,out}' \tag{9}$$

and $\qquad T_{c,out} = Y\, T_{c,in} + (1-Y)T_{c,out}' \tag{10}$

The outlet temperatures from the heat exchanger, $T_{h,out}'$ and $T_{c,out}'$, are for the reduced heat capacity rates, $(1-X)C_h'$ and $(1-Y)C_c'$, respectively, and obviously, for new values of Ntu' and $C^{*'}$, and, consequently, ϵ'. With these values, the temperatures $T_{h,out}'$ and $T_{c,out}'$ can be obtained from equations (3) and (4) in terms of the inlet temperatures. Substitution of these equations into equations (9) and (10) provides the following expressions for the temperatures after by-pass:

$$T_{h,out} = \left[1 - \epsilon' \frac{C_{min}'}{C_h'} \right] T_{h,in} + \epsilon' \frac{C_{min}'}{C_h'} T_{c,in} \tag{11}$$

and $\qquad T_{c,out} = \epsilon' \dfrac{C_{min}'}{C_c'} T_{h,in} + \left[1 - \epsilon' \dfrac{C_{min}'}{C_c'} \right] T_{c,in} \tag{12}$

where $\quad \epsilon' = fn(Ntu',\ C^{*'}) \tag{13}$

$\qquad Ntu' = U'A/C_{min}' \tag{14}$

$\qquad C^{*'} = C_{min}'/C_{max}' \tag{15}$

and $\qquad C_{min}' = min[(1-X)C_h',\ (1-Y)C_c'] \tag{16}$

Hence if by-pass can achieve the specified outlet temperatures of the original design, the following equalities must be satisfied:

$$\epsilon \, \frac{C_{min}}{C_h} = \epsilon' \, \frac{C'_{min}}{C'_h} \tag{17}$$

$$\text{and} \quad \epsilon \, \frac{C_{min}}{C_c} = \epsilon' \, \frac{C'_{min}}{C'_c} \tag{18}$$

An Algorithm for Determining the By-Pass for Various Changes in Process Conditions

An algorithm has been developed to calculate the by-pass flow of either one or both streams to satisfy specified outlet conditions for a previously designed heat exchanger [3]. A flow diagram of the main program is shown in Figure 3.

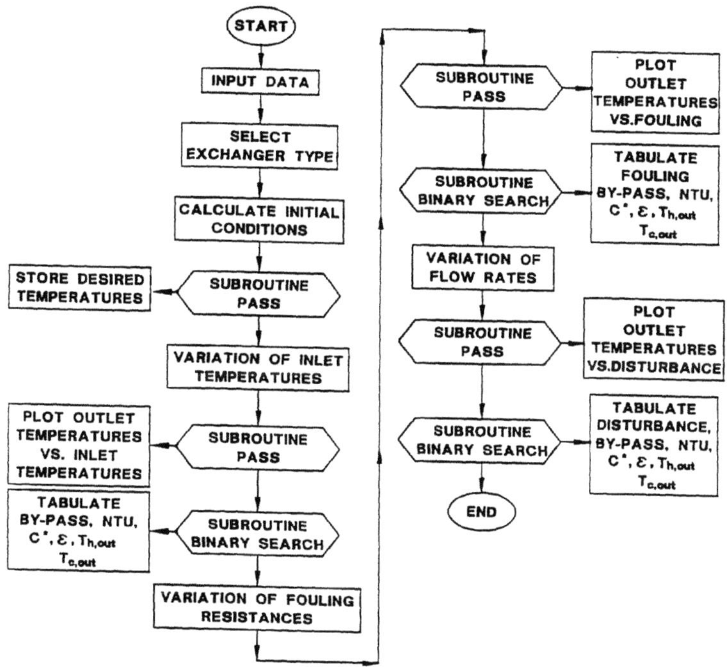

Fig. 3 Main Program Flow Diagram

The exchanger type is initially chosen, and the normal construction information required to predict the outlet temperatures from the specifications of flowrates and inlet temperatures are used in subroutine PASS. Appropriate correlations for predicting the heat transfer coefficients on the tube and shell sides must be included in PASS. The various changes in process conditions are then chosen: inlet temperatures, fouling coefficients or flowrates. For each type of process condition change, the resulting outlet temperatures of the particular exchanger are evaluated using subroutine PASS. These may be plotted against the process change. Subroutine BINARY SEARCH is then used to find whether by-pass flow can provide the required outlet temperatures. All possible conditions that can be controlled using a by-pass flow are tabulated along with the by-pass flow, various thermal parameters and the final outlet temperatures after combination of the by-pass flow and the flow through the exchanger. More detailed description of the algorithm and the various subroutines may be found in reference [3].

<u>Example Illustrating the Use of the By-Pass Algorithm</u>

The example has been taken from Holland et al [4], and it is the design of a 1:1 double pipe countercurrent heat exchanger in which it is required to cool 1.512 kg/s of ethylene glycol from 355.4 K down to 341.5 K and to heat 0.867 kg/s of toluene from 300 K up to 336 K. The ethylene glycol is on the tube side. The details of the design data are listed in Table 1. The relevant ϵ-Ntu parameters are as follows: $C_{min} = C_c = 1553.5$ J/s K, $C_{max} = C_h = 4038.6$ J/s K, $C^* = 0.385$, $U = 617.8$ W/m^2 K, Ntu $= 1.239$, $\epsilon = 0.65$. The by-pass requirements for the control of the outlet temperatures at $T_H = 341.5$ K and $T_C = 336$ K are $\epsilon C_{min}/C_h = 0.25$ and $\epsilon C_{min}/C_c = 0.65$ respectively.

Space precludes investigation of all possible process condition disturbances but reference [3] includes the following: variation of inlet temperatures, stream flowrates and fouling conditions.

One of the more complicated investigations is reported here: variation of both stream flowrates concurrently. The outlet temperatures $T''_{h,out}$ and $T''_{c,out}$, from the exchanger without by-pass but subject to the changing flowrates are shown in Figure 4 along with the desired outlet temperatures. The stream flowrate variations $(1+Z)\dot{M}$ cover the range from $Z = -0.8$ to $+0.8$ of the specified flowrates.

At values of disturbances more than zero, the outlet temperature of the cold stream decreases below the desired value, while the outlet temperature of the hot stream exceeds the desired value, therefore by-passing neither one nor both streams will help the streams to achieve their target temperatures.

Design Parameters	Tube Side Ethylene Glycol	Shell Side Toluene
Mass flow rate, kg/s	1.512	0.8669
Inlet temperature, K	355.4	299.9
Specific heat, J/kg K	2671	1792
Viscosity, Ns/m^2	0.00334	4.42×10^{-4}
Density, kg/m^3	1071	842.4
Thermal conductivity, $W/[m^2(K/m)]$	0.248	0.147
Fouling coefficient, W/m^2 K	5678	5678

Length = 27.28 m

Area = 3.115 m^2

Thermal conductivity = 45 $W/[m^2(K/m)]$

Inside diameter of tube = 0.03636 m

Outside diameter of tube = 0.0429 m

Inside diameter of shell = 0.05383 m

Outside diameter of shell = 0.06032 m

Tube flow area = 0.001038 m^2

Annular flow area = 8.33×10^{-4} m^2

Table 1 Design data

At values of disturbances less than zero, the outlet temperature of the cold stream is higher than the desired value, while the outlet temperature of the hot stream is below the desired value, therefore, by-passing either one or both streams could help to maintain the outlet temperatures at the desired values. These regions are shaded in Figure 4.

The flow of the hot stream is shown to be changed from turbulent to transition and this is clearly shown on the plot after the point of intersection of the two curves. Table 2 lists the values of the Reynolds numbers of both streams at various values of the stream disturbances together with the ϵ, Ntu and C^* values. C^* remains constant for all values of disturbances considered since both streams are disturbed by the

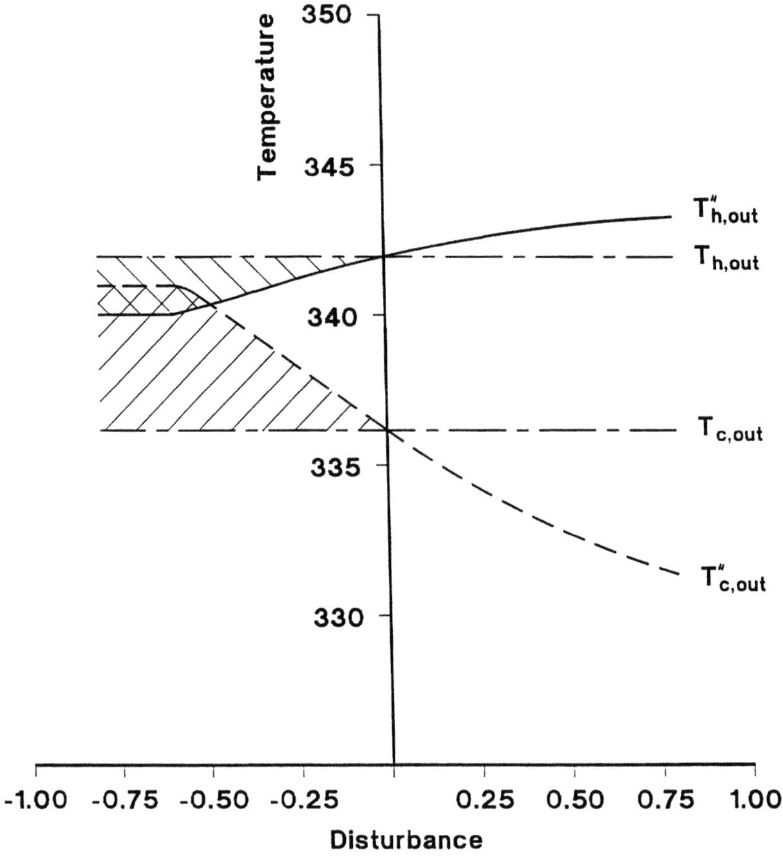

Fig. 4 Temperature vs disturbance of both streams

same amount. Ntu and ϵ are less than the desired values for positive disturbances and higher for negative disturbances.

In Table 3 the results are depicted for control of the cold outlet temperature using by-pass of the cold stream for four values of disturbance. For this arrangement, as the size of the negative disturbance increases, the amount of by-pass flow also becomes larger. The values of $C^{*'}$ decrease because the cold stream has the minimum flowing heat capacity. Although the overall heat transfer coefficient will become smaller, the flowing heat capacity falls more and hence the value of Ntu' increase, and so do the values of ϵ'. At the disturbance $Z = -0.8$, the trend has changed because the flow of the hot stream has fallen into the transition region (see Table 2 and also the effect on the outlet temperatures in Figure 4). The product $\epsilon' \, C'_{min}/C_c$ is always close to 0.65 and the outlet temperatues are both at their desired values. The hot outlet temperature from the exchanger is at the desired value, because the disturbance to both streams is identical. For unequal disturbances this would not be so.

Dist. Z	Re_h	Re_c	C^*	Ntu	ϵ
0.8	28534	104936	0.385	0.962	0.568
0.6	25363	93276	0.385	1.016	0.585
0.4	22193	81616	0.385	1.078	0.605
0.2	19022	69957	0.385	1.151	0.626
0.0	15852	58297	0.385	1.239	0.650
-0.2	12681	46638	0.385	1.349	0.678
-0.4	9511	34978	0.385	1.467	0.704
-0.6	6340	23319	0.385	1.606	0.733
-0.8	3170	11659	0.385	1.607	0.733

Table 2   Disturbance to both streams

Dist. Z	By-pass Y	$C^{*'}$	Ntu'	ϵ'	$T_{c,out}$ K	$T_{h,out}$ K	$\epsilon' \frac{C'_{min}}{C_c}$
-0.2	0.073	0.357	1.429	0.701	335.97	341.52	0.649
-0.4	0.125	0.337	1.619	0.744	336.03	341.49	0.651
-0.6	0.175	0.317	1.851	0.788	335.98	341.53	0.650
-0.8	0.173	0.318	1.840	0.786	335.98	341.54	0.650

Table 3   Disturbance to both streams   By-pass on cold stream

Table 4 lists the results for control of the hot outlet temperature using by-pass of the hot stream for the same four values of disturbance used in the previous discussion for the cold by-pass. For this arrangement, the hot stream has the larger flowing heat capacity rate. The values of $C^{*'}$ now increase for larger negative values of the disturbance, Z, because with by-pass the value of C_{max} will be less whereas C_{min} remains fixed. The values of Ntu' also increase and the values of ϵ' remain fixed at 0.65. At disturbance $Z = -0.8$, the by-pass flow X is less because again the tube side flow has become transitional. Again, both the desired outlet temperatures have been obtained and the by-pass criterion of $\epsilon' C'_{min}/C_h = 0.25$ has been achieved.

Hence for the case of equal disturbances to each stream it is possible to control both outlet temperatures at their desired values using by-pass on either the hot or cold streams for negative disturbances. For positive disturbances, by-pass cannot be employed to control the outlet temperatures.

Dist. Z	By-pass X	$C^{*'}$	Ntu'	ϵ'	$T_{h,out}$	$T_{c,out}$	$\epsilon'\dfrac{C'_{min}}{C_h}$
-0.2	0.130	0.442	1.275	0.650	341.53	335.98	0.250
-0.4	0.200	0.481	1.304	0.651	341.51	336.03	0.250
-0.6	0.250	0.513	1.323	0.650	341.53	335.98	0.250
-0.8	0.170	0.464	1.287	0.649	341.54	335.92	0.250

Table 4   Disturbance to both streams   By-pass on hot stream

Conclusions

By-pass of either stream around a heat exchanger may be used to control the outlet temperatures at desired values. The ϵ-Ntu methodology is an ideal tool for the implementation of an algorithm for investigation of the potential for employing by-pass control. If by-pass is possible then following equalities must be satisfied:

$$\epsilon \frac{C_{min}}{C_h} = \epsilon' \frac{C'_{min}}{C'_h} \quad \text{and} \quad \epsilon \frac{C_{min}}{C_c} = \epsilon' \frac{C'_{min}}{C'_c}$$

It is impossible to predict *a priori* whether by-pass control will be able to achieve the desired values of outlet temperatures and each exchanger must be investigated individually. The algorithm discussed in this presentation will be very helpful in these investigations.

Nomenclature

C	flowing heat capacity rate, $\dot{M}C_p$	$W/m^2\ K$
C^*	ratio of flowing heat capacities, C_{min}/C_{max}	
X	by-pass flow for hot stream, see figure 2	
Y	by-pass flow for cold stream, see figure 2	
Z	flow disturbance, $C' = (1+Z)C$	
ϵ	thermal effectiveness, equations (1) and (2)	

Subscripts

c	cold
h	hot
in	inlet
max	maximum
min	minimum
out	outlet

References

1. Shinskey, F.G.: Process Control Systems, McGraw-Hill, New York, 1976.

2. ESDU, Heat Transfer Sub-Series 5a: Heat exchangers: Effectiveness–Ntu relationships. Item 89016 – August 1989, Item 86018 – July 1986, Item 88021 – November 1988. ESDU International plc, London, N1 6UA.

3. Abid, I.M.: Mathematical Modelling of the Operation and Control of Heat Exchange Equipment. PhD. Thesis, University of Bradford, UK, 1991.

4. Holland, R.A.; Moores, R.M.; Watson, F.A.; Wilkinson, J.K.: Heat Transfer. Heinemann Educational Books, London, 1970.

Shell and Tube Heat Exchangers

Effects of Unequal Transfer Area
in Multi-Pass Heat Exchangers

P.J. HEGGS and I.M. ABID
Department of Chemical Engineering
University of Bradford, BRADFORD, BD7 1DP, U.K.

Summary

This investigation has revealed that unequal multi-pass 1:2 and 1:3 heat exchanger arrangements with the same sized tubes in each pass and the same total surface area are less efficient thermally than equal pass systems. In addition, the tube side pressure drop is least for the equal pass arrangements.

Introduction

Multi-pass heat exchangers are frequently encountered in the process industries, either as single units or in a series of exchangers. Thermodynamically a multi-pass heat exchanger can never better a 1:1 countercurrent exchanger with the same number of transfer units (Ntu). Using multi-pass configurations does provide higher tube-side velocities and, hence, larger tube side heat transfer coefficient and overall coefficients. This will lead to larger values of Ntu for the same transfer area and this compensates for the lower thermodynamic performance of the multi-pass configuration. The increased multi-pass velocity and the longer tube-side flow path will create larger pressure drops than the 1:1 arrangement.

The majority of design charts/formulae for multi-pass arrangement, whether ϵ-Ntu or F-factor, assume that the transfer area is equally distributed between each pass. Actual exchangers will have slightly unequal area for each pass due to mechanical construction restrictions, although for U-tube bundles, the area is equally distributed. Any multi-pass arrangement will have at least one pass operating in co-current flow and thus this pass will be less thermally efficient than an equivalent countercurrent pass. Intuitively one would expect that unequal transfer area between passes with more countercurrent flow area would be thermally more effective than equal area distribution. For instance, a 1:2 configuration has 50% countercurrent flow and 50% co-current flow, whereas a 1:3 configuration can have 67% countercurrent flow and 33% co-current flow. Present design charts show that the 1:3 arrangement is thermodynamically the best for either even numbers of passes or odd numbers of passes (greater than 3) at the same value of

NTU. What is of interest, is the effect of more countercurrent flow area in either of the 1:2 and 1:3 arrangements.

Gardner [1] considered the mean temperature difference in unbalanced-pass 1:2 and 1:4 exchangers and presented his results in terms of correction factors for use with the logarithmic mean temperature difference of a truly countercurrent 1:1 exchanger. This is an ideal approach for sizing exchangers but requires iteration for performance type calculations. He did present examples that showed an unbalanced-pass unit was less expensive than either a 2:2 pass exchanger or two even balanced 1:2 exchangers in series, even though the unbalanced-pass unit required more surface area.

Roetzel [2] and Roetzel and Spang [3,4] have more recently published results for the thermal calculation of single pass shell and multi-pass (even or odd) tube exchangers. They have solved the sets of ordinary differential equations and presented charts in the form of single curves of Ntu values plotted against the thermal ratios, Φ_1 and Φ_2, and the ratios of the flowing heat capacities, C_1 and C_2. They have proposed [4] an unbalanced construction with a large single tube for the co-current pass(es). This greatly reduces the Ntu for the co-current pass(es) and thus provides more countercurrent Ntu.

This present work solves the multi-pass system in the ϵ-Ntu-C^* methodology and considers the effects on both thermal and pressure drop performance of a fixed tube bundle with 1, 2 and 3 tube passes and a single shell pass.

<u>Mathematical Representation</u>

The following equations provide a generalised formulation for 1 pass shell and multi-pass tube exchanger. The normal assumptions are: well mixed fluids, constant heat transfer coefficients per pass, constant physical properties, no axial conduction effects either in the fluids or the walls and no heat loss or gain from the ambient.

$$\text{Shell side:} \quad \pm \frac{C_s}{C_{min}} \frac{d\theta}{dy} = \theta \sum_{i=1}^{n} Ntu_i - \sum_{i=1}^{n} Ntu_i \emptyset_i \tag{1}$$

$$\text{Tube side:} \quad \frac{C_t}{C_{min}} \frac{d\emptyset_i}{dy} = (-1)^{i+1} Ntu_i \, (\theta - \emptyset_i) \, , \, i = 1 \text{ to } n \tag{2}$$

where the positive sign in equation (1) implies the shell side fluid enters the exchanger at y = 1.0, referred to as opposite end arrangement, and the negative sign in equation (1) implies the shell side fluid enters at y = 0.0, referred to as same end arrangement, see Figure 1. The tube side fluid always enters at y = 0.0. Equations (1) and (2) hold for all positive values of n. If n = 1, the equations represent 1:1 arrangements; n = 2, 1:2 arrangements, etc.

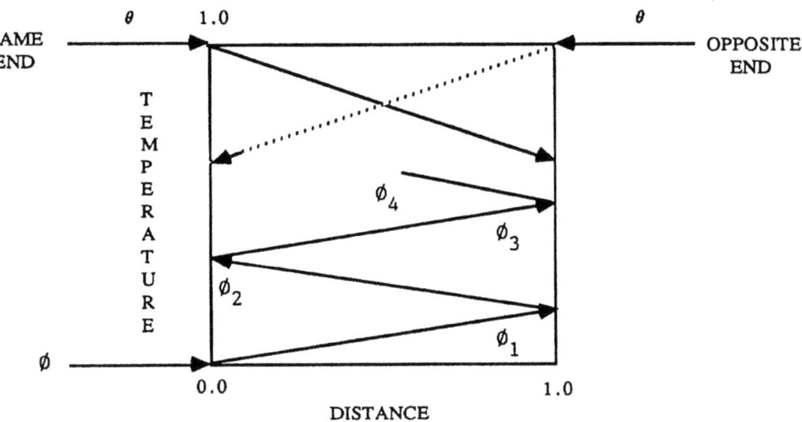

Fig. 1 Schematic diagram of a multipass exchanger

The boundary conditions are as follows:

Inlet temperatures:

      Opposite end: at y = 0.0, ϕ_1 = 0.0 and at y = 1.0, θ = 1.0   (3)

      Same end: at y = 0.0, ϕ_1 = 0.0 and θ = 1.0   (4)

Header temperatures:

      at y = 0.0 and for n > 3, ϕ_{i-1} = ϕ_i for i = 3,5,7,...,n   (5)

      at y = 1.0 and for n > 2, ϕ_{i-1} = ϕ_i for n = 2,4,6,...,n   (6)

The dimensionless variables and groups in equations (1) and (2) are defined as follows:

      y = 1/L, normalised length   (7)

      θ = $(T_s - T_{t,in})/(T_{s,in} - T_{t,in})$, normalised shell side temperature,   (8)

      ϕ_i = $(T_{t,i} - T_{t,in})/(T_{s,in} - T_{t,in})$, normalised tube side temperature   (9)

56

$$\text{Ntu}_i = (UA)_i/C_{min}, \text{ number of transfer units for pass } i, \qquad (10)$$

$$C = \dot{M}C_p, \text{ flowing heat capacity}, \qquad (11)$$

$$\text{and} \quad C_{min} = \min[C_s, C_t], \text{ minimum flowing heat capacity}. \qquad (12)$$

The solutions to the equations (1) to (6) will provide the outlet temperatures of the shell side and tube side fluids, which then allows the thermal effectiveness to be evaluated, i.e.

$$\epsilon = \text{Heat transferred/Thermodynamic maximum transfer} \qquad (13)$$

$$= \frac{C_t}{C_{min}} \frac{(T_{t,out} - T_{t,in})}{(T_{s,in} - T_{t,in})} = \frac{C_s}{C_{min}} \frac{(T_{s,in} - T_{s,out})}{(T_{s,in} - T_{t,in})} \qquad (14)$$

$$= \frac{C_t}{C_{min}} \varnothing_{n,out} = \frac{C_s}{C_{min}} 1 - \theta_{out} \qquad (15)$$

Space precludes the details of the solutions. Analytical solutions were obtained where possible using the characteristic equation methodology for sets of ordinary differential equations. Numerical solutions employing central difference approximations for the derivatives were obtained for all unbalanced systems up to $n = 3$. These were shown to be stable, compatible and convergent by comparison with the analytical solutions for equal pass arrangements. Full details may be found in I.M. Abid's thesis [5].

Effectiveness-Ntu Charts for Unequal Pass 1:2 and 1:3 Exchangers

The most complete set of effectiveness-Ntu charts has been published by ESDU [6], but does not include any information on unequal pass arrangements.

For 1:2 equal pass exchangers theory dictates that the results are identical for opposite and same end arrangements, and irrespective whether C_{min} occurs on either the shell or tube side. The same has been found for unequal 1:2 pass exchangers.

Figures 2, 3 and 4 are plots of effectiveness for unequal 1:2 pass exchangers against Ntu relationships for fixed values of C^* over the range 1.0 (0.1) 0.0 for $2/3$ counter current flow on the 1st pass and $1/3$ co-current flow on the second pass, 80% and 20%, and finally, 90% and

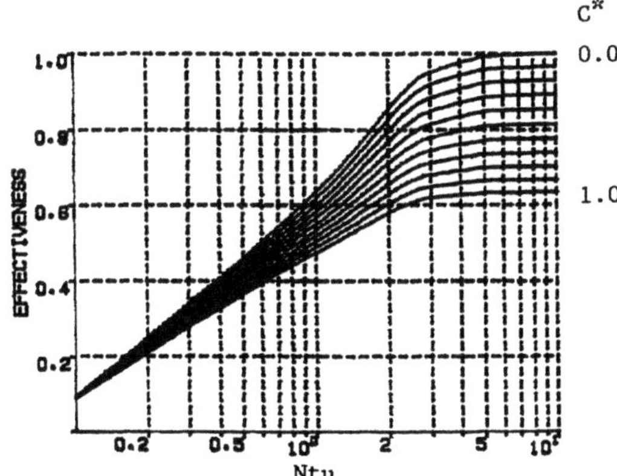

Fig. 2 Opposite end 1:2 exchanger with $Ntu_1 = {}^2/3\ Ntu$ and $Ntu_2 = {}^1/3\ Ntu$

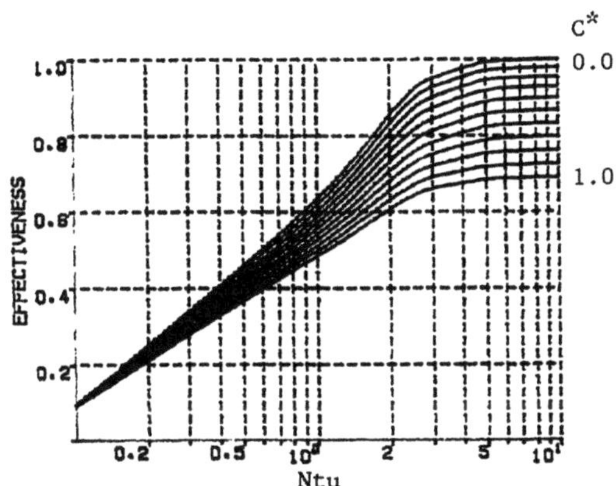

Fig. 3 Opposite end 1:2 exchanger with $Ntu_1 = 0.8\ Ntu$ and $Ntu_2 = 0.2\ Ntu$

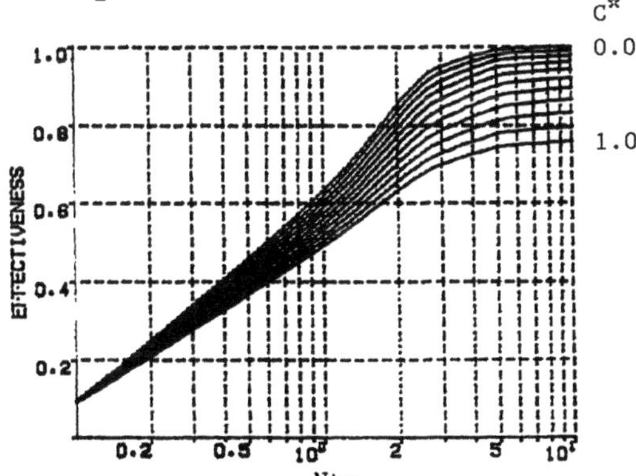

Fig. 4 Opposite end 1:2 exchanger with $Ntu_1 = 0.9\ Ntu$ and $Ntu_2 = 0.1\ Ntu$

58

10%, respectively. For fixed values of Ntu, the effectiveness increases
for larger values of countercurrent flow, as is to be expected.

Figures 5, 6 and 7 are plots of effectiveness against Ntu for unequal
pass 1:3 exchangers for the opposite end arrangements. Same end
arrangements are not considered, because this would have two passes in
co-current flow and the configuration is not thermodynamically efficient.
Figure 5 is for equal passes and it is interesting to note that for
values of Ntu greater than 3, larger values of effectiveness are obtained
if C_{min} is on the tube side. The same phenomenon is also found for the
unequal pass configurations. The effectiveness of the unequal pass
arrangements for the same value of Ntu is higher than the equal pass
systems.

It is important however to remember that for fixed flowrates of the tube
side fluid, variation of the transfer area between passes will have
corresponding effects on the cross-sectional area of flow for each pass.
Hence the tube side heat transfer coefficients and pressure drops will
change also, and it is impossible to hypothesise whether unequal pass
arrangements will be better than equal pass systems by simply considering
plots similar to Figures 2 to 7.

<u>Investigation of a Single Tube Bundle for Various Exchanger Arrangements</u>
The investigation is centred around a design problem presented by Holland
[7]. The design considers a 1:2 equal pass exchanger for the cooling of
18.9 kg/s of ethylene glycol from 394 K to 378 K on the tube side and the
heating of 13.9 kg/s of toluene from 300 K to 336 K on the shell side.
The final design required 196 steel tubes of 14SWG thickness having an
outside diameter of $\frac{3}{4}$ in. The tubes are 8 ft. long and are laid out on a
1 in. triangular pitch. The shell contains 25 per cent cut segmental
baffles spaced 6 in. apart.

Table 1 lists the design parameters for the 1:2 arrangement with equal
passes on the tube side. The thermal effectiveness for the required duty
is 0.481 and with the ratio of the flowing heat capacities, C^* being
0.462. This is a relatively easy duty for a 1:2 exchanger, because
maximum thermal effectiveness is 0.78 for this arrangement and C^* equal
to 0.462.

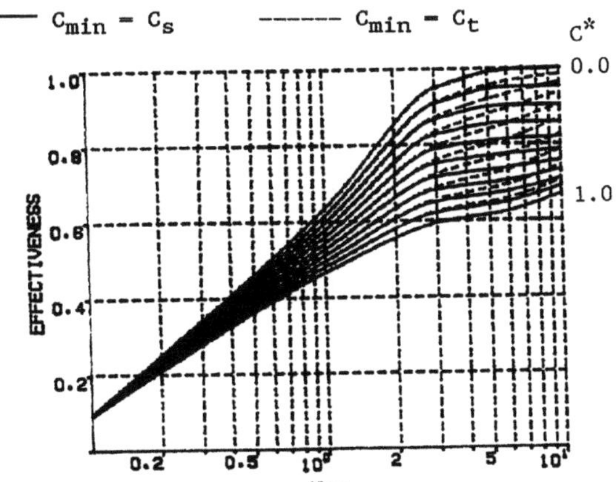

Fig. 5 Opposite end 1:3 exchanger with equal passes
$Ntu_1 = Ntu_2 = Ntu_3 = {}^1\!/_3\,Ntu$

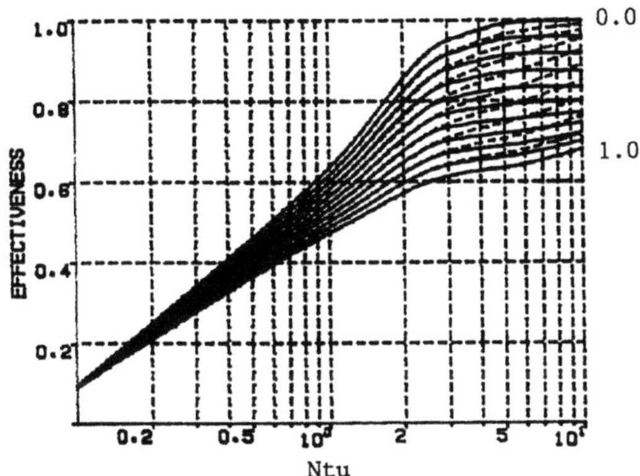

Fig. 6 Opposite end 1:3 exchanger with $Ntu_1 = Ntu_3 = 0.4\,Ntu$
and $Ntu_2 = 0.2\,Ntu$

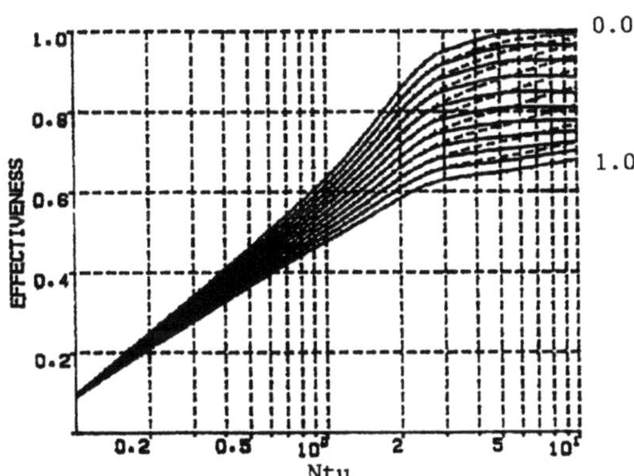

Fig. 7 Opposite end 1:3 exchanger with $Ntu_1 = 0.4\,Ntu$,
$Ntu_2 = 0.1\,Ntu$ and $Ntu_3 = 0.5\,Ntu$

A_t , m^2	22.5
$(\dot{M}c_p)_s$, J/s K	24908.8
$(\dot{M}c_p)_t$, J/s K	53959.5
c^*	0.462
α_t , W/m^2 K	1601
R_w , $(W/m^2$ K$)^{-1}$	0.00004
α_s , W/m^2 K	1630
U , W/m^2 K	871.6
Ntu	0.787
ϵ	0.481
Δp_t , N/m^2	8234

Table 1. Design data from Holland for 1:2
shell and tube heat exchanger

The tube bundle of the design has been used throughout the rest of the
investigation, and only the flow paths through the bundles have been
altered. The shell side conditions have remained fixed and it is assumed
that the headers and the pass partition plates can be changed to provide
various flows through the tubes. ,For each arrangement, the tube side
coefficient and pressure drop have been recalculated for the fluid flow
through each pass in accordance with the design correlations in Holland
[4].

The arrangements considered are as follows: 1:1 countercurrent, 1:2 same
end arrangement with the surface area divided equally, $2/3$ counter- and
$1/3$ co-current, 80% counter- and 20% co-current, 90% counter- and 10% co-
current, and 1:3 opposite end arrangement with the surface area divided
equally, 40% 1st counter-, 20% 2nd co- and 40% 3rd counter-current pass,
then 60%, 20% and 20%; 45%, 10% and 45%, and finally 30%, 10% and 60%.
The final results of the performance calculations for the various flow
arrangement are listed in Table 2. Included are the values for the
individual pass Ntu values, the total value of Ntu, the thermal
effectiveness and the tube side pressure drop.

The 1:1 countercurrent exchanger effectiveness of 0.373 is much lower than design value for the 1:2 equal pass system. This is to be expected because the velocity in the tubes will only be half that of the base case. The Ntu value is only 0.491 due to the lower tube side coefficient, but it is important to note the dramatic reduction in the pressure drop.

Exchanger type	Area distribution	Ntu_i	Ntu	ϵ	Δp_t N/m^2
1:1	countercurrent	–	0.491	0.373	1231
1:2	0.5 counter 0.5 co	0.394 0.394	0.787	0.481	8234
1:2	2/3 counter 1/3 co	0.436 0.309	0.745	0.466	10811
1:2	0.8 counter 0.2 co	0.465 0.220	0.685	0.442	22039
1:2	0.9 counter 0.1 co	0.480 0.127	0.607	0.410	68990
1:3	1/3 counter 1/3 co 1/3 counter	0.309 0.309 0.309	0.927	0.528	24958
1:3	0.4 counter 0.2 co 0.4 counter	0.346 0.219 0.346	0.911	0.526	32324
1:3	0.6 counter 0.2 co 0.2 counter	0.417 0.220 0.220	0.857	0.508	43443
1:3	0.45 counter 0.10 co 0.45 counter	0.371 0.129 0.371	0.871	0.515	77384
1:3	0.3 counter 0.1 co 0.6 counter	0.289 0.129 0.418	0.836	0.503	80685

Table 2. Ntu, effectiveness and tube-side pressure drop for the various shell and tube exchanger arrangements

None of the unequal 1:2 configurements are better than the equal pass case. The total number of transfer units, Ntu, is always less than the equal pass value. The transfer areas have been proportioned in a linear manner, but the overall heat transfer coefficients for each pass do not vary linearly. The tube side coefficient varies according to $u^{0.8}$, and the evaluation of the overall coefficient includes the unaltered resistances for the tube wall and the shell side convection. The greater the difference between the countercurrent area and co-current area, the worse the situation becomes. The pressure drop on the tube side also increases considerably for unequal passes. The pressure drop is roughly dependent upon the square of the tube-side velocity, and thus the smaller cross-sectional area of flow dominates the total pressure drop. The equal pass 1:3 exchanger has an effectiveness value approximately 10% greater than the base case, but the tube side pressure drop has increased by 3 times. All of the unequal 1:3 arrangements are worse than the equal 1:3 system both thermally and with respect to the pressure drop on the tube-side. The reasons for these results are the same as those for the 1:2 exchanger investigation.

Conclusions

The investigation has revealed that multi-pass exchangers with the same sized tubes in each pass should be constructed with equal transfer areas and, consequently, equal cross-sectional areas of flow. Unequal pass areas result in lower thermal effectiveness values and larger tube side pressure drops. Perhaps the only way to achieve better thermal performance and lower pressure drops is to use larger sized tubes on the co-current pass(s), or even a large single tube, as proposed by Roetzel and Spang [4].

Nomenclature

C	flowing heat capacity, $\dot{M}C_p$,	W/m^2 K
F	correction factor for non 1:1 heat exchangers,	–
n	number of tube side passes,	–
Δp	pressure drop	N/m^2
y	normalised exchanger length,	–
e	voidage,	–
θ	normalised shell side temperature,	–
$\emptyset$	normalised tube side temperature,	–
Φ	normalised thermal ratio,	–

References

1. Gardner, K.A.: Mean temperature difference in unbalanced pass exchangers. Ind. Eng. Chem. 33 (1941) 1215–1223.

2. Roetzel, W., Thermische Berechnung von dreigängigen Rohrbündelwärmeübertragern mit zwei Gegenstromdurchgängen gleicher Grösse. Wärme- und Stoffübertragung 22 (1988) 3–11.

3. Roetzel, W.; Spang, B.: Analytisches verfahren zur Thermischen berechnung mehrgängiger Rohrbündel-Wärmeübertrager. Fortschr.-Ber VDI 19 (18) (1987).

4. Roetzel, W.; Spang, B.: Thermal calculation of multipass shell and tube exchangers. Chem. Eng. Res. Des. 67 (1989) 115–120.
5. Abid, I.M.: Mathematical Modelling of the Operation and Control of Heat Exchange Equipment. PhD. Thesis, University of Bradford, UK, 1991.

6. ESDU, Heat Transfer Sub-Series 5a: Heat exchangers: Effectiveness-Ntu relationships. Item 89016 – August 1989, Item 86018 – July 1986, Item 88021 – November 1988. ESDU International plc, London, N1 6UA.

7. Holland, R.A.; Moores, R.M.; Watson, F.A.; Wilkinson, J.K.: Heat Transfer. Heinemann Educational Books, London, 1970.

Variable Pitch Tube Layout Concept for Shell and Tube Heat Exchanger

S. KRISHNAN & G.K. SADEKAR

Larsen & Toubro Limited
Powai Works
Bombay 400 072

Summary

A special tube layout configuration has been developed to achieve a relatively uniform heat transfer rate over the cross section on the shellside of a shell and tube heat exchanger. For large size heat exchangers, use of disc and doughnut baffles will result in wide velocity variation of 4.5:1 and heat transfer coefficient variation of 2.5:1 at any cross section. With the variable pitch layout, where tubes are arranged at reducing pitch as one moves from inner zone to outer periphery, heat transfer coefficient variation is reduced to 1.5:1 at any cross section of exchanger. The flow distribution has been studied using network analysis and is used to evaluate different configurations.

Introduction

Shell and tube heat exchanger configurations and design methods have undergone several improvements to cater to varied range of applications, specific process requirements and larger size of equipments. The shellside configuration, a major design aspect, is governed by parameters such as desirable heat transfer rate, minimisation of pressure losses and the need to obtain uniform flow distribution. In addition to these parameters, certain applications call for specific process requirements to be satisfied. One such parameter is uniform heat transfer rate at any cross section of the heat exchanger. With large size heat exchangers, such requirements cannot be satisfied by usual configurations.

A number of alternative design configurations can be considered, but the selection of final configuration is governed by parameters such as relative effectiveness, ease of manufacture, accuracy of performance prediction etc.

Design Strategy

The need to obtain more uniform heat transfer rate across any cross section of heat exchanger implies that all tubes should be in the zone of similar flow characteristics. Thus the exchanger should be designed with no tubes in the window region of baffles. This will have effect on the heat exchanger shell diameter. Also the window flow areas should be sufficiently large to avoid excessive pressure losses.

The baffle configuration to be chosen is a function of the various parameters discussed earlier. The desired heat transfer rate and pressure losses can be obtained with appropriate baffle spacing for a given baffle type[1]. Selection of baffle type depends on the desired flow pattern and extent of uniform flow distribution giving rise to less stagnant zones in addition to required heat transfer rate and pressure losses. This aspect becomes more significant when the shell diameter exceeds about 2500 mm resulting in a larger cross flow length compared to baffle pitch.

Considering these factors, it has been found that the use of disc and doughnut baffles gives better flow distribution and less stagnant areas as compared to single and double segmental baffles. In a few cases one can resort to axial flow with the use of grid baffles to obtain good flow distribution. However, this may not be suitable for all applications as heat transfer rate becomes independent of baffle configuration.

The disc and doughnut baffle configuration has radial flow as compared to one dimensional flow for single and double segmental baffle configuration[2]. When the flow traverses from one end to the other end of single or double segmental baffles the flow area does not change significantly. But with disc and doughnut baffles flow area changes continuously as the fluid moves from inner to outer zone. This variation in flow area will result in velocity and heat transfer rate variation across the cross section. Typically for an exchanger with a diameter of 5000 mm, the variation in velocity is of the order of 4.5:1 and heat transfer rate, 2.5:1. Thus, the process requirement of minimising variation in heat transfer rate at any cross section of heat exchanger will not be satisfied when disc and doughnut baffles are used.

Various alternatives can be considered in order to reduce the variation in velocity and heat transfer with disc and doughnut baffles, e.g.
a) Varying tube hole clearance throughout the bundle
b) Providing additional flow areas at selected regions by drilling holes in baffles

c) Variable tube pitch throughout the bundle thereby directly
   reducing the flow area variation.

The above mentioned ideas were evaluated and the concept of
variable tube pitch layout was developed in order to minimise
velocity variation at any cross section while using disc and
doughnut baffles.

<u>Concept of Variable Pitch Layout</u>

The tubes in the exchanger are arranged with a pitch varying
from inner to outer radius. The tube pitch would be maximum
at the center and minimum at the outer edge. The pitch can
vary continuously at every row in radial direction to obtain
uniform flow area. Considering the manufacturing feasibility
and allowing for certain variation in heat transfer rate,
tubes can be arranged at variable pitch in certain
bands/regions in the exchanger. The tube pitch and the number
of variable pitch bands are to be evaluated based on the
application, complexity of manufacturing and permissible
variation in heat transfer rate.

This concept is illustrated for a typical problem in Fig.1,
wherein Fig.1A indicates the tubes arranged with uniform pitch
and Fig.1B, the tubes arranged with variable pitch in five
bands.

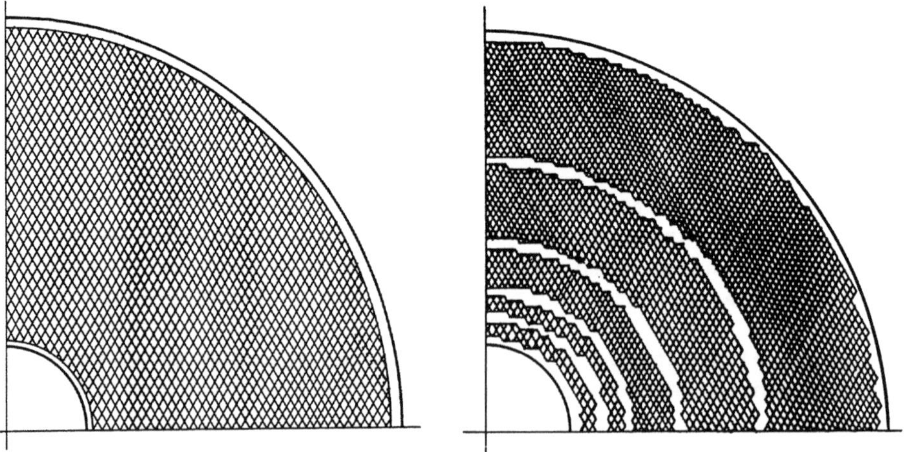

Fig.1A. Uniform pitch layout   Fig.1B. Variable pitch layout

As the exchanger has tubes arranged in variable pitch the standard programs available cannot be used for predicting flow distribution in the exchanger. A new computer program is therefore developed using 'Network Technique'.

<u>Network Analysis Program</u>

The flow pattern on the shellside of an exchanger with the presence of baffles gives rise to various flow streams in different paths e.g.

leakage stream in the orifice formed by the clearance between the baffle tube hole and the tube wall,
leakage stream between the baffle edge and shell wall,
main effective crossflow stream,
tube bundle bypass stream in the gap between the bundle and the shell wall and
bypass stream in pass partition lanes.

The shell and tube heat exchanger using variable pitch layout has disc and doughnut baffles. The flow pattern of disc and doughnut baffle eliminates the bundle bypass stream. The pass partition lane flow is absent due to single pass configuration on the tubeside. Furthermore, the process requirement calls for either welding the baffle to shell or minimising the clearance by other means and thereby eliminating the baffle to shell leakage.

Thus, the flow stream on the shellside consists of crossflow stream and flow in tube to baffle leakage path only. The analysis involves writing mass balance and pressure balance equations in the flow paths and solving the same to obtain flow and pressure distribution.

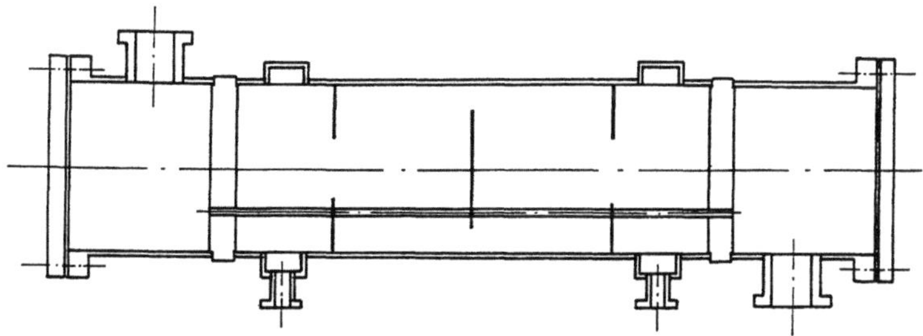

Fig.2. Shell and tube exchanger

The exchanger that has been modeled is shown in Fig.2. The complete exchanger has been considered for analysis with four flow passes on shellside. To account for variable pitch effect, the exchanger is divided into five bands in the tube region and two in the window regions. The centre point of each band forms a node. Fig.3 shows the nodal diagram representing the shellside of the exchanger with 25 nodes and 39 streams.

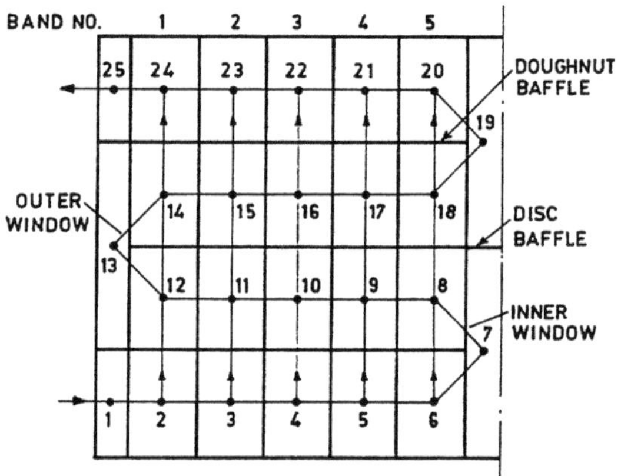

Fig.3. Nodal Diagram

The mass balance and pressure difference equations are written as follows:

Mass Balance Equations

Flow from any node i to i+1 is represented as F_iF_{i+1}. If we consider node 2, for example, this node receives flow from node 1 and sends flow to node 12 and 3. Mass balance equation can be therefore written as

$$F_1F_2 = F_2F_{12} + F_2F_3 \qquad \qquad \ldots(1)$$

Similar mass balance equations are written for every node with $F_{24}F_{25}$ = FLOW IN as boundary condition.

Pressure Difference Equations

Fluid flow is either in cross path, in leakage path or in a mixture of cross and window paths. Pressure difference equations based on the flow path can be written as follows

Case 1

Consider the pressure difference P_1-P_2. The flow path 1-2 is a crossflow path.

Pressure difference, P_1-P_2

$$= CF_1(L_1/d_o\rho s)(F_1F_2/A_1)^2 \qquad \ldots(2)$$

where CF_1 : Loss coefficient for zone between 1 and 2
      L_1 : Path traveled from 1 to 2
      F_1F_2 : Mass flow from 1 to 2
      A_1 : Mean superficial crossflow area between 1 and 2
      ρs : Fluid density
      do : Tube outside diameter

Case-2:

Consider pressure difference P_2-P_{12}. The flow 2-12 is a leakage path in tube hole.

Pressure difference, P_2-P_{12}

$$= (CN_1)(0.5)(F_2F_{12}/A_{L_1})^2/\rho s \qquad \ldots(3)$$

where CN_1 : No. of velocity heads
      F_2F_{12}: Mass flow from 2 to 12
      A_{L_1} : Leakage area in band 1

Case-3:

Consider pressure difference P_6-P_7. The flow path 6 to 7 consists of resistance in crossflow and window flow. The window flow resistance is divided as 50 % for path 6 to 7 and 50% for path 7 to 8.

Pressure difference, P_6-P_7

$$= (CF_6)(L_6/d_o\ \rho s)(F_6F_7/A_5)^2+(CN_{W1})(F_7F_8)^2/2 \qquad \ldots(4)$$

where CN_{W1} : Loss coefficient for inner window.

Similarly, pressure difference equations are written for every flow path depending on type of flow in the path.

These equations are non linear in nature and are linearised and solved by iteration method using Gaussian elimination technique for linearised equation at each iteration. The computer code has been developed and various configurations have been evaluated using the same. The flow chart of the program is given in Fig.4.

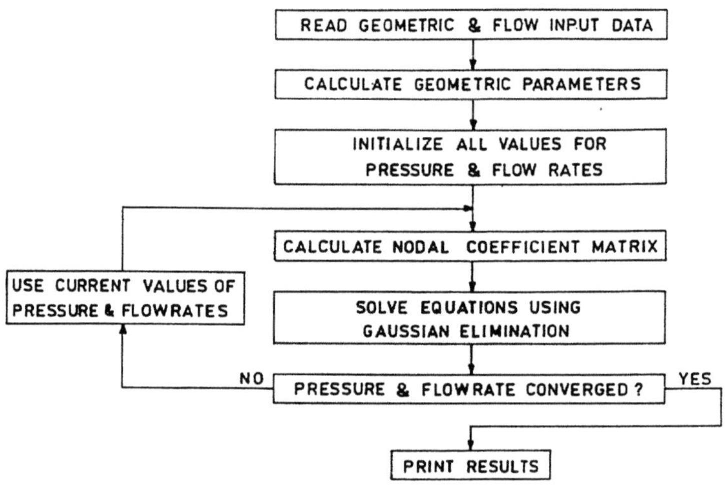

Fig.4. Flow Chart

Discussions

The network analysis program developed for the flow distribution has been used to analyze the variable pitch and uniform pitch layout. The velocity and the heat transfer coefficient variation has been estimated based on the flow distribution. The velocity at any point is calculated using the flow fraction at that point and crossflow area. The heat transfer coefficient variation is evaluated using standard correlations with only crossflow as effective stream[3].

Case-I: Flow distribution in variable pitch layout

The geometric and process data assumed is given in Table-1. The flow fraction in various paths is given in Fig.5 and the pressure distribution is shown in Fig.6.

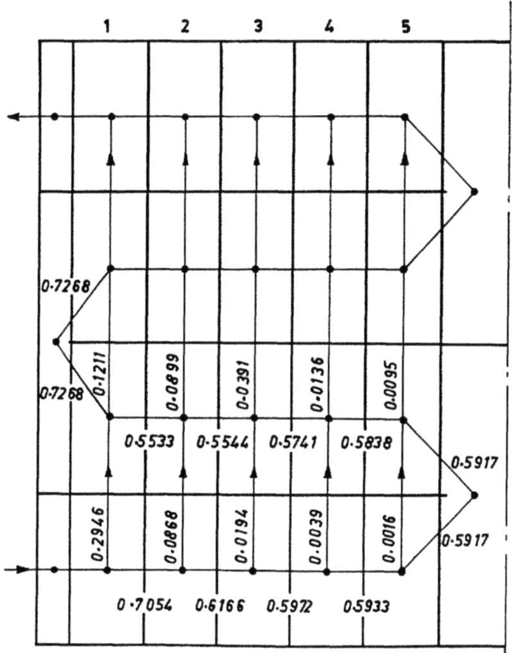

Fig.5. Flow Fraction

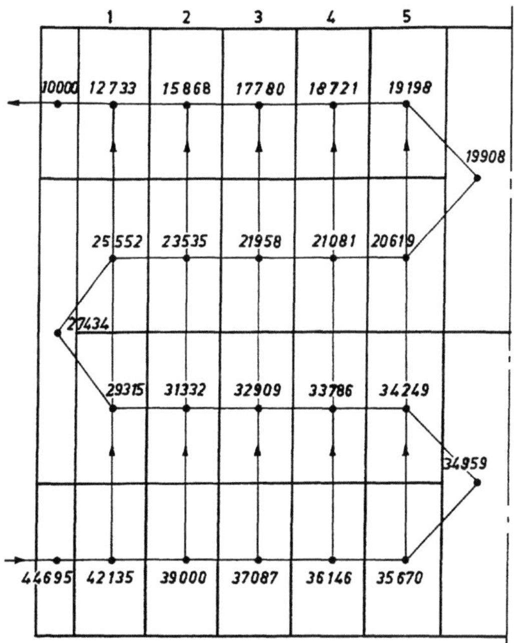

Fig.6. Pressure Distribution

Case-II: Flow distribution in uniform pitch layout

Similar exercise has been carried out for uniform pitch keeping all other parameters similar to case-I.

The velocity and heat transfer coefficient variation at a cross section as function of radius for Case-I and II is shown in Fig.7   and Fig.8.   The ratio of maximum to minimum heat transfer coefficient at any cross section is 2.5 for case I and 1.5 for case-II.

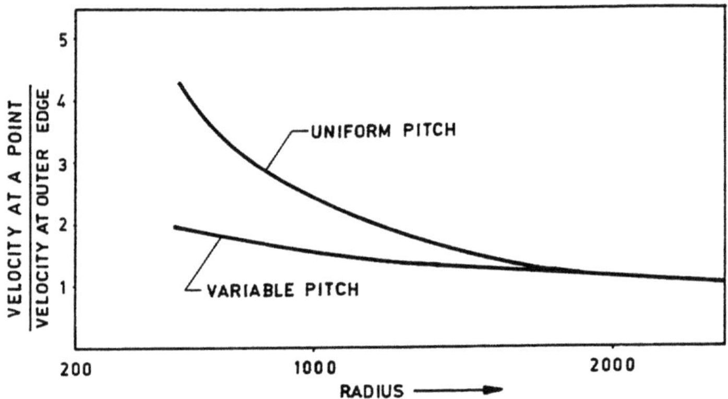

Fig.7. Velocity Variation

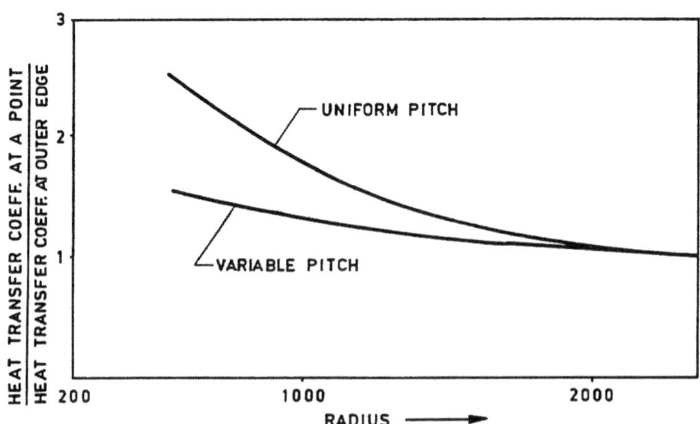

Fig.8. Heat Transfer Coefficient Variation

The average heat transfer coefficient in uniform pitch case is lower than that in the variable pitch configuration in addition to higher variation at a particular cross section. Thus, the choice of variable pitch layout is better than uniform pitch layout for application requiring more uniform heat transfer rate. The variations can be further reduced by selecting different combinations of the pitch and the number of bands. The results shown here are for a typical case and are given to illustrate the effectiveness of variable pitch layout.

Conclusion

1. The concept of variable pitch layout is useful to minimise heat transfer rate variation at any cross section in shell and tube heat exchanger using disc and doughnut baffles.

2. Network analysis technique is useful for prediction of flow distribution in heat exchanger and can be applied to account for variable flow geometries in the exchanger.

References

1. Haseler, L.E.; Hirst, C.; Murray, P.W.; Diaper, A.D.: Flow velocities on the shell and tube heat exchanger: Second UK National Conference on heat transfer. Vol.1 Sessions 1A-3D: Mechanical Engineering Publications (1988).
2. Founti, M.A.; Vafidis, C.; Whitelaw, J.H.: Shellside distribution and the influence of inlet conditions in a model of a disc and doughnut heat exchanger: Experiments in fluids. 3(1985) 293-300.
3. Schlunder, E.U. (Ed). Heat exchanger design handbook: Vol.3. Thermal hydraulic design of heat exchangers. Hemisphere Publishing Corporation (1983).

TABLE-1: HEAT EXCHANGER DATA

Shell diameter	4900 mm
Number of tubes	18000
Tube diameter	25 mm
Tube pitch - Uniform (Case I)	32 mm
- Variable (Case II)	31-41 mm
Process fluid flow rate	3000 m^3/h

Design Improvements of a Shell and Tube Heat Exchanger Based on Practical Experience and Numerical Analysis

C.W.M. van der Geld and J.M.W.M. Schoonen

Eindhoven University of Technology
Faculty of Mechanical Engineering P.O. Box 513 5600 MB Eindhoven
The Netherlands Telefax +(31) (40) 441749

Summary

It is shown that the performance, maintenance and lifespan of a vertical shell and tube evaporator critically depends on the geometry near the outlet. Several different geometries are examined both theoretically and experimentally, utilizing some heat exchangers of Shell Chemie B.V. Wear mechanisms are located, recognized and quantified with the aid of an eddy current measuring technique. In situ measurements during operation are performed in order to verify the 2–D numerical modelling.

I Introduction

This study aims at facilitating the judgement of (re)design propositions of shell–and–tube heat exchangers by affording an experimentally verified modelling approach of a general kind. Prevailing wear mechanisms will be examined in order to define the requirements of the modelling approach.

As a spin–off this study will provide with a discussion of advantages and drawbacks of some exchanger geometries.

This study has been initiated and sponsored by SHELL Moerdijk in the Netherland.

II Geometries studied both theoretically and experimentally

Vertical shell–and–tube heat exchangers that are operational at the SHELL chemical facility at Moerdijk exhibit two different baffle configurations. One will be denoted as EXCH–102, the other as EXCH–103 (see Fig. 1). In 1988 the sole shell–sided steam outlets of one particular 102 and one particular 103 was enlarged, from 14" (35 cm) to 20" (50 cm) along the same centerline (see Fig. 2). The rim of the 20" outlet just

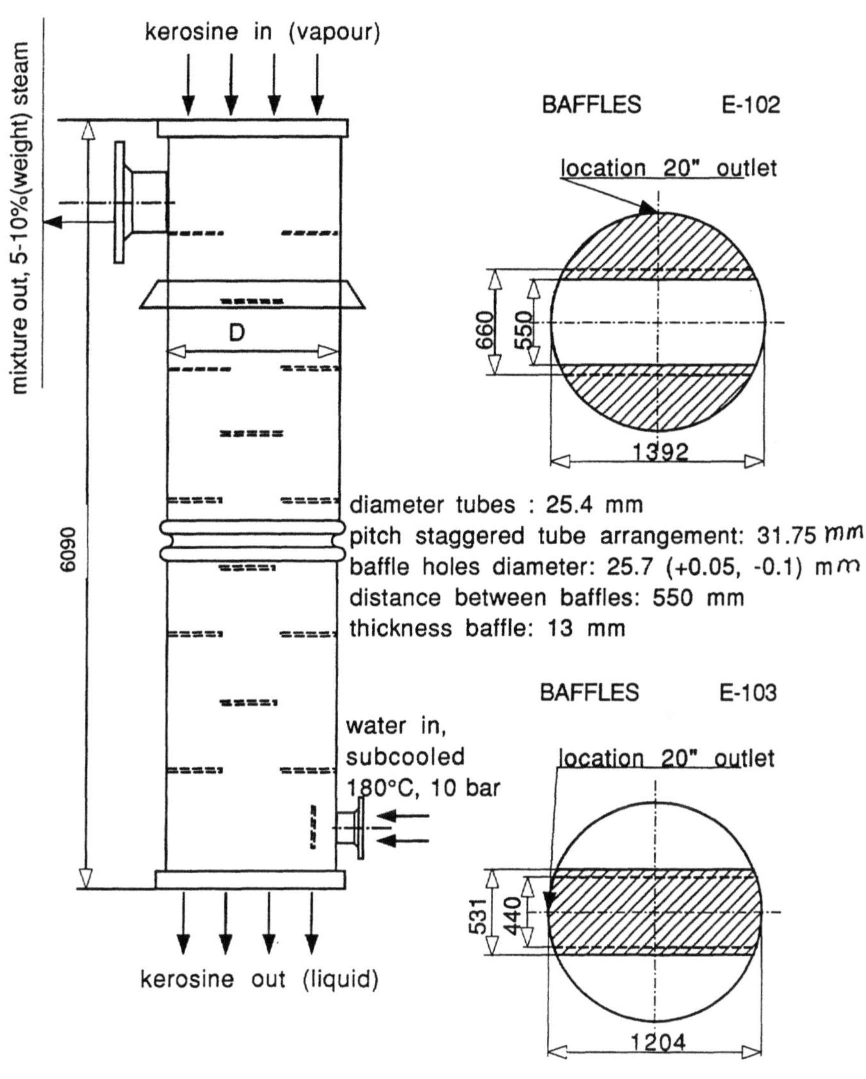

Figure 1 Schematic and dimensions of the EXCH heat exchangers.

reached the upper baffle plate.

In oktober 1990 this particular EXCH–103 outlet diameter was decreased to 14"
again, retaining the same centerline and placing an identical outlet just opposite of

the first one. In 1992 other 103 and 102 geometries will be adapted similarly.

The following numbering will be used henceforth (meas. ≡ measured; ΔP ≡ pressure drop over the upper baffle plate) :

Nr.	Geometry Description	Wear meas.?	ΔP meas.?
1	EXCH–103 before first adaptation outlet	Yes	No
2	EXCH–103 after enlarging outlet	Yes	Yes
3	EXCH–102 with a single outlet	Yes	No
4	EXCH–103 with the 1990 design improvement	No	Yes

All these geometries, but also other ones such as an EXCH–102 with two 14" outlets will be studied theoretically.

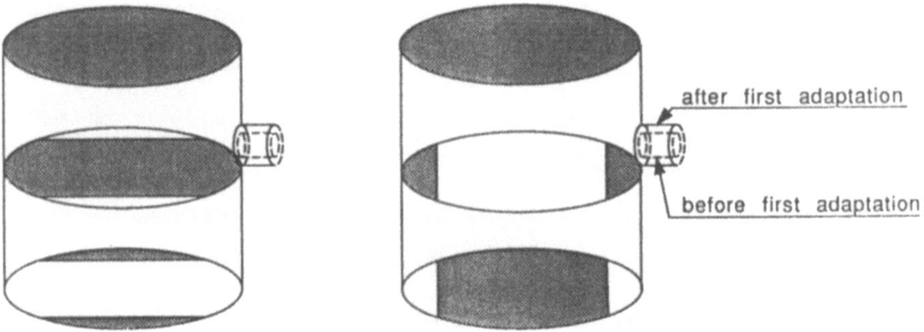

Figure 2 The upper part baffle and outlet geometries of the EXCH E–103 (left figure) and EXCH E–102 (to the right).

Typical operating conditions will be presented in Ch. VI (see also Fig. 1). Demineralised water (pH = 8,5) enters shell sided at the bottom with a few degrees subcooling. Kerosine is totally condensed in downflow inside the carbon–steel tubes. The 'duty', the rate of kerosine condensation, is gradually increased over a period of ca. two years in order to compensate for increasing catalyst imperfections in the process that heats the kerosine.

The flow pattern of the steam–water mixture near the outlet is some type of churn–spray–mist flow with a high ($\sim$ 70 – 95 %) steam void fraction.

III Selection criteria for optimal design

In order to optimize a heat exchanger several parameters can be varied : the number of baffle cuts, the dimensions of baffles plates, outlet geometry, type of process fluid, inclination of the exchanger, total mass flow rate, etc. The main decisive parameter in design is cost. All cost aspects should normally be addressed, from primary investment, preventive and corrective maintenance to eventual profits. The determining of the annual cost requires knowledge of lifetime, investment, depreciation factor and interest rate. However, the translation into cost of future maintenance is often cumbersome. It requires predictive tools for the wear rate in all parts of the heat exchanger.

This problem will be addressed in the following chapters.

IV Design improvement by preventing wear

IV 1   Possible wear mechanisms.

Malfunctioning of machinery must be analyzed from the history and current state of the device in relation to operating conditions. If wear is observed this might be due to several causes :

- vibration [1, 2]
- impingement [4]
- erosion/corrosion [7, 8]

- cavitation [3]
- corrosion [5, 6]

In this paper attention is focussed on the geometries described in Ch. II.

IV.2   Assessment of the condition of some heat exchangers.

From visual observations two types of wear were found to occur in the EXCH heat exchangers :

Type A   Upstream severe thinning of the tube wall on both sides of the stagnation point with a axial height of circa 15 cm. This damage merely occurred just in front of the outlet and manifested itself especially in geometry 1 (see Ch. II) with the smaller outlet.

Type B   Circumferential material loss of pipes just in and a few centimeters downstream the near–exit upper baffle plate, see photographs 3. This damage occurred in geometry 2 (see Ch. II) with the rim of the outlet just reaching the baffle plate. Very near the baffle plate pits are observed on the surface, see Fig. 3.

Tube color at other places is dark grey to black, indicating a magnetite skin of $Fe_3 O_4$ protecting against corrosion.

Measurements with an eddy current measurement technique, to be further discussed in Ch. VI, showed an increase of type B material loss towards the exit (see Fig. 4). Some tubes just in front of the exit even had to be plugged off because of excessive

Figure 3 Photographs of wear damage type B, near the baffle plate.

wear in order to guarantee safe operation of the plant.

IV.3 Wear mechanisms for some heat exchangers.
Vibration might cause fretting of the tubes in the co–annular spacing inbetween upper baffle and tubes (see also Ch. 2). This might initiate some flaws of the magnetite layer of vibrating tubes.

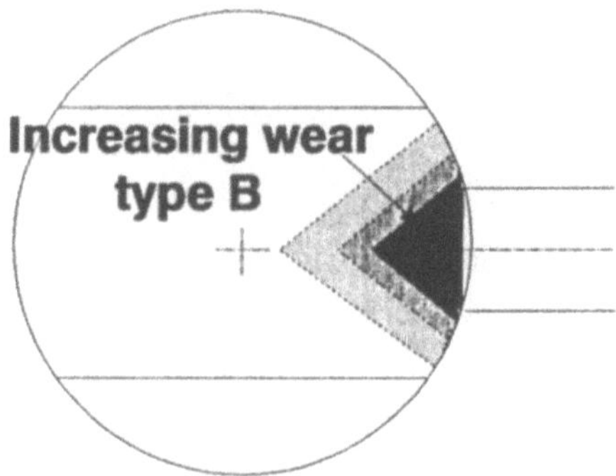

Figure 4 Wear damage intensities at the height of the baffle plate.

Cavitation is a high speed phenomenon, typically occurring at velocities exceeding

200 m/s [4,3]. Under normal operating conditions maximum local velocity is estimated to be less than 40 m/s. Cavitation is therefore considered to be unimportant.

A rough estimate of impingement *erosion* caused by 0,5 mm diameter droplets with
the aid of Springer's formulae [4] yields an incubation time of 3.10^{11} years for configuration 1. Although the uncertainty of several parameters is large impingement as a
sole process may therefore safely be disregarded.

Corrosion itself is an unimportant wear mechanism if a magnetite layer is present.
Observations (see IV.2) and measurements of pH (value 8,3) and temperature
(180°C) indicate the presence of a stable magnetite layer [9].

The synergetic effect of impingement and corrosion is called "*erosion/corrosion*".
The mechanism basically comprises the repeated withdrawel of a protective layer by
impingement and subsequent layer reparation by corrosion [7,8]. This process depends on temperature and is fostered by fretting and tube vibrations. Maximum
possible wear rate occurs at temperatures around 180 ^{O}C [9].

If corrosion is dominating, the dissolution model [7] yields for the EXCH–103 an
estimated wear rate of 0,1 mm/year utilizing a typical mixture mass flux of 420
kg/m^2s (total mass flow rate, $\dot{m}_{tot}$, is 85 kg/s). However, if erosion is dominating
through the local droplet velocity normal to the wall, v_d, the droplet impact model
yields the wear rate [8]

$$m^* = C_{pf} \dot{m}_{tot} (1-x) v_d^4 F_e F_h / \{A_c P_{hd}^2\} \tag{1}$$

for steam quality x and other parameters that hardly depend on $\dot{m}_{tot}$. The accuracy
of this equation is circa 10%, but improves considerably if merely used for assessing
the dependency of m^* on the steam quality for a specific configuration.

For geometry 2 and the same mass flow rate as above a typical droplet velocity is 35
m/s as will be seen in Ch. V. This yields 1,05 mm/year wear rate, which corresponds
to the actually observed wear rate (see Ch.2).

Erosion dominated corrosion is therefore the most probable cause for the wear of type
B observed in geometry 2. This will be further examined in the chapter V, where
differences with configuration 1 will be elucidated.

IV.4 The importance of velocity prediction.

Eq. (1) shows that an accurate assessment of the local (droplet) velocity is of vital
importance for wear rate computation. Note the power four of v_d and the occurrence

of $\dot{m}_{tot}$ in Eq. (1). For optimal design of heat exchangers it is therefore mandatory to reduce local velocities as much as possible. Temperature and pH control of the coolant are also very important if erosion–corrosion wear should be avoided.

The next chapter is merely devoted to the computation of local velocities in the geometries described in Ch. II in the context of wear rate prediction.

V Theoretical predictions

The prediction of the dependance of local velocities on geometry was the main issue of the theoretical analysis, since in Ch. IV local velocities were found to be essential for wear rate prediction. Focussing attention on the predictability of a trend rather than accurate values allowed the simplification of actual 3–D geometries to corresponding 2–D ones (see Fig. 5). The observations of Fig. 4 verify to some extend that it is sufficient to study flow characteristics after projection to the plane through the vertical and the outlet axis.

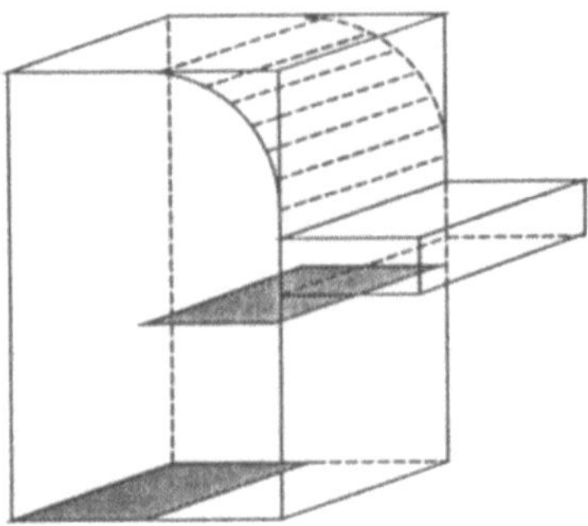

Figure 5 Two–dimensional model representing the upper part geometry of the EXCH–103.

Following Patankar and Spalding [10] the tube array was modelled as an anisotropic homogeneous flow resistance (see Fig. 6).

The average void fraction downstream of the upper baffle plate always exceeding 70 % the two–phase flow mixture on shell side was modelled as a homogeneous with averaged mixture properties. The slip strongly depends on flow regime and geometry, and was varied in the range 1 – 4.

Computations were carried out with PHOENIX after adapting the ground file. Porosity in the baffle plate was taken to be 0,028 which was about 10 % of the porosity of the tube array. More details are given by Houtermans [11].

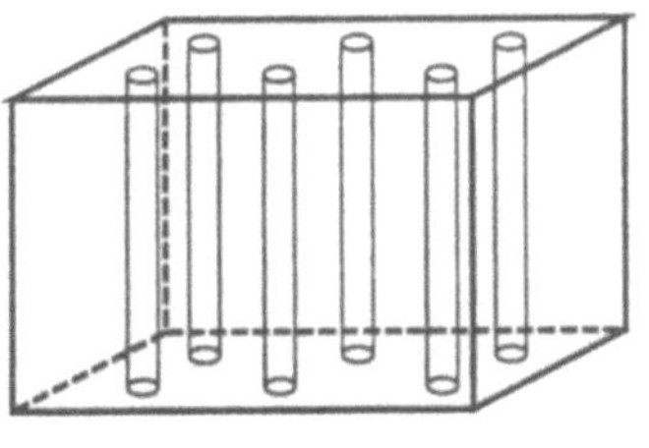 

a) actual cell filled with tubes   b) mock-up,'porous', cell

<u>Figure 6</u> Porous cell representation of a staggered tube array.

Some typical results exhibiting the dependancy on geometry of some important local velocities are shown in Figures 7 and 8. It is seen in Fig. 7 that the highest velocities occur just below the centerline at exactly the same place where the biggest damage of type A was observed (see IV.2).

Some computational results are summarized below. Here ϵ denotes the global void fraction and $v_\perp$ the velocity component normal to the tubes just downstream of the baffle plate right in front of the exit (see also Figs. 7 and 8); $v_\perp = |v|.\sin(\alpha)$.

Geometry	$\dot{m}_{tot}$	x	ϵ	$|v|$	$v_\perp$	Figure
	kg/s	%		m/s	m/s	
1	165	10	0,95	62	14	7 →
1	165	5	0,69	33	5	
2	165	10	0,95	70	22	8
2	165	5	0,69	36	11	
3 (1 x 20")	82	5	0,69	63	14	
3 (2 x 14")	82	5	0,69	42	5	

Note that the mass flow rate for geometry 3 was choosen to be half of the one for geometry 2. In addition the heat exchanger diameter is larger for geometry 3 than for geometry 2, while the EXCH–102 baffle plate occupies more space near the outlet than the one in geometry 3. For these reasons the part of the flow that passes through the baffle plate is much bigger for geometry 2 than for geometry 3, leading to relatively high baffle plate velocities.

Some important conclusions can be drawn from the above table :

→Computed velocity in the outlet, slightly off the centerline : $|v| = 78$ m/s, $v_\perp = 42$ m/s.

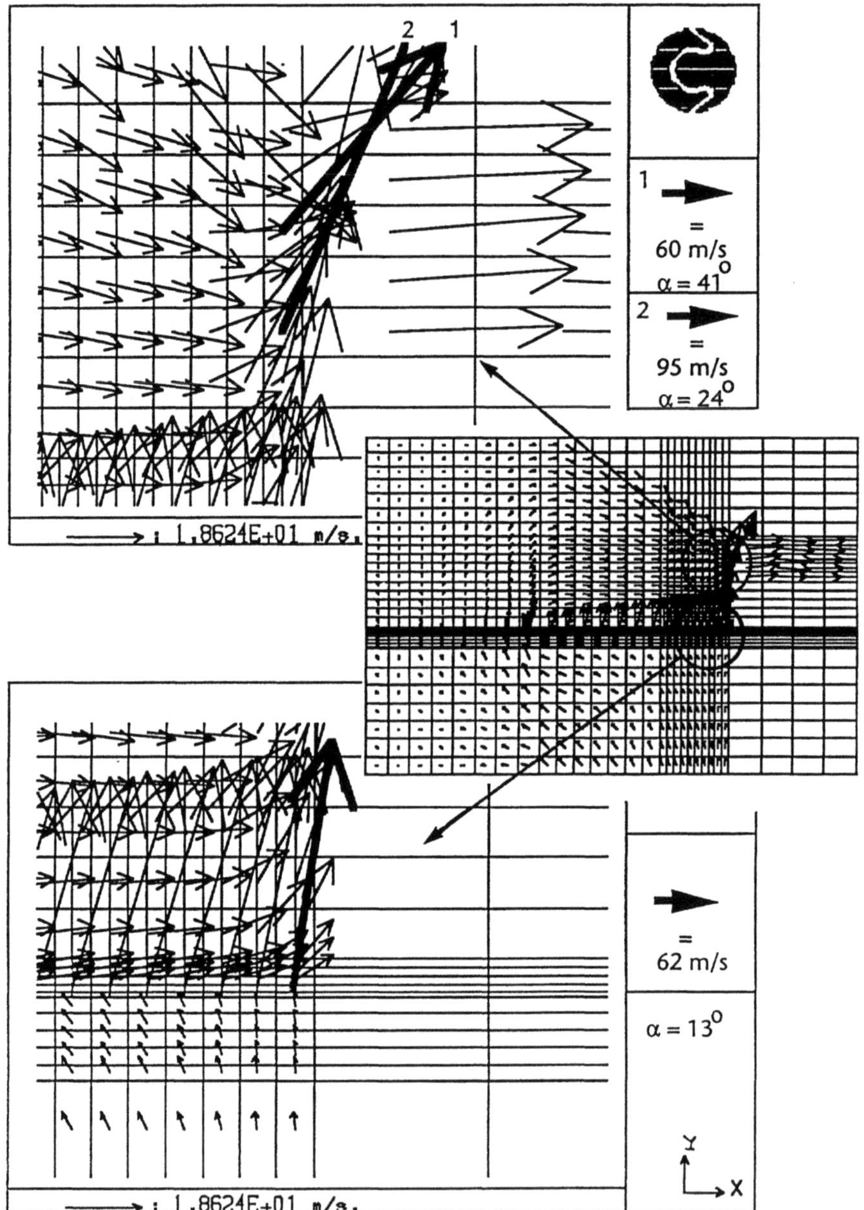

Figure 7 Computational results for geometry 1; $\dot{m}$ = 165 kg/s; x = 10 %; ϵ = 0,95.

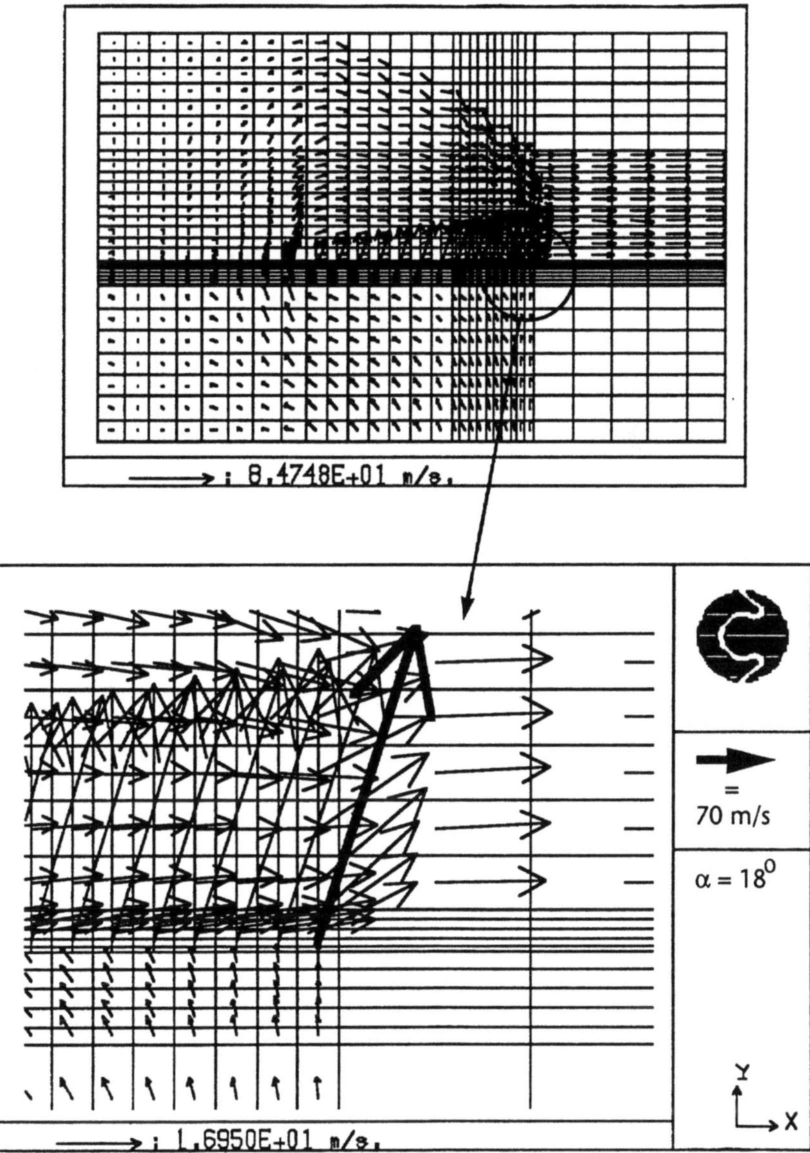

Figure 8 Computational results for geometry 2. Flow conditions the same as those in Figure 7.

a The enlarging of the outlet diameter as effectuated by selecting geometry 2 in stead of 1 diminishes the velocities in the outlet but increases the velocities in the baffle plate clearances.

b   The trend observed in *a* is even more distinct at low mass flow rates.

c   The trend observed in *a* is even more clear if only $v_\perp$ is considered. This should be done in view of v_d in Eq. 1. Since $v_\perp$ represents an average over the co–annular spacing inbetween the baffle plate and a tube, the effect might locally be even more pronounced. The instationary character of the two–phase mixture might compensate for circumferential differences but not for fluctuations and high peak values.

d   Based on the geometry 3 computations it should be expected that geometry 4 restores the low baffle plate velocities of geometry 1 while retaining the low outlet velocities of geometry 2.

Observation *a* must be attributed to the only difference between configurations 1 and 2 : the position of the outlet relative to the baffle plate. This is what we name the "wistle effect".

The importance of conclusion *c* for wear rate prediction should be obvious from the power 4 in Eq. 1.

VI <u>Measurements and comparison with theoretical predictions.</u>

In the EXCH–103 of the SHELL plant in Moerdijk, the Netherlands, measurements were carried out for the validation of the predictions of ch.'s IV and V. The results for two geometries are discussed below.

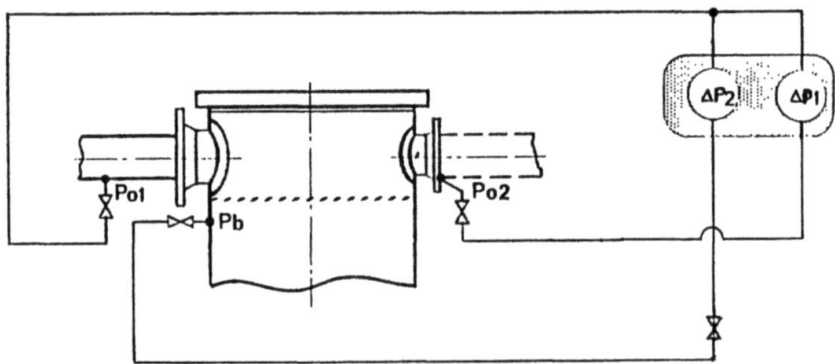

Figure 9 Location of pressure taps for configuration 2.

VI.1 In situ measurements of flow parameters.

Figure 9 shows the locations of the additional pressure drop sensors mounted on the EXCH–103 and the definitions of $\Delta P1$ and $\Delta P2$. Pressure drops were corrected for hydrostatic pressure drops of the tapping lines.

Other flow parameters, temperatures, mass flow rate and pressures, were measured elsewhere as a routine procedure furnishing daily averages only.

Some averaged results over periods of quasi–stationary operation for geometries 2 (period 4–19 august 1990) and 3 (26 okt.–1 nov. 1990) are listed below. Here T_{ko} denotes the temperature at the outlet for the kerosine in the tubes, and T_{ki} and the pressure P_{ki} similarly at the inlet. P_{so} denotes the pressure of the steam near the outlet, $\dot{m}_s$ the mass flow rate of steam coming from the heat exchanger.

	Geometry 2	*Geometry 4*			*Geometry 2*	*Geometry 4*
P_{ko}	1,3 bar	0,64 bar		T_{ko}	181 ^{o}C	192 ^{o}C
$\dot{m}_{ko}$	20 kg/s	34 kg/s		T_{ki}	220 ^{o}C	202 ^{o}C
P_{so}	9,2 bar	8,8 bar		T_{so}	181 ^{o}C	181 ^{o}C
$\dot{m}_{so}$	2,7 kg/s	4,3 kg/s		x	3 %	5 %
$\Delta P1$	4700 Pa			$\Delta P2$	5900 Pa	1875 Pa

Note that the ratio $\dot{m}_{so}/\dot{m}_{ko}$ is mainly determined by phase change enthalpies. The total mass flow rate on shell side, - 85 kg/s, is based on an average over several heat exchangers of the same type.

Employing homogeneous 1–D flow equations, a slip factor of 3 and the above experimental values, for geometry 4 a mixture density, ρ_m, of 95 kg/m^3 is computed as well as an outlet steam velocity of ca. 5,5 m/s.

Most important experimental result is the strong pressure drop decrease over the upper baffle plate as a consequence of the adaptations leading to geometry 4, despite the increase of the duty.

A rough estimate for the average leakage velocity in the baffle plate is obtained from this pressure drop measurement (v - $\sqrt{2\,\Delta P\,/\,\rho_m}$), yielding 6,3 m/s for geometry 4 and 11,1 m/s for geometry 2. Despite the fact that these values are not expected to be accurate, their ratio must approximately be correct. It is therefore safe to conclude that the leakage flow through the upper baffle plate is strongly decreased by the adaptations leading to configuration 4.

Exactly the same trend was already predicted in Ch. V.

More model validation is obtained from pressure drop computations for the flow conditions of the above table. These computations were performed for geometry 3, a EXCH–102 with a 50" outlet, yielding $\Delta P1$ - 3000 Pa and $\Delta P2$ - 6000 Pa. Although the geometry differs, these results can qualitatively be compared† with the experimental ones for geometry 2. In view of the approximations involved the results agree remarkably well.

The exit mixture velocity was for this geometry 3 predicted to be 6 m/s, again quite close to the above estimate based on experiments (5,5 m/s).

VI.2 Eddy current inspection measurements.

The tubing of several heat exchanger configurations was inspected with the aid of an eddy current method. This method is especially suited for non ferromagnetic materi

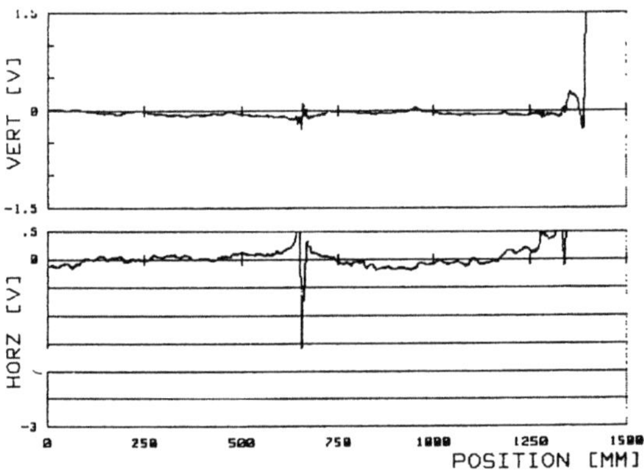

Figure 10 Specimen of eddy current measurement in the EXCH–103. The pike near the centre indicates outer pipe damage at the height of the baffle plate.

als and consists of an internal diameter probe connected to automatic pulling and registration equipment. The circumferentially averaged material loss is quantified after careful calibration. Also is indicated whether the damage occurs inside or outside the tube.

A typical signal obtained with a measurement of a EXCH–103 tubing is shown in Fig. 10.

The results of the measurements for geometry 2 (see ch. II) have already been summarized in Fig. 4. Average wall thickness loss increases gradually from less than 0,5 mm to more than 1,5 mm to even more, as indicated in Fig. 4. Since some tubes had to be plugged off no new information was inferred after early 1990. Since

---

†Computations were performed before the experiments, leading to some difficulty in matching both flow conditions and geometry at the same time.

geometry 4 was realized in oktober 1990 and the next eddy current inspection is due for 1992, the eddy current analysis of the design improvements is yet to come.

The authors wish to express their gratitude to G. Jansen (Gemco), R. van de Ploeg, S. Bosch, P. Geenen (Shell) and R. Bakker, M. Houtermans (graduate students) for their help in this study.

<u>References</u>

1   W.J. Heilker and R.Q. Vincent
    Vibration in nuclear heat exchangers due to liquid and two—pase flow.
    Trans. of ASME, Vol. 103, 1981

2   Pettigrew M.J., Y. Sylvestre and A.O. Campagna
    Flow—induced vibration analysis of heat exchanger and steam generator designs.
    4th Int. Conf. on structural mechanics in reactor techn., San Fransisco, part F6/1, 1977

3   R. Bakker
    Wear mechanisms and design aspects.
    Report WOC—WET 90.031, Eindhoven University of Technology, 1990

4   G.S. Springer
    Erosion by liquid impact.
    J. Wiley & Sons, New York, 1976

5   S.J. Green
    Thermal, hydraulic and corrosion aspects of PWR steam generator problems.
    Heat Transfer Eng., Vol. 9, no. 1, 1988

6   Several authors
    Corrosion and other types of damage.
    Hemisphere Publication Corporation, New York, 1983

7   R.G. Keck and P. Griffith
    Prediction and mitigation of erosive/corrosive wear in secondary piping systems of nuclear
    power plants.
    Nureg/CR—5007, R5; MIT; Cambridge, MA 0222139, 1987

8   J.G. Hines and F. Neufeld.
    Corrosion of mild steel due to impingement in the make—gas stream.
    In "Materials technology in steam reforming processes", ed. C. Edeleanu, Pergamon, Oxford,
    pp. 357—369, 1966.

9   Several authors
    Metals handbook; ninth edition
    Volume 13, "Corrosion"; American Soc. of Mech.Eng. pp. 964—996, 1989

10 S.V. Patankar and D.B. Spalding
    Computer analysis of the three—dimensional flow and heat transfer in a steam generator.
    In "Forschung im Ingenieurwesen", Vol. 44, no. 2, pp. 47—52, 1978.

11 M.P.A. Houtermans
    Numerical analysis of the two—phase flow on shell side of "shell—and—tube" heat exchangers.
    Report WOC—WET 90.029, Eindhoven University of Technology, 1990.

Simple Algorithms for Optimization of Shell and Tube Heat Exchangers

K RAMANANDA RAO, U SHRINIVASA & J SRINIVASAN
Department of Mechanical Engineering
Indian Institute of Science
Bangalore-560 012, INDIA

Summary

A simple and reliable algorithm for optimization of shell and tube heat exchangers is discussed. The algorithm is based on the observation of weak coupling between the geometry and the heat transfer in the near-optimal region of the feasible design space irrespective of the objective function used. Hence, the algorithm is quite general in nature and can handle various objective functions like weight, cost and volume. A decision table is provided which summarizes the strategies to obtain velocity and pressure drop constrained designs from the above optimum by systematic variation of the geometry parameters. Sub-algorithms based on this decision table are developed. Finally, the use of these algorithms is illustrated by an example.

Introduction

Heat exchangers are considered as essential equipment in many process industries, power plants, heat recovery systems and other related industries. Many types of heat exchangers are used in these industries. Among these, the shell and tube heat exchangers are extensively used. Therefore, optimizing these heat exchangers could reduce the initial investment and operating costs of the plant considerably. We propose here a simple and general algorithm for optimal synthesis of shell and tube heat exchangers for objective functions such as weight, cost and volume. Several sub-algorithms are developed which handle practical constraints such as those on velocity and pressure drops commonly enforced by the customer on the design of shell and tube heat exchangers.

A number of investigators have proposed many different methods for optimization of heat exchangers. Most of the methods require explicit algebraic expressions for the objective function and constraints. Moreover, the synthesis problem of shell and tube heat exchangers involves lengthy and complex procedures. Also, iterative calculations are involved. Therefore, the use of these methods necessitate the conversion of these procedural calculations into algebraic forms which amounts to making a number of assumptions. The methods which are capable of handling such procedures are the search methods, case study methods and the non-linear programming methods. The search methods are more effective when a unique optimum exists. The case study methods which are similar to search methods need good engineering judgement to properly direct the convergence of the optimization process. The non-linear programming methods do not ensure convergence on to a global optimum and if the interrelationship of the parameters are complex they may not even give a feasible design.

Structure of Feasible Designs

We have studied a large number of feasible designs obtained for different process specifications and by considering different objective functions such as weight, cost and volume. From this study, we have developed useful insight into the structure of feasible designs and the importance of geometry and heat transfer parameters in the design process[1]. The most interesting result from this study is the observation of weak coupling between the heat transfer and the geometry in the near-optimal region of the feasible design space. This is shown in Figs. 1, 2 and 3. Fig. 1 shows the variation of the objective function, weight of the heat exchanger, with the overall heat transfer coefficient Uo. Figs. 2 and 3 are similar plots for the cost functions obtained from the data of Purohit[2] and Saunders[3] for USA and UK respectively. Feasible designs are obtained for the process specifications given in Table 1. From these figures we find that, for a particular value of Uo, there are a number of feasible designs of varying objective function values. These exchangers also have the same exchanger area since for a particular tube layout and for a given number of tube passes, the product of the overall heat transfer coefficient and the exchanger surface

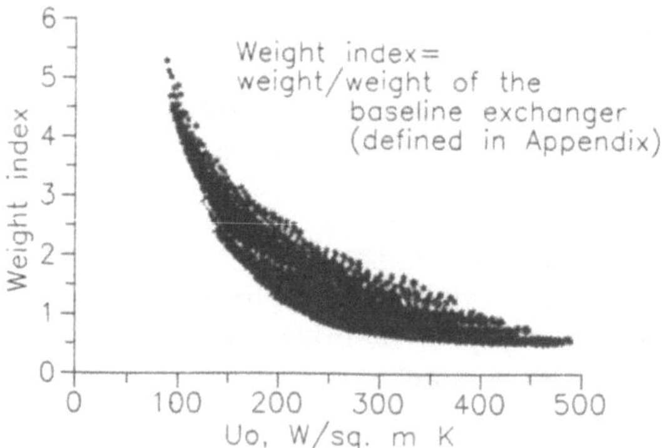

Fig. 1. Variation of weight with Uo

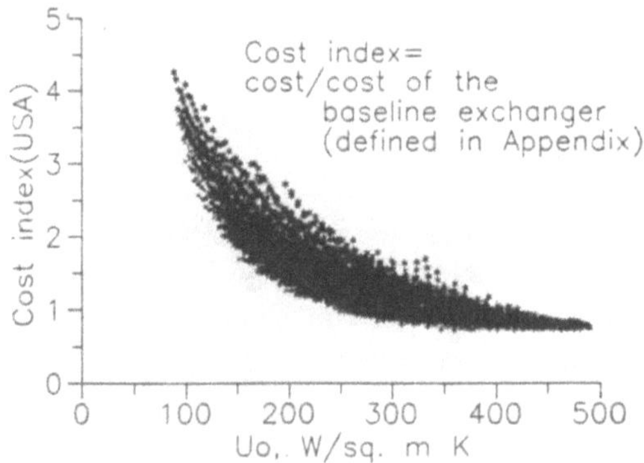

Fig. 2. Variation of cost(USA) with Uo

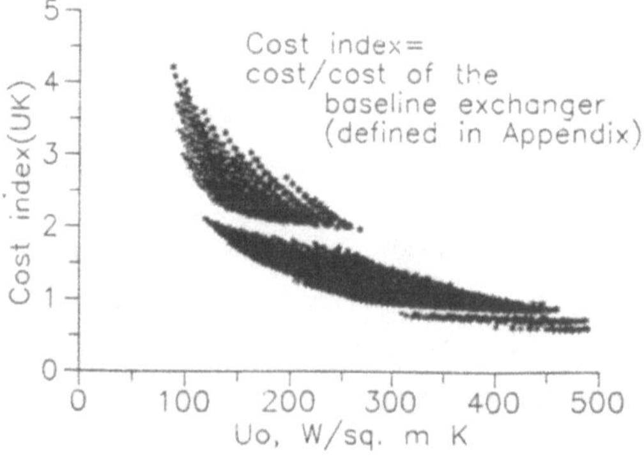

Fig. 3. Variation of cost(UK) with Uo

area Ao is a constant. On the other hand, we find in the near-optimal region a large number of feasible designs having nearly the same value for the objective function but, widely varying values of Uo. This shows that, the objective function is weakly dependent on Uo in this region. Thus, the above plots indicate that, as one approaches the optimum region, the dependence of the objective function on Uo is likely to be weak and hence, the geometric optimum design will also be one of the near optimum designs satisfying the heat duty requirements. Thus, the above trend suggests the possibility of decoupling of the geometry and the heat transfer aspects while looking for optimal solutions. This makes the optimization process much simpler since one has to handle now only the geometry optimization problem. Therefore, we first concentrate our attention on geometry optimization.(We have also obtained analytical support for the weak coupling between the heat transfer and geometry in the near-optimal region[4].)

Geometry Optimization

The geometry optimization involves determination of the optimum geometry for a given exchanger surface area subject to geometric constraints alone. We have based the geometry optimization on the exchanger area since, every heat exchanger

Table 1. Specifications of the shell and tube heat exchanger

Heat duty kW	643
Exchanger type	BEM class R
Tube layout	Triangular
Tube diameters used	19.05 mm and above
Number of tube passes	4

	Shell side	Tube side
Fluid type	organic liquid	water
Mass flow rate kg/s	8.05	21.94
Temperature (in/out) K	372/313	306/313
Density (bulk) kg/m^3	1165	1000
Viscosity (bulk/wall) kg/ms	$4.88 \times 10^{-4}/6.6 \times 10^{-4}$	$8 \times 10^{-4}/8 \times 10^{-4}$
Specific heat (bulk) J/kg K	1340	4187
Thermal conductivity W/m K	0.116	0.625
Fouling resistance m^{2}K/W	4.65×10^{-4}	4.65×10^{-4}

is specified by the surface area packed. However, we have seen in the previous section that, there can be a large number of alternative geometries satisfying a given surface area requirement. The geometry optimization will find the optimum geometry among these alternatives. The geometry optimization is still a non-linear optimization problem subject to non-linear, procedural and iterative constraints. However, it is much simplified as compared to the original problem with heat duty constraints. We have also shown that, knowing the nature of the objective function used, the geometry optimization problem can be greatly simplified. From this, we can even obtain approximate analytical solutions for the geometry optimization problem[4].

The Basic Algorithm

The geometry optimization module explained in the previous section should be properly linked to the thermal rating module to obtain the optimum design satisfying the heat duty constraints. We have developed an algorithm to interact with these two modules and find an optimum design satisfying the heat duty demanded by the process specifications. The algorithm is explained by the flowchart given in Fig. 4. The algorithm works as follows. First the thermal rating module estimates the value of the overall heat transfer coefficient Uo for the given process specifications. From this value of Uo, the required exchanger surface area Ao to satisfy the specified heat duty is determined. The optimum geometry for this Ao is determined by the geometry optimization module. This geometry is again rated by the thermal rating module and the above procedure is repeated till convergence is achieved.

Consideration of Velocity and Pressure Drop Restrictions

The optimum design obtained by the use of the above algorithm may not satisfy the limitations on velocity and pressure drops. In this section, we propose sub-algorithms which modify this design to take into consideration the velocity and pressure drop limitations. The velocities on the shell and tube sides

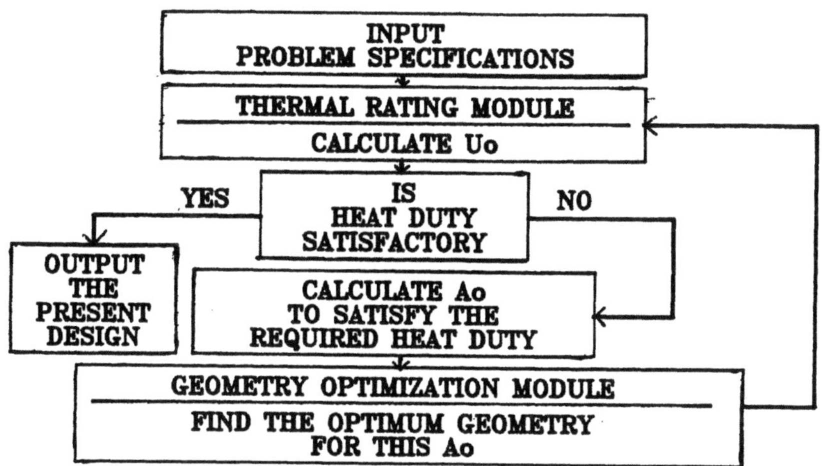

Fig. 4. Basic algorithm for optimization

can be controlled by the variation of the flow areas. These can be varied by systematic variation of the major geometry parameters such as shell diameter, tube pitch, baffle spacing etc. If the pressure drop exceeds the allowable pressure drop on any side, the equivalent velocities for these allowable pressure drops can be determined from the relation,

$$PD \propto V^n$$

The Heat Exchanger Design Handbook[5] gives the value of n as,

$n=1.6 - 1.8$ for turbulent flow

$n=1.0$ for laminar flow

Once the allowable pressure drop is converted into equivalent velocity, this velocity is treated as the limiting velocity and the design is modified as discussed earlier. In some cases, a geometry modification to contain a velocity on one side might violate the velocity limitations on the other side. In such a situation, it may not be possible to find a feasible solution. We have developed a decision table which helps us to know whether a solution to the problem exists and if it exists what are the possible steps to be taken to get the required design. The outline of the decision table is given in Fig. 5. The boxes which are crossed out are the cases where no solution is possible. In other cases, the solution is possible by a proper variation of major geometric parameters. We have developed sub-algorithms based on the above decision table to obtain velocity and pressure drop constrained optimal designs.

<u>Example</u>

We illustrate by an example the above algorithms developed for the optimal synthesis of shell and tube heat exchangers. Here we have considered weight as the objective function. The process specifications are given in Table 1. The velocity and pressure drop restrictions are given in Table 2. The details of iterations are given in Table 3. The convergence process is illustrated in Fig. 6. The optimum design from the above methodology has been compared with the optimum design obtained

SHELL SIDE \ TUBE SIDE	VT LOW PRT HIGH	VT HIGH PRT HIGH	VT WITHIN LIMITS PRT HIGH	VT LOW PRT WITHIN LIMITS	VT HIGH PRT WITHIN LIMITS	VT AND PRT WITHIN LIMITS
VS LOW PRS HIGH	X	X	X	X	X	X
VS HIGH PRS HIGH	X					
VS WITHIN LIMITS PRS HIGH	X					
VS LOW PRS WITHIN LIMITS	X	X			X	
VS HIGH PRS WITHIN LIMITS	X					
VS AND PRS WITHIN LIMITS	X					

Fig. 5. Outline of the decision table

Table 2. Velocity and pressure drop restrictions

Velocity restrictions	Shell side	Tube side
Minimum	0.6 m/s	1.0 m/s
Maximum	1.5 m/s	3.0 m/s
Pressure drop restrictions	0.69 Mpa	0.98 MPa

Table 3. Details of iterations

Optimum from the basic algorithm

				Geometric optimum							
Itn.	Uo W/m^2K	Ao m^2	DT mm	LTP mm	DS mm	LBC mm	LTO mm	VT m/s	VS m/s	PDT MPa	PDS MPa
1	578	55	19.05	23.81	520	104	2750	1.51	0.59	0.026	0.010
2	373	86	19.05	23.81	610	122	3000	1.05	0.43	0.012	0.007
3	338	94	19.05	23.81	610	122	3350	1.05	0.43	0.013	0.007
4	338	94	19.05	23.81	610	122	3350	1.05	0.43	0.013	0.007

Modifications using sub-algorithms to meet velocity and pressure drop requirements

Itn.	Uo	Ao	DT	LTP	DS	LBC	LTO	VT	VS	PDT	PDS
5	–	–	19.05	23.81	517	103	–	1.53	0.59	–	–
6	376	85	19.05	23.81	514	103	4500	1.57	0.60	0.046	0.016

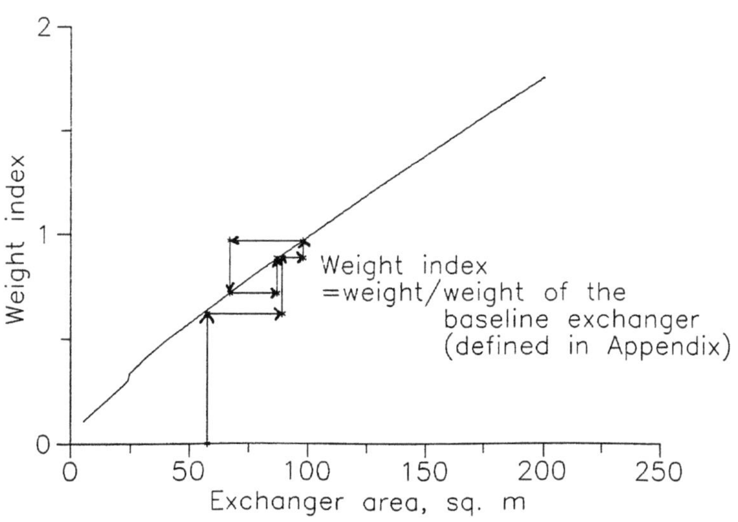

Fig. 6. Convergence of the optimization process

Table 4. Comparison of optima from the present method and
exhaustive search

Parameter	Present method	Exhaustive search
DT, mm	19.05	19.05
LTP, mm	23.81	23.81
DS, mm	514	495
LBC, mm	103	99
LTO, mm	4500	4850
Ao, m^2	85	83
Weight, kg	2547	2534

by an exhaustive search of the feasible designs generated for
the given process specification. These two designs agree very
well as can be seen from Table 4.

Conclusions

A generalized algorithm for optimal synthesis of shell and tube
heat exchangers is proposed. This algorithm is developed based
on the observation of weak coupling of geometry and heat
transfer in the near-optimal region of feasible designs. The
algorithm is simple and can adapt itself to manual calculations
as well as sophisticated computing facilities. Several sub-
algorithms are discussed which aid in systematically modifying
the above design to satisfy the velocity and pressure drop
limitations set forth by the customer. The proposed method is
the simplest among the various methods used for optimization of
shell and tube heat exchangers.

Nomenclature

Uo   Overall heat transfer coefficient, W/m^2 K
Ao   Exchanger surface area, m^2
V   Velocity, m/s
VT   Tube side velocity, m/s
VS   Shell side velocity, m/s
PD   Pressure drop, Mpa
PDT Tube side pressure drop, MPa
PDS Shell side pressure drop, MPa

References

1. Rao, K.R.; Shrinivasa, U.; Srinivasan, J.: Feasible designs
    of shell and tube heat exchangers - A study. Report 90VAR3.
    Department of Mechanical Engineering. Indian Institute of
    Science. Bangalore. India (1990).

2. Purohit, G.P.: Estimating costs of shell and tube heat
    exchangers. Chem. Engg. 90 (1983) 56-67.

3. Saunders, E.A.D.(ed.) Heat exchangers - selection, design
    and construction. London: Longman Scientific and Technical
    (1988).

4. Rao, K.R.; Shrinivasa, U.; Srinivasan, J.: Synthesis of
    optimal shell and tube heat exchangers. Lecture notes.
    Short term course on compact and process heat exchangers.
    Indian Institute of Science. Bangalore. INDIA (1990).

5. Schlunder, E.U. (ed.-in-chief) Heat exchanger design
    handbook. Dusseldorf: VDI Verlag GmbH (1983).

Appendix

The base line exchanger is defined as follows.

Exchanger type	BEM(No expansion joint), all carbon steel
Tubes	19.05 mm, welded, 14 BWG, average wall
Tube layout	triangular, single pass
Tube pitch	1.25 x outside tube diameter
Exchanger length	6096 mm
Exchanger area	100 m^2
Design pressures	<=1 MPa

Dispersion Model for Divided-Flow Heat Exchanger

Wilfried Roetzel and Yimin Xuan
Institute of Thermodynamics, University of the Federal Armed Forces
Hamburg, FRG

The axial dispersion model is used to analyse the thermal performance of divided-flow heat exchangers with N tube passes, arbitrary distribution of mass flow rate on the shellside, variable entrance location of the shellside flow and piecewise variation of heat transfer coefficient (or NTU), considering shellside maldistribution such as bypassing and leakage. Especially, the explicit formulas for calculating thermal effectivenesses as well as temperature profiles are derived for the 1-1 divided-flow heat exchanger. Measurements are carried out on a segmentally baffled 1-1 divided-flow heat exchanger with various clearances between segmental baffles and shell, in order to investigate the relationship between the dispersion factor and the flow pattern.

1. INTRODUCTION

Up to now, the rating and design of shell and tube heat exchangers are mostly based on the assumption that the shellside flow pattern is ideal, i.e., the plug flow always occurs on the shellside in spite of the complicated geometrical configuration of heat exchangers. In other words, all parts of the shellside fluid move parallel to the tubes at the same speed and are completely mixed in one cross-section. There is no maldistribution and no residence time distribution. Obviously, this assumption deviates from the real shellside flow pattern in segmentally baffled shell and tube heat exchangers mainly due to the cross-flow between the baffles and the existence of several different shellside flow paths with different thermal effects, as analysed by Tinker [1, 2] and by Palen and Taborek [3]. The shellside maldistribution may result from bypassing and leakage influenced by mechanical design or manufacturing tolerances. Mueller and Chiou [4] extensively discussed the causes of maldistributions. Furthermore, there exist regions of stagnation and the local velocities may change as the fluid flows across a bundle of tubes [5].

In the situation that the shellside maldistribution is strong, the result will be not satisfactory, if the idealized plug-flow model is still applied to describing the thermal performance of shell and tube heat exchangers. In fact, this maldistribution will deteriorate heat transfer process and result in a lower effective mean temperature difference, which is beyond the conventional plug-flow model. To make up this inherent deficiency of the plug-flow model, a few methods [3, 6, 7] were developed. Since they are either proprietary or complicated, these methods are not easy to use. Diaz and Aguayo [8] proposed an axial dispersion model for the countercurrent flow in the 1-1 shell and tube heat exchanger. However, the relationship between shellside Peclet number and flow pattern is still unknown. Roetzel and Spang [9, 10] developed a most generally valid analytical solution and calculation programme for the 1-N shell and tube heat exchanger using the dispersion model, and the Peclet number was fitted to the experimental data with countercurrent and cocurrent flow (N = 1).

2. THE AXIAL DISPERSION MODEL

Figure 1 shows schematic representations of the heat transfer in 1-N divided-flow heat exchangers. As shown in Fig. 1, the total heat transfer region is divided into two subregions, which are designated as subregions e and f, respectively.

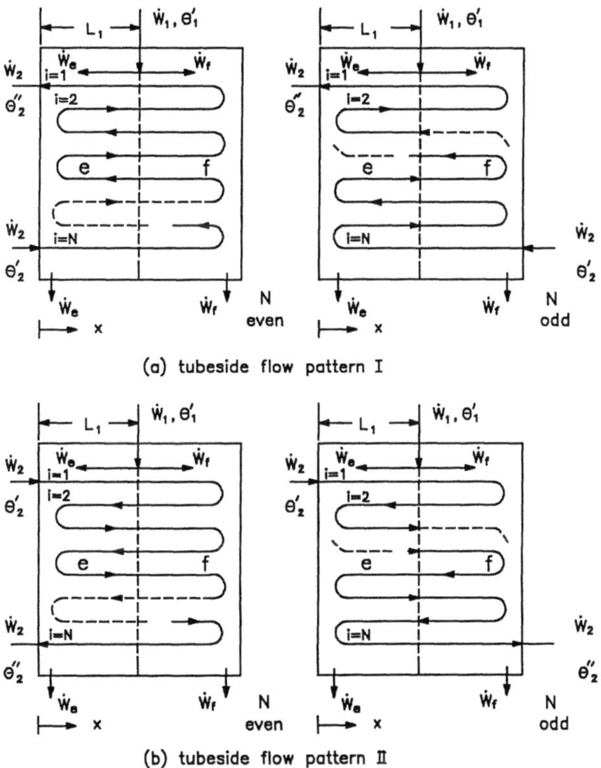

(a) tubeside flow pattern I

(b) tubeside flow pattern II

Fig. 1: Schematic representation of 1-N divided-flow heat exchangers

After the shellside fluid has flowed into the 1-N divided-flow heat exchanger via the entrance, it is immediately divided into two parts which flow across subregions e and f. These two parts of the shellside fluid exit from the apparatus via two nozzels at opposite ends. The thermal analysis of 1-N divided-flow heat exchangers was finished by means of the ideal plug-flow model [11], in which no cross-flow and no bypassing and leakage of the shellside flow were presumed besides other necessary assumptions. In the axial dispersion model, however, the Pecelt number is introduced to deal with the shellside maldistribution, such as bypassing and leakage, which really occur during the heat transfer process in heat exchangers.

The essence of the axial dispersion model is that a Fourier-type axial conduction term or a Fick-type diffusion term is introduced to describe the energy balance during the heat transfer process with a dispersion coefficient, so that one can correct the deviation of an ideal flow model from the real flow pattern. This kind of axial dispersion model is often used for the description of heat and mass transfer processes in chemical reactors and in packed beds [12, 13]. Here it will be applied to the derivation of the formulas for calculating temperature profiles and thermal effectivenesses of 1-N divided-flow heat exchangers with arbitrary division and arbitrary entrance location of the shellside fluid as well as piecewise constant heat transfer coefficient (or NTU) which may vary with tube pass and subregion. The dimensionless governing equations for subregion e ($0 \le x \le x_1$) are derived as follows:

$$\frac{dt_{ei}}{dx} = \pm(-1)^i NTU_2\, \varepsilon_i\, (T_e - t_{ei}) \qquad (i=1,\ 2,\ ...,\ N) \qquad (1)$$

$$\frac{1}{Pe_e}\frac{d^2 T_e}{dx^2} + \frac{dT_e}{dx} = \frac{NTU_1}{\beta_e} T_e - \sum_{i=1}^{N} \frac{NTU_1}{\beta_e} \varepsilon_i\, t_{ei} \qquad (2)$$

Similarly, the following equations are given for subregion f ($x_1 < x \le 1$):

$$\frac{dt_{fi}}{dx} = \pm(-1)^i NTU_2\, \varepsilon_i\, (T_f - t_{fi}) \qquad (i=1,\ 2,\ ...,\ N) \qquad (3)$$

$$\frac{1}{Pe_f}\frac{d^2 T_f}{dx^2} - \frac{dT_f}{dx} = \frac{NTU_1}{\beta_f} T_f - \sum_{i=1}^{N} \frac{NTU_1}{\beta_f} \varepsilon_i\, t_{fi} \qquad (4)$$

where the positive sign $(+)$ and the negative sign $(-)$ of sign $(\pm)$ in the above equations are valid for tube flow pattern I and for tube flow pattern II, respectively, which are shown in Fig. 1. The complete system consisting of second-order and first-order linear differential Eqs. (1), (2), (3) and (4) can be solved by means of matrix theory [14]. With eigenvalues λ_{ej} ($j=1, 2, ..., N+2$)

of E and corresponding eigenvectors $C_j = (c_{1j}, c_{2j}, ..., c_{N+2,j})^T$ as well as eigenvalues λ_{fj} of F and eigenvectors $D_j = (d_{1j}, d_{2j}, ..., d_{N+2,j})$ the general solutions to Eqs. (1) and (2) as well as Eqs. (3) and (4) can be shown as follows, respectively [15]:

$$T_E = \sum_{j=1}^{N+2} g_j C_j \exp(\lambda_{ej} x) \tag{5}$$

$$T_F = \sum_{j=1}^{N+2} h_j D_j \exp(\lambda_{fj} x) \tag{6}$$

where g_j and h_j ($j=1, 2, ..., N+2$) are unknown coefficients, which must be determined according to boundary and interface conditions. These necessary $(2N+4)$ determinant conditions are given in table 1 and Eqs. (7) and (8):

Table 1: Boundary and interface conditions for $t_{ej}(x)$ and $t_{fj}(x)$

| | $x = 0$ | $x = 1$ | tubeside flow pattern | |
			I	II
N even	$t_{ei} = t_{ei+1} = t_{ei,i+1}$ $i=2, 4, ..., N-2$	$t_{fi} = t_{fi+1} = t_{fi,i+1}$ $i=1, 3, ..., N-1$	$t_{eN}(0) = 0$	$t_{e1}(0) = 0$
N odd	$t_{ei} = t_{ei+1} = t_{ei,i+1}$ $i=2, 4, ..., N-1$	$t_{fi} = t_{fi+1} = t_{fi,i+1}$ $i=1, 3, ..., N-2$	$t_{fN}(1) = 0$	$t_{e1}(0) = 0$

The following boundary conditions pertinent to the dispersion model for $T_e(x)$ and $T_f(x)$ are derived:

$$T_e + \frac{1}{Pe_e} \frac{dT_e}{dx} = 1 \quad \text{and} \quad T_f - \frac{1}{Pe_f} \frac{dT_f}{dx} = 1 \quad \text{at} \quad x = x_1 \tag{7}$$

$$\frac{dT_e}{dx}\bigg|_{x=0} = 0 \quad \text{and} \quad \frac{dT_f}{dx}\bigg|_{x=1} = 0 \tag{8}$$

The temperature steps at the inlet according to Eq. (7) take the " conductive " axial heat flow through the inlet cross-section into account and fulfil the total energy balance. Making use of these conditions, one can obtain $(2N+4)$ unknown coefficients according to the following equation:

$$Y = W^{-1} G \tag{9}$$

where $Y = (g_1, g_2, ..., g_{N+2}, h_1, h_2, ..., h_{N+2})^T$ and $G = (0, 0, ..., 0, 1, 1)^T$, if the boundary conditions described by Eq. (7) are placed into the last two equations in Eq. (9). W is $(2N+4) \times (2N+4)$ matrix, elements of which depend on different factors, such as the number of tube passes, the tubeside flow pattern and the multiplicity of eigenvalues [15]. Thus, the dimensionless temperature changes P_1 and P_2 can be determined:

$$P_1 = 1 - \sum_{j=1}^{N+2} [\beta_e \, g_j \, c_{N+1,j} + \beta_f \, h_j \, d_{N+1,j} \exp(\lambda_{fj})] \tag{10}$$

$$P_2 = [1 - \sum_{j=1}^{N+2} [\beta_e \, g_j \, c_{N+1,j} + \beta_f \, h_j \, d_{N+1,j} \exp(\lambda_{fj})]] \, R_1 \tag{11}$$

The correction factor F of the logarithmic mean temperature difference (LMTD) can be calculated according to the following equations:

$$F = \begin{cases} \dfrac{\ln \dfrac{1 - P_2}{1 - P_2 R_2}}{NTU_2 (R_2 - 1)} & R_2 \neq 1 \\[4ex] \dfrac{P_2}{NTU_2 (1 - P_2)} & R_2 = 1 \end{cases} \tag{12}$$

Especially, for a 1-1 divided-flow heat exchanger the temperature profiles and the formulas for calculating P_1 and P_2 can be explicitly given. In subregion e:

$$\begin{vmatrix} t_e \\ T_e \\ \dfrac{dT_e}{dx} \end{vmatrix} = g_1 \begin{vmatrix} c_{11} \\ \lambda_{e1}^{-1} \\ 1 \end{vmatrix} \exp(\lambda_{e1}x) + g_2 \begin{vmatrix} c_{12} \\ \lambda_{e2}^{-1} \\ 1 \end{vmatrix} \exp(\lambda_{e2}x) + g_3 \begin{vmatrix} 1 \\ 1 \\ 0 \end{vmatrix} \tag{13}$$

where $c_{1j} = \dfrac{NTU_2}{\lambda_{ej}(NTU_2 - \lambda_{cj})}$, $(j=1, 2)$ and eigenvalues λ_{e1} and λ_{e2} are:

$$\lambda_{e1,2} = \frac{NTU_2 - Pe_e \pm \sqrt{(NTU_2 - Pe_e)^2 + 4(NTU_2 \, Pe_e + NTU_1 \, Pe_e/\beta_e)}}{2}$$

in subregion f:

$$\begin{vmatrix} t_f \\ T_f \\ \dfrac{dT_f}{dx} \end{vmatrix} = h_1 \begin{vmatrix} d_{11} \\ \lambda_{f1}^{-1} \\ 1 \end{vmatrix} \exp(\lambda_{f1}x) + h_2 \begin{vmatrix} d_{12} \\ \lambda_{f2}^{-1} \\ 1 \end{vmatrix} \exp(\lambda_{f2}x) + h_3 \begin{vmatrix} 1 \\ 1 \\ 0 \end{vmatrix} \tag{14}$$

where $d_{1j} = \dfrac{NTU_2}{\lambda_{fj}(NTU_2 - \lambda_{fj})}$, $(j=1, 2)$ and λ_{f1} and λ_{f2} are:

$$\lambda_{f1,2} = \frac{NTU_2 + Pe_f \pm \sqrt{(NTU_2 + Pe_f)^2 - 4(NTU_2\,Pe_f - NTU_1\,Pe_f/\beta_f)}}{2}$$

The 6 coefficients in Eqs. (13) and (14) are as follows:

$$h_1 = \frac{1}{(\dfrac{1}{\lambda_{f1}} - \dfrac{1}{Pe_f})\exp(\lambda_{f1}x_1) - \exp(\lambda_{f1} - \lambda_{f2} + \lambda_{f2}x_1)(\dfrac{1}{\lambda_{f2}} - \dfrac{1}{Pe_f}) - (d_{11} - d_{12})\exp(\lambda_{f1})}$$

$$h_2 = - \exp(\lambda_{f1} - \lambda_{f2})\,h_1 \quad \text{and} \quad h_3 = - (d_{11} - d_{12})\exp(\lambda_{f1})\,h_1$$

$$g_1 = \frac{1 - h_1\,[d_{11}(\exp(\lambda_{f1}x_1) - \exp(\lambda_{f1})) + d_{12}\exp(\lambda_{f1})\,(\exp(\lambda_{f2}x_1 - \lambda_{f2}) - 1]}{(\dfrac{1}{\lambda_{e1}} + \dfrac{1}{Pe_e} - c_{11})\exp(\lambda_{e1}x_1) - (\dfrac{1}{\lambda_{e2}} + \dfrac{1}{Pe_e} - c_{12})\exp(\lambda_{e2}x_1)}$$

$$g_2 = - g_1 \quad \text{and} \quad g_3 = 1 - g_1\,[(\dfrac{1}{\lambda_{e1}} + \dfrac{1}{Pe_e})\exp(\lambda_{e1}x_1) - (\dfrac{1}{\lambda_{e2}} + \dfrac{1}{Pe_e})\exp(\lambda_{e2}x_1)]$$

The values of P_1 and P_2 are determined by Eqs. (15) and (16), respectively:

$$P_1 = 1 - \beta_e\,(\frac{g_1}{\lambda_{e1}} + \frac{g_2}{\lambda_{e2}} + g_3) - \beta_f\,(\frac{\exp(\lambda_{f1})}{\lambda_{f1}}\,h_1 + \frac{\exp(\lambda_{f2})}{\lambda_{f2}}\,h_2 + h_3) \tag{15}$$

$$P_2 = 1 - g_1\,[(\frac{1}{\lambda_{e1}} + \frac{1}{Pe_e})\exp(\lambda_{e1}x_1) - (\frac{1}{\lambda_{e2}} + \frac{1}{Pe_e})\exp(\lambda_{e2}x_1) - c_{11} + c_{12}] \tag{16}$$

One example is plotted in Fig. 2 with the assumption $Pe_e = Pe_f = Pe$. Compared with the results in [11], the values of P_1 and P_2 according to the dispersion model decrease obviously at the same NTU and R_1, which reveals the fact that heat transfer degenerates because of maldistribution. The Peclet number Pe is an important criterion to describe the effect of maldistribution. Theoretically, the dispersion model is converted into the plug-flow model if $Pe \to \infty$. In fact, both models are almost identical if $Pe > 60$.

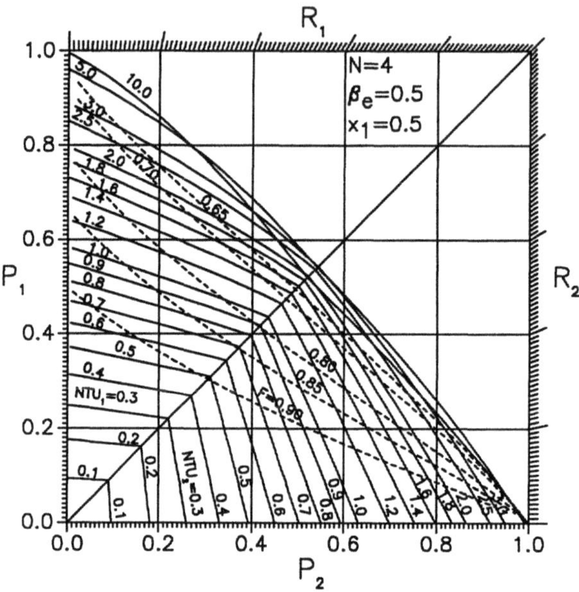

Fig. 2: P_1, P_2-chart for a 1-4 divided-flow heat exchanger with tubeside flow pattern I ($\varepsilon_1 = \varepsilon_2 = \varepsilon_3 = \varepsilon_4 = 0.25$, Pe = 10)

3. EXPERIMENT

The application of the dispersion model involves knowledge of the Peclet number, which must be determined in advance. For a simple flow its value can be obtained theoretically [16, 17]. However, for a complicated flow such as the shellside flow in shell and tube heat exchangers it is impossible so far to predict Pe theoretically and it must be determined experimentally. The involved experiments were executed on a 1-1 divided-flow heat exchanger, in which the clearance between baffles and shell could be changed by means of polyamide rings which were attached to the edge of baffles. Thus, one is able to study the influence of the leakage stream between baffles and shell, which plays the greatest role in deteriorating the mean temperature difference [3]. During the experiment inlet temperatures as well as temperature differences between inlet and outlet of each fluid (water) were measured by thermocouples, flow rates by turbine flowmeters and pressure drops by strain-gauge pressure transducers. The apparatus and the experimental procedures are described in detail in ref. [15].

The Reynolds numbers for fluids 1 and 2 are defined as follows:

$$Re_1 = \frac{u_1 \, l}{\psi \, \nu_1} \quad \text{and} \quad Re_2 = \frac{u_2 \, d_i}{\nu_2} \tag{17}$$

where $u_1 = \dot{V}_1/(D_s\,S)$. During the experiments Re_1 varied from 1500 to 12000, Re_2 from 3000 to 7000. The correlations for the shellside Nusselt number Nu_1 and the Nusselt number Nu_2 pertinent to turbulent and transitional flow through circular tubes are separately given as follows:

$$Nu_1 = c_1\,Re_2^{0.6}\,Pr_1^{0.36} \quad \text{and} \quad Nu_2 = c_2\,Re_2^{m_2}\,Pr_2^{1/3} \tag{18}$$

Using these correlations, the product of heat transfer surface and overall heat transfer coefficient is expressed as:

$$\frac{1}{UA} = \frac{1}{c_1\,Re_1^{0.6}\,Pr_1^{0.36}\,\dfrac{k_1}{l}\,A_1} + \frac{s}{k\,A_m} + \frac{1}{c_2\,Re_2^{m_2}\,Pr_2^{1/3}\,\dfrac{k_2}{d_i}\,A_2} \tag{19}$$

where $A_1 = N\,\pi\,d_a\,L$, $A_2 = N\,\pi\,d_i\,L$ and $A_m = (A_1 - A_2)/\ln(A_1/A_2)$. The Peclet number as well as the two unknown coefficients c_1 and c_2 and one exponent m_2 are determined by fitting experimental data with the dispersion model. For this purpose, the objective function to be minimized is given by:

$$\sum_{i=1}^{M} (P_{2,i,calc} - P_{2,i,exp})^2 \quad \rightarrow \quad \min \tag{20}$$

where $P_{2,i,calc}$ are the effectivenesses calculated from Eq. (16) and $P_{2,i,exp}$ are the corresponding experimental values. Nelder and Mead's technique was used to implement the minimization according to Eq. (20). The final results show that the values of c_1, c_2 and m_2 are approximately equalled among different groups of experimental data related to the different clearances between baffles and shell, respectively, and only the Peclet number varies with the clearances. The average of each parameter is given by Eq. (21) and table 2 as follows:

Table 2: Variation of Pe with the clearance δ

δ (mm)	0.00	0.25	0.50	1.00	2.45
Pe	52.19	42.09	27.67	18.47	4.11

$$c_1 = 0.34895, \quad c_2 = 0.01277 \quad \text{and} \quad m_2 = 0.84911 \tag{21}$$

Thus, the shellside and tubeside Nusselt numbers are obtained:

$$\text{Nu}_1 = 0.349 \, \text{Re}_1^{0.6} \, \text{Pr}_1^{0.36} \quad \text{and} \quad \text{Nu}_2 = 0.0128 \, \text{Re}_2^{0.85} \, \text{Pr}_2^{1/3} \qquad (22)$$

The heat transfer correlations given in Eq. (22) are suitable for the shell and tube heat exchangers with different clearances. Combined with the corresponding Peclet number, they can be used to predict the thermal perfomance of similar heat exchangers. Because of the limited space, only a few results pertinent to the smallest and biggest clearances are plotted in Fig. 3 - Fig. 4. It is obvious that the results calculated from the dispersion model agree satisfactorily with the experimental data. The calculation has shown that the effect of Pe on the heat transfer process has almost nothing to do with the number of tube passes, if NTU is smaller than 3.0. As a result, the correlations in Eq. (22) can also be extended to shell and tube heat exchangers with more than one tube pass.

The smallest leakage stream between baffles and shell will occur at the nought clearance ($\delta = 0.00$ mm). The difference between the dispersion and plug-flow models reaches the minimum and one can substitute the latter for the first at this situation. According to the experimental data corresponding to the nought clearance and the nonlinear regression method three unknowns c_1, c_2 and m_2 related to the plug-flow model were determined as follows:

$$c_1 = 0.325, \quad c_2 = 0.01284 \quad \text{and} \quad m_2 = 0.85079 \qquad (23)$$

As a result, Nu_1 and Nu_2 for the nought clearance can be expressed as:

$$\text{Nu}_1 = 0.325 \, \text{Re}_1^{0.6} \, \text{Pr}_1^{0.36} \quad \text{and} \quad \text{Nu}_2 = 0.0128 \, \text{Re}_2^{0.85} \, \text{Pr}_2^{1/3} \qquad (24)$$

Comparing Eq. (24) with Eq. (22), one can easily find that the tubeside Nusselt numbers are identical and there is a slight deviation of coefficient c_1 for both models. The reason for this deviation is that there occurs really the leakage stream between baffles and shell to some extent even at the numerical nought clearance. The bypassing around the tube bundle, the cross-flow effect and other forms of maldistribution exist always in the heat transfer process and are neglected by the plug-model. Therefore, one obtains a bigger apparent mean temperature difference and a lower apparent heat transfer coefficient from the plug-flow model. In fact, coefficient c_1 decreases further with increasing clearance (but the tubeside Nusselt number Nu_2 can retain the same), if this model is used to determine the three parameters c_1, c_2 and m_2 [15]. This means that according to the experimental data on a specified shell and tube heat exchanger one cannot expect to obtain universal heat transfer correlations with the conventional plug-flow model. In its applicable range the Nusselt number Nu_1 in Eq. (24) agrees well with Gnielinski's equation [19], the maximal relative deviation is 5%. Because there exists no reliable correlation for the transitional flow range, the tubeside Nu_2 is compared with Hausen's equation [20] and the maximal relative deviation between both correlations reaches 10%.

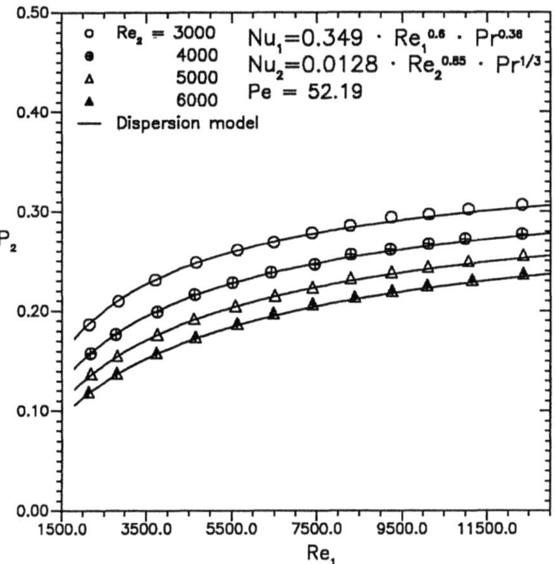

Fig. 3: Thermal effectiveness P_2 related to the nought clearance $\delta = 0.00$ mm

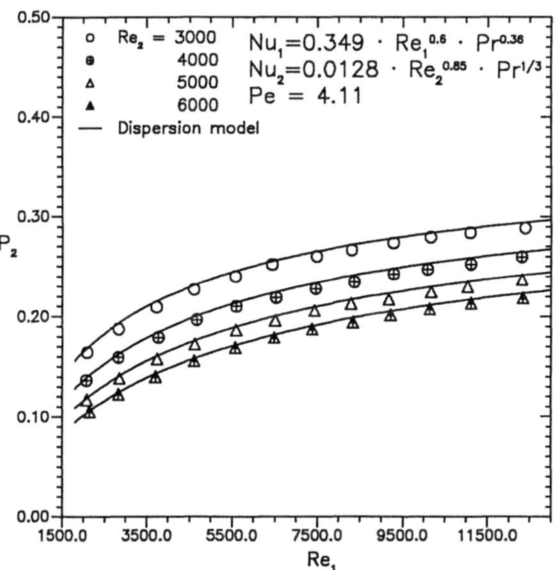

Fig. 4: Thermal effectiveness P_2 related to the clearance $\delta = 2.45$ mm

4. CONCLUSION

The dispersion model can be used to predict the thermal performance of 1-N divided-flow heat exchangers with arbitrary division ratios of the shellside flow rate, arbitrary inlet locations of the shellside flow and piecewise variable heat transfer coefficients (or NTU). The heat transfer correlations (or NTU) obtained from the experimental data on a 1-1 divided-flow heat exchanger agree well with some existing correlations. Combined with the corresponding Peclet number, these correlations are recommended to execute the thermal rating and design of multipass shell and tube heat exchangers.

ACKNOWLEDGEMENT

The authors would like to express many thanks to the " Deutsche Forschungs-gemeinschaft " for the financial support of this research project.

NOMENCLATURE

A	heat transfer surface
a	pitch ratio
d_i, d_a	inside and outside diameters of a single tube
D_s	inside shell diameter
E	matrix pertinent to subregion e
F	matrix pertinent to subregion f
F	LMTD correction factor
L	the whole length of divided-flow heat exchanger
l	characteristical length $l = \pi\, d_a/2$
M	number of measurement readings
N	number of tube passes
NTU	number of transfer units $NTU_{1,2} = (UA)/\dot{W}_{1,2}$
P	dimensionless temperature change through heat exchangers
Pe	Peclet number for axial dispersion
R	thermal flow rate ratio $R_1 = 1/R_2 = \dot{W}_1/\dot{W}_2 = P_2/P_1$
s	wall thickness of a single tube
S	baffle span
t	tubeside dimensionless temperature $t = (\theta_2 - \theta'_2)/(\theta'_1 - \theta'_2)$
T	shellside dimensionless temperature $T = (\theta_1 - \theta'_2)/(\theta'_1 - \theta'_2)$
U	overall heat transfer coefficient
$\dot{W}$	thermal flow rate
x	dimensionless flow path coordinate

β	division ratio of shellside flow rate $\beta_{e,f} = \dot{W}_{e,f}/\dot{W}_1$
ε_i	ratio of NTU_i in tube pass i to overall NTU of heat exchanger
θ	temperature
λ	eigenvalue
ψ	bundle porosity $\psi = 1 - \pi/(4a)$

Subscripts

e, E	subregion e
f, F	subregion f
1	shellside
2	tubeside

Superscripts

'	inlet
"	outlet

REFERENCES

1. Tinker, T.: Shell Side Characteristics of Shell and Tube Heat Excahngers, Part I, II, III.
   General Discussion on Heat Transfer, IMechE, Lodon (1951), 89-116.

2. Tinker, T.: Shell-Side Characteristics of Shell-and-Tube Heat Exchangers - A Simplified Rating System for Commercial Heat Exchangers.
   Trans. ASME 80 (1958), 36-52.

3. Palen, J.W. and J. Taborek: Solution of Shell Side Flow Pressure Drop and Heat Transfer by Stream Analysis Method.
   Chem. Eng. Progr. Symp. Ser. 92, 65 (1969), 53-63.

4. Mueller, A. C. and J. P. Chiou: Review of Various Types of Flow Maldistribution in Heat Exchangers.
   Heat Transfer Engineering, vol. 9 no. 2 (1988), 36-50.

5. Berner, C., F. Durst and D. M. McEligot: Flow Around Baffles.
   J. Heat Transfer 106 (1984), 743-649.

6. Short, B.E.: Shell-and-Tube Heat Exchangers: Effect of Bypass and Clearance Stream on the Main Stream Temperature. ASME Paper 60-HT-16 (1960).

7. Bell, K.J. and W.H. Kegler: Analysis of Bypass Flow Effects in Tube Banks and Heat Exchangers. AIChE Symp. Ser. 174, 74 (1978), 47-52.

8. Diaz, M. and A.T. Aguayo: How Flow Dispersion Affects Exchanger Performance. Hydrocarbon Processing 66, 4 (1987), 57-60.

9. Roetzel, W. and B. Spang: Effective Mean Temperature Difference in Segmentally Baffled Shell and Tube Heat Exchangers. Proceedings of the Ninth Int. Heat Transfer Conf., Jerusalem, vol. 5 (1990), 79-84.

10. Spang, B.: Über das thermische Verhalten von Rohrbündelwärmeübertragern mit Segmentumlenkblechen. Fortschr.-Ber. VDI Reihe 19, 1991. Düsseldorf: VDI-Verlag.

11. Xuan, Y., B. Spang and W. Roetzel: Thermal Analysis of Shell and Tube Exchangers with Divided-Flow pattern. Int. J. Heat & Mass Transfer 34 (1991), 853-861.

12. Wen, C.Y. and L.T. Fan: Models for Flow Systems and Chemical Reactors. New York: Marcel Dekker, Inc. 1975.

13. Dixon, A. G. and D. L. Cresswell: Effective Heat Transfer Parameters for Transient Packed-Bed Models. AIChE Journal 32 (1986), 809-819.

14. Hochstadt, H.: Differential Equations - A Modern Approach. New York: Dover Publications, Inc. 1964.

15. Xuan, Y.: Thermische Modellierung mehrgängiger Rohrbündelwärmeübertrager mit Umlenkblechen und geteiltem Mantelstrom. Fortschr.-Ber. VDI Reihe 19 Nr. 52. Düsseldorf: VDI-Verlag, 1991.

16. Taylor, S.G.: The Dispersion of Matter in Turbulent Flow through A Pipe. Proc. Royal Society of Lodon, Ser. A 223 (1954), 446-468.

17. Backman, L.V. and V.J. Law, etc: Axial Dispersion for Turbulent Flow with A Large Radial Heat Flux. AUChE Journal 36 (1990), 598-604.

18. Nelder, J.A. and R. Mead: A Simplex Method for Function Minimization. Computer Journal 7 (1965), 308-313.

19. Gnielinski, V.: Wärmeübertragung im Außenraum von Rohrbündelwärmeübertragern mit Umlenkblechen. VDI-Wärmeatlas, 4. Aufl., Düsseldorf: VDI-Verlag (1984), Abschnitt Gg.

20. Hausen, H.: Neue Gleichungen für die Wärmeübertrgung bei freier oder erzwungener Strömung. Allg. Wärmetechnik, Band 9 H. 4/5 (1959), 75-79.

Heat Exchanger in Transient Conditions

by

G.RATEL, P.MERCIER, G.ICART
(Groupement pour la Recherche sur les Echangeurs Thermiques
CENG/GRETh-85X- 38041 GRENOBLE CEDEX- FRANCE)

1/ INTRODUCTION

In many industrial applications, heat exchangers (HX) are generally a part of a system. They are therefore exposed to many transient regimes when a part of the system changes its operating conditions.
The first step in determining the behaviour patterns of a complex network is to study the basic elementary component which is a heat exchanger. In order to predict the response of the HX at a variation of the process, a software has been developped to predict the time and the way to reach the new steady state regime. This software has been built with part of the shell and tube heat exchangers design sofware CETUC /REF 1/
The different types of time dependant inlet conditions which can be studied are :
 - any temperature transient
 - any mass flow rate transient
 - a change in fluid (which can be a mixture of fluids) without chemical reaction.
These different transient conditions can be mixed together and applied to one or two sides of the HX.
The main limitations are :
 - the fluid is in single phase (either liquid or gas) ;
 - the shell must be of the TEMA type E with single segmental baffle ;
 - a maximum of 12 tube passes;
 The prior step of the modelisation consists in studying all the different parts of the HX to determine their thermal behaviour and to treat each necessary part of the HX separately. The different components of the HX are : the shell side flow, the tube side flow, the shell, the tubes, the tubular plates, the baffles, the headers and the turnarounds if the pass number is greater than 1.
Those HX parts have a specific thermal behaviour and must be modelled.

2/ ENERGY EQUATIONS FOR THE COMPONENTS

Two kinds of energy equations can be written in temperature terms :

for a flow (shell side or tube side) :

$$vol.\rho.Cp\,\frac{\partial T}{\partial t} + u.vol.\rho.Cp\,\frac{\partial T}{\partial x} = \sum_{As} \dot{q}.As$$

where As are the heat exchange areas

for a solid region :

$$vol.\rho.Cp.\frac{\partial T}{\partial t} = \sum_{As} \dot{q}.As$$

And the heat flux is :

$$\dot{q}= U.(Text- Tb)$$

Where U is the overall heat transfer coefficient between the region with the bulk temperature Tb and the outside region with a bulk temperature Text.

For a solid component two different approaches may be used either by solving a specific equation in the material or using a lumped model in connecting the component with the outside flow. The first method is used for tubes and tube plates. The second method is used for the baffles and for the metallic parts of the headers and the turnarounds which are in contact with the outside.
The lumped model performed a perfect heat exchange between the flow and the solid region and can be used when the biot number and the fourier number of the component are smaller than those number for the flow.

If s subscript is used for the solid region and f for the flow region, a lumped model modifies the energy equation of the flow :

$$(\rho_f.Cp_f.vol_f+\rho_s.Cp_s.vol_s)\frac{\partial T}{\partial t} + u.vol_f.\rho_f.Cp_f\,\frac{\partial T}{\partial x} =\sum_{As}\dot{q}.As'$$

Where As' are the heat exchange areas with the exception of the area between the flow and the lumped solid.

The tubes

As the tubes are thermally dependent on both flows, the thermal flux rate is :

$$\dot{Q} = U_o.Am.(T_{ss} - T_t) + U_i.Am(T_{ts} - T_t)$$

The overall coefficients of the inside tube Ui and outside tube Uo are :

$$\frac{1}{U_i} = \frac{1}{h_{ss}}.\frac{Di}{Dm} + \frac{Dm}{2\,\lambda} \log \left(\frac{Dm}{Di}\right)$$

$$\frac{1}{U_o} = \frac{1}{h_{ts}}.\frac{Dm}{Do} + \frac{Dm}{2\,\lambda} \log \left(\frac{Do}{Dm}\right)$$

The shell

The shell is thermally insulated from the outside environment, and the heat flux is between the shell side flow and the shell. As there is no particular study on the heat transfer coefficient between the shell and the flow inside, the shell side coefficient law was used.

The tubular plates

As in the former case, the heat transfer coefficient between the tube plates and the flows around them is difficult to obtain due to the flow's complexity. The heat rate is :

$$\dot{Q} = U_{tps}.A_{tps}.(T_{ss} - T_{tp}) + U_{tpt}.A_{tpt}.(T_{ts} - T_{tp})$$

T_{ts} is the flow temperature inside the header or the turnaround;
T_{ss} is the temperature inside the first or the last shell compartment.

The overall coefficients of shell side U_{tps} and header or turnaroud side U_{tpt} are

$$\frac{1}{U_{tps}} = \frac{1}{h_{ss}}.+ \frac{e_{tp}}{2\,\lambda}$$

$$\frac{1}{U_{tpt}} = \frac{1}{h_{ts}}.+ \frac{e_{tp}}{2\,\lambda}$$

The baffles

The baffles are related to the shell flow as previously stated.

The shell side heat transfer coefficient : the Bell method

The Bell method /REF2/ is used to ev luate the he transfer coefficients for the shell side flow. this method gives a good heat transfer prediction for classical shells. The ideal crossflow heat transfer coefficient law is modified using correction factors to take into account all the bypass or leakage effects due to geometrical clearances.

$$h_{ss} = h_{ideal}\ P(J_i)$$

where

h_{ss} is the shell side heat transfer coefficient

h_{ideal} is the ideal crossflow heat transfer coefficient

$P(Ji)$ is the product of all the correction factors (REF/2/)

Furthermore the Bell method allows us to consider the shell stream and the tube stream as cocurrent or countercurrent flow.

The tube side heat transfer coefficient

For the tube side flow, classical correlations are used to evaluate the heat transfer coefficient:
    Hausen for the laminar flow /REF 3/
    Gnielinsky for the turbulent flow regime /REF 4/
In the header and turnaround, the same laws are used to calculate the heat fluxes between the solid components of the hx and the flows.

3/ CONCENTRATION EQUATION FOR THE FLOWS

It is possible to describe a transient change of fluid without chemical reaction in one or two parts of the hx. The concentration C of the initial fluid is treated with a passive scalar transport equation.

$$vol.\rho.\ \frac{\partial C}{\partial t} + u.vol.\rho\ \frac{\partial C}{\partial x} = 0$$

The concentration C acts on physical properties of the fluids.

4/ PHYSICAL PROPERTIES

The solid physical properties are supposed constant with temperature variation.

The fluid physical properties are supposed variable with temperature and concentration

- ρ, Cp : linear variations with temperature and concentration;
- μ : an Arhenus type law for temperature variation with an unlinear variation for concentration;
- λ : linear variation with temperature and unlinear variation with concentration.

5/ RESOLUTION

The model is based on a one dimensional finite volume method in which the shell and tube HX is divided into an adequate number of control volumes. The figures 1 and 2 show the different meshes for the components in the case of two pass HX.
The initial steady state temperature fields for flows and solid parts are set up with the results of the steady state calculated with the CETUC code. Because of differences in modelization, it is necessary to obtain a stabilisation by the transient algorithm on a steady state solution.
The transient equations are then solved using a time step Δt. For a flow the equation is :

$$vol^* \; \rho^{n*} \; cp^{n*} \; \frac{T^{n+1} - T^n}{\Delta t} \; + M.Cp.(T_{am}^{n+1} - T^{n+1})$$

$$= \sum_{As} U.As.(Text^n - T^{n+1})$$

For the solid regions the equation is :

$$vol.\rho^n.cp^n. \; \frac{T^{n+1} - T^n}{\Delta t} \; = \sum_{As} U.As.(Text^n - T^{n+1})$$

The subscript n is used if the term is evaluated at the time step n and the subscript n+1 is used if the term is calculated at the new time step. The * sign is used to point out that solid regions can be lumped.
For each time step Δt the software first evaluates the HX coefficient using thermal fields at the previous time , and the tube side flow temperatures (inside tubes, header and turnaround), the shell side flow temperatures and at last the temperatures of the solid parts.

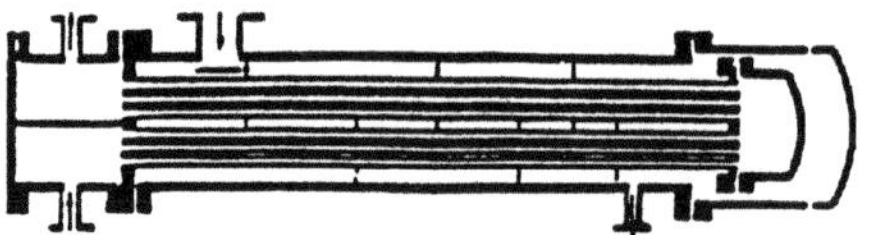

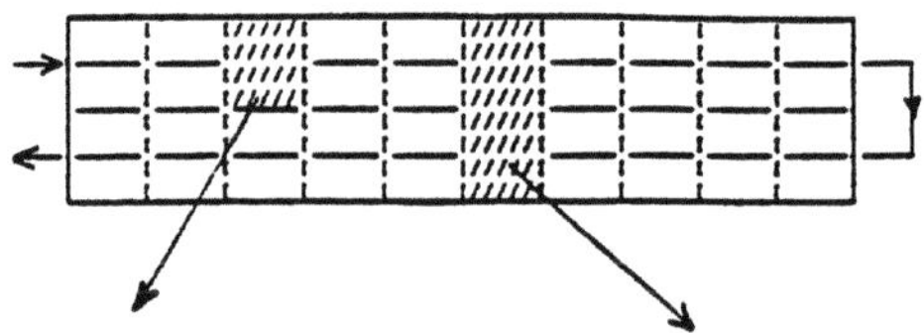

Figure 1 : Thermal modelisation of a heat exchanger

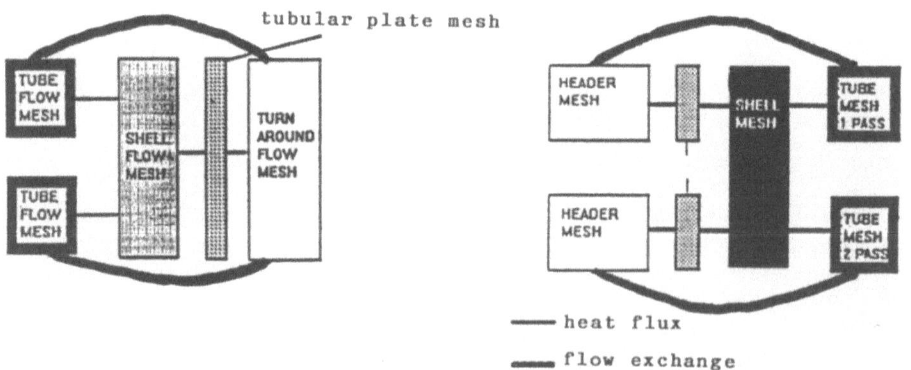

Figure 2 : Flow and heat flux exchanges for the header meshes (bottom) and the turnaround mesh (top)

6/ <u>RESULTS</u>

When the new steady state is reached , the software provides :
- the dynamic characteristics (temperature, heat transfer coefficient) of the HX in each cell at each moment,
- the evolution of the inlet and outlet thermal characteristics. these characteristics can be plotted on the screen or on a printer.

7/ <u>VALIDATION</u>

7.1 <u>Double-pipe HX</u>

A comparison between the software results and experimental results for a double pipe hx is shown on figure 3b. Figure 3a describes the HX geometry. Mass flowrate and temperature transient can be produced on an experimental test loop /REF 5/.
  The figure 3b shows the inside tube outlet temperature evolution after a step variation of the outside tube inlet temperature while other process characteristics remain constant.
  The numerical calculations and experiments show similar results but since the temperature is plotted in a nondimensional form, the influence of the temperature variations on physical properties and on the heat transfer coeffficents can not be well described.

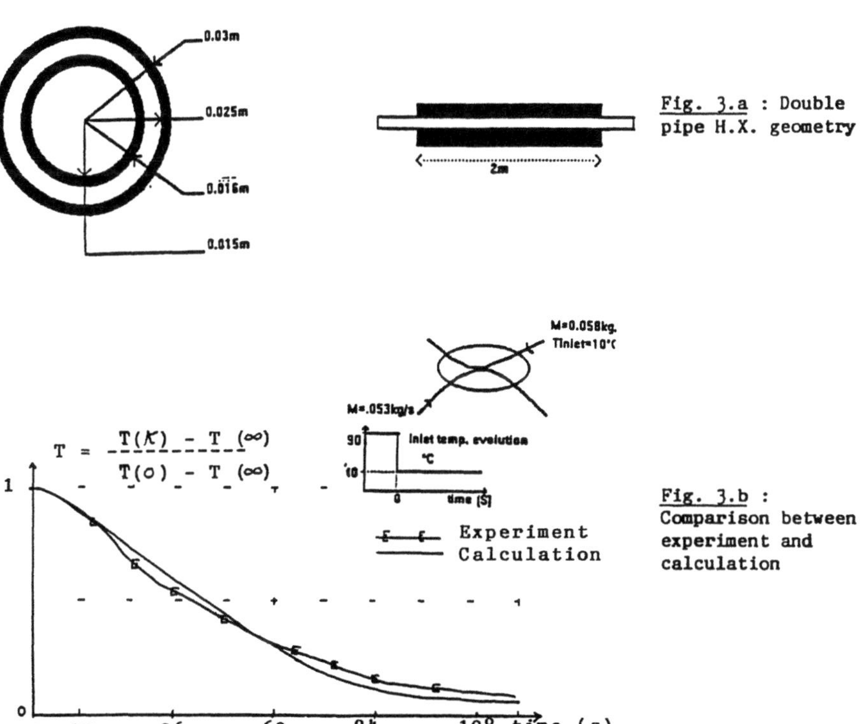

Fig. 3.a : Double pipe H.X. geometry

$$T = \frac{T(K) - T(\infty)}{T(o) - T(\infty)}$$

Fig. 3.b : Comparison between experiment and calculation

7.2 Industrial HX

An industrial HX has been tested on the esther test facilty
under transient conditions.
The geometrical characteristics of the HX are :
    Length : 1.4 m,
    100 tubes with an outside diameter 0.018m and an inside
      diameter 0.016m,
    Pitch angle 90 degrees with tube pitch at 0.0225m,
    Two tube passes,
    Shell side diameter : 0.26 m,
The instrumentation allows for the measurement of inlet and
outlet temperatures and mass flow rates on both sides of the
HX. Inlet temperature or mass flow rate variations can be
produced on one or two sides of the HX. From these experiments
the results are :

<u>Steady state</u>

The HX has been first tested with steady state conditions and
the CETUC code has been used to calculate the steady state. The
experiment shows a maximum deviation on the heat duty of 5% and
about 10% on the pressure drop.

<u>Mass flowrate transient</u>

Figure 4 shows numerical and experimental comparisons for a
sudden variation of the tube side mass flowrate. The left curve
shows the mass flow rate evolutions and on the other curve the
inlet-outlet temperature differences for both streams.

<u>Inlet temperature transient</u>

Figure 5 shows numerical and experimental comparisons for a
sudden variation of the shell side inlet temperature.

The comparisons show similar results between the experiment and
the software.

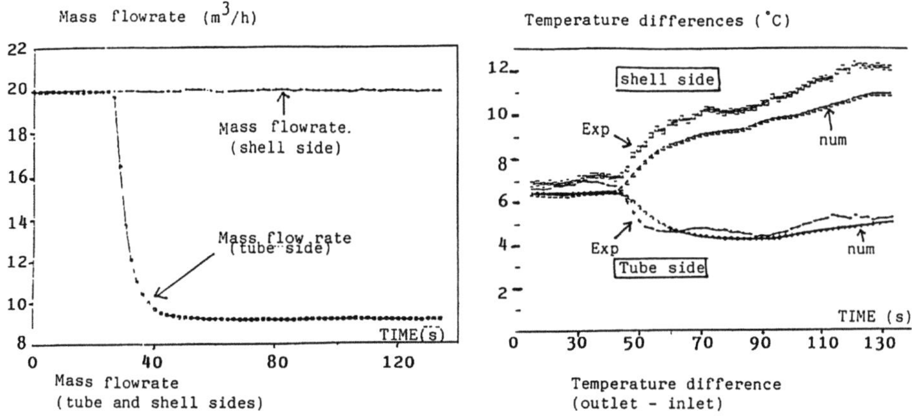

**Figure 4 : Numerical and experimental calculation
for mass flowrate transient**

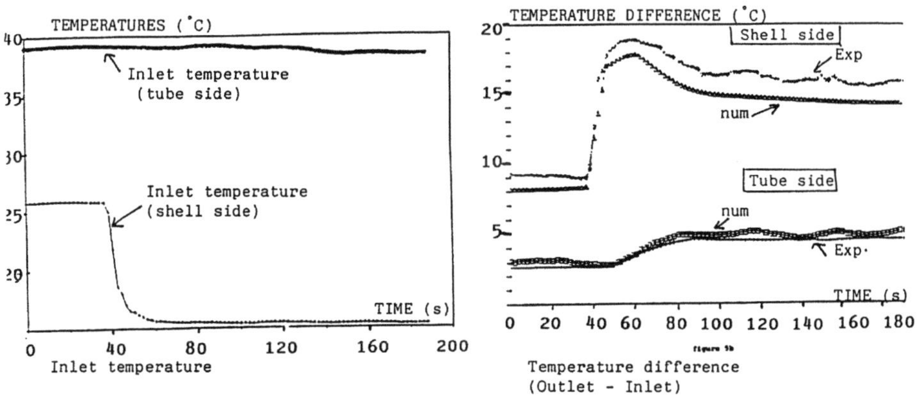

**Figure 5 : Numerical and experimental calculation
for temperature transient**

8/ USING THE SOFTWARE

The software has been written with the cetuc environments and is currently an option of the cetuc code. The software allows for the calculation of the initial and final steady states using the cetuc calculation and postprocessor units. The main characteristics for the input data unit are in common. The sofware is written in fortran 77 and can run on pc computers.
In addition to the cetuc software, during transient calculation a synoptic (figure 6) allows the user to look at the intlet and outlet processes and main geometrical characteristics. A graphic postprocessor unit can be used to visualise the evolution of the process.

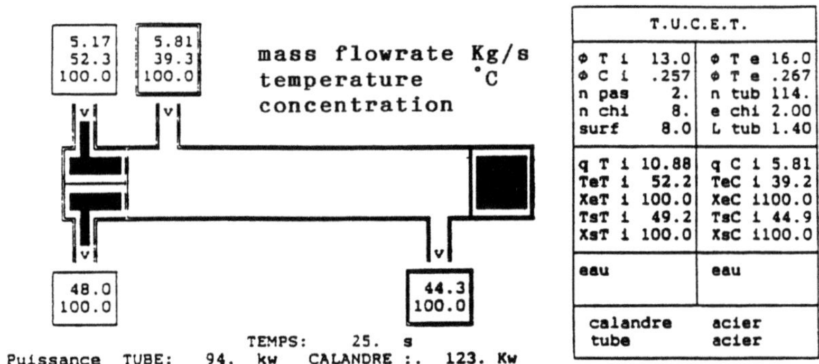

Figure 6 : Synoptic of the heat exchanger during calculation

9/ CONCLUSION

The Software is a practical validated tool to study the dynamic behaviour of an HX under transient conditions. Further validations will deal the changing fluids transient.Future developments will treat a complex network of different heat exchangers.

REFERENCE

ref 1 : P.MERCIER and G.RATEL : Dimensionnement des échangeurs à tubes et calandre, Le logiciel CETUC. Revue générale de thermique N°313 janvier 1988.

ref 2 : S.KAKAC, A.E.BERGLES, F.MAYINGER : Heat Exchangers thermal hydraulic fundamentals and design. Mac graw - Hill

ref 3 : GNIELINSKI Heat exchangers design handbook ; HEmisphère publishing corporation P 2.5.1.2

ref 4 : GNIELINSKI Heat exchangers design handbook ; Hemisphère publishing corporation P 2.5.1.52

ref 5 : D.AZILINON, P.PIERSON, J.PADET : constante de temps dans les échangeurs thermiques. Revue générale de thermique N°338 janvier 1990.

ref 6 : R.VIDIL , G.ICART : La plateforme d'essai ESTHER du GRETh: un outil de qualification des échangeurs à la disposition des constructeurs. Journées MEI90, Echangeurs et récupérateurs de chaleur,1990.

NOMENCLATURE

ρ mass Density
μ dynamic Viscosity
λ thermal conductibility

Am   heat exchange area for the tube calculated with Dm
As   heat exchange areas
C   mass concentration
Cp   heat capacity t constant pressure
D   tube diameter
Dm   average tube diameter
e   thickness
h   heat transfer coefficient

$\dot{Q}$ heat rate (power)

$\dot{q}$ heat flux
t   time
u   velocity
U   overall heat transfer coefficient
vol   volume of the component

Subscript

am   upwind
b   bulk
ext   external
f   flow region
i   inside tube
o   outside tube
s   solid region
ss   for the shell side flow
t   for the tubes (component)
ts   for the tube side flow
tp   tubular plate
tps   shell side of the tubular plate
tpt   tube side(header or turnaround) of the tubular plate

Cross-Flow Heat Exchangers

Approximate Equations for the Design
of Cross-Flow Heat Exchangers

B. Spang and W. Roetzel

Institute of Thermodynamics
University of the Federal Armed Forces Hamburg, Germany

Summary

Cross-counterflow configurations with one fluid unmixed throughout are important
flow arrangements for the design of air-coolers and extended surface exchangers.
Approximate equations for the thermal calculation of these flow arrangements with
arbitrary numbers of passes are proposed. Their accuracy and limitations are
discussed. The accuracy of the given equations is more than sufficient for practical
design.

1 Introduction

In the thermal theory of cross-flow heat exchangers, fluids are treated as either fully
mixed or unmixed at every cross section of a single pass. In multi-pass
arrangements the thermal performance depends additionally on the overall
arrangement (either cross-counter or cross-parallel flow) and on the degree of
mixing between passes (either mixed or unmixed). If one fluid is unmixed
throughout, a further criterion is whether the other fluid has alternate flow
directions in consecutive passes or the same flow direction in each pass.

Cross-counterflow arrangements with one fluid unmixed throughout represent
important industrial exchanger designs. Figure 1 shows schematic sketches of three
different arrangements. Following the usual notation, fluid 2 is unmixed throughout
the exchanger. The arrangement shown in Fig. 1a with fluid 1 unmixed within
passes, mixed between passes, and with alternate flow directions models fin-tube
exchangers containing a large number of tube rows per pass or by plate-fin
exchangers, both with header connections between passes. The arrangements shown
in Fig. 1b and 1c with fluid 1 mixed throughout differ only by the flow direction
of fluid 1 in subsequent passes. Examples of the arrangement shown in Fig. 1b are
air-fin coolers with one tube row per pass and U-tube connections. The arrangement shown in Fig. 1c models exchangers with helically coiled tubes (Fig. 2).

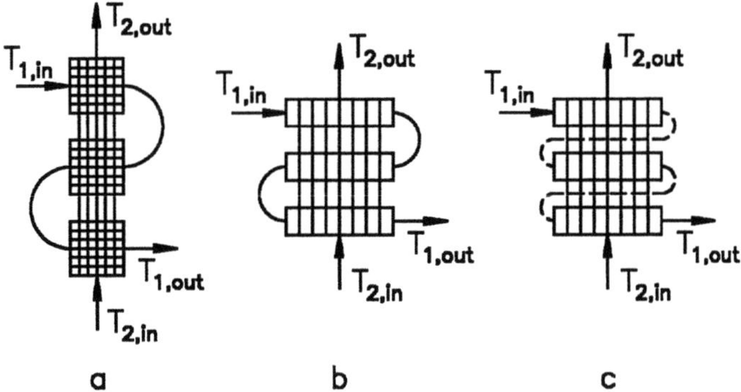

Fig. 1 Types of cross-counterflow arrangements (schematic with three passes).

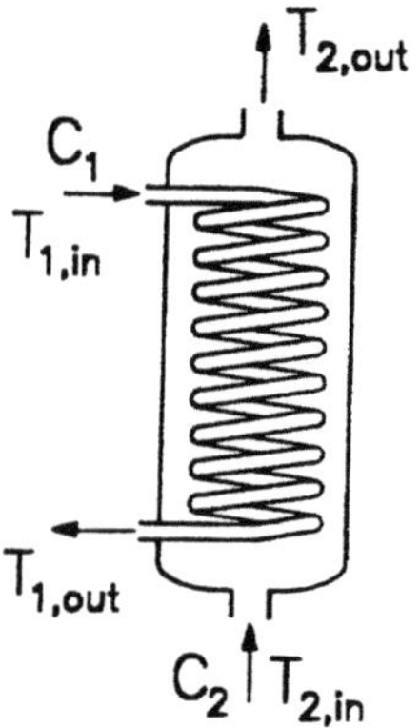

Fig. 2 Exchanger with helically coiled tubes.

With one pass, the solution for the unmixed-unmixed arrangement in Fig. 1a was derived by Nusselt [1], later brought into a more mangeable form by Mason [2]. Closed-form equations for the arrangements in Fig. 1b and 1c with two and three passes were given by Stevens et al. [3] and with up to six passes by Nicole [4]. In general, equations for the MTD of these arrangements with arbitrary number of passes can be obtained using an analytical method given by Braun [5]. However, this procedure yields rather complex expressions even with relatively small numbers of passes. For the arrangement shown in Fig. 1a with two and three passes Baclic and Gvozdenac (cited after [6]) have derived complex expressions using Bessel functions.

Obviously, closed-form equations for the calculation of the MTD of these important flow arrangements are available only for small numbers of passes. In this paper approximate equations for arbitrary numbers of passes will be developed and their limitations and accuracy will be discussed. The analytical and numerical methods used to derive and test the approximate equations are based on the usual idealizations for heat exchanger analysis [7].

2 Development of the approximate equations

2.1 Fluid 1 unmixed within and mixed between passes

This is the arrangement shown in Fig. 1a. Exact values of the MTD or the log mean temperature (LMTD) correction factor F were determined for a single pass ($N=1$) from the equation given by Mason [2] and for multi-pass units ($N=2$, 3, ...) by a numerical finite difference technique similar to that described by Stevens et al. [3]. Each pass was divided into 100 x 100 elements. A more precise subdivision does not change the results for NTU-values up to 100.

The starting point for the development of the approximate equation are equations for the LMTD correction factor $F_1(NTU_1)$ for $N=1$ and $F_\infty(NTU_1)$ for $N \to \infty$, both for a heat capacity rate ratio $R_1=1$. Correction factors F for arrangements with arbitrary number N of passes are then calculated according to

$$F = \frac{1}{N} F_1 + \frac{N-1}{N} F_\infty \tag{1}$$

The equations for $R_1=1$ are extended to arbitrary values of R_1 ($0 \le R_1 < \infty$) by replacing NTU_1 by the geometric mean value ξ of NTU_1 and NTU_2 in one pass

$$\xi = \frac{1}{N} \sqrt{NTU_1 NTU_2} = \sqrt{R_1} \frac{NTU_1}{N} \tag{2}$$

The equations for F_1 and F_∞, which can be written immediately as functions of ξ, were determined as follows. The exact equation for F_1 with $R_1=1$ can be written in terms of modified Bessel functions [6]. For large values of ξ the exact equation can be approximated by asymptotic expansions of these modified Bessel functions for large arguments. Inclusion of the first two terms of these expansions yields

$$F_1 = \frac{\sqrt{\pi \cdot \xi}}{\xi - 0.0625} - \frac{1}{\xi} \tag{3}$$

Use of eq. (3) is recommended for $\xi > 2$. For smaller values of ξ the use of the limiting forms of modified Bessel functions for small arguments does not prove accurate enough. Instead, the empirical equation

$$F_1 = (1 + 0.9\,\xi^2)^{-0.15} \tag{4}$$

is recommended for $\xi \leq 2$.

Values of F_∞ were generated from exact values of F for finite $N \geq 2$ according to

$$F_\infty = \frac{N \cdot F - F_1}{N-1} \tag{5}$$

A parameter fit using exact values (from numerical calculations) of F for $R_1 = 1$ and N up to 10 yielded

$$F_\infty = (1 + 0.63\,\xi^2)^{-0.24} \tag{6}$$

Equations (1) – (6) were primarly developed for the thermal calculation of multi-pass arrangements. As a side result eqs. (3) or (4) emerged for the calculation of the single-pass unmixed-unmixed cross-flow configuration. However, even more accurate results can be obtained for $N=1$ if the variable

$$\tilde{\xi} = NTU_1 \left(0.6\,\sqrt{R_1} + \frac{0.8\,R_1}{1 + R_1} \right) \tag{7}$$

is used in eqs. (3) or (4) instead of ξ according to eq. (2).

2.2 Fluid 1 mixed throughout, alternate flow directions in alternate passes

This is the arrangement shown in Fig. 1b. Exact values of F were generated using the procedure given by Braun [5]. The starting point for the development of an approximate equation was a formula derived by Hausen [8] for $R_1 = 1$ under the assumption that the outlet temperature profiles of fluid 2 of all passes are similar and that they are not affected by the uniform inlet temperature profile of the first pass. This equation was extended to arbitrary values of R_1 by replacing the expression NTU_1/N by the variable ξ according to eq. (2). The resulting approximate equation for the LMTD correction factor is

$$F = \frac{1}{\xi} \cdot \frac{3\sinh\xi}{1 + 2\cosh\xi} \tag{8}$$

2.3 Fluid 1 mixed throughout, same flow direction in each pass

This is the arrangement shown in Fig. 1c. Exact values of F were generated using an analytical calculation method which is described in the unpublished ref. [9]. Hausen [8] has also derived an equation for this arrangement with $R_1 = 1$ under the above mentioned assumption. For arbitrary R_1 it follows that

$$F = \frac{1}{\xi} \cdot \frac{2(1-e^{-\xi})}{1 + e^{-\xi}} \tag{9}$$

So far only equations have been given which are directly suitable for rating of heat exchangers (given NTU-values) but have to be solved iteratively for sizing problems (given thermal effectivenesses). For the arrangement under consideration an approximate equation explicit in NTU was also derived. For this it was assumed that the thermal effectiveness P_{1i} of pass i is equal for all passes $i = 1 \dots N$. Then, according to the rules for series coupling of identical exchangers in overall counterflow [6, 7], P_{1i} is related to the overall effectiveness P_1 as

$$\Psi = \left(\frac{1 - R_1 P_1}{1 - P_1} \right)^{1/N} = \frac{1 - R_1 P_{1i}}{1 - P_{1i}} \qquad \text{for } R_1 \neq 1 \tag{10}$$

Using an equation of Hausen [8] for the relation between P_{1i} and the NTU-value NTU_{1i} of an individual pass, the overall NTU can be calculated according to

$$NTU_1 = N \cdot NTU_{1i} = \frac{N}{R_1} \cdot \ln \frac{1 - \Psi + R_1 \cdot \ln \Psi}{1 - (1 - R_1 \cdot \ln \Psi) \cdot \Psi} \qquad \text{for } R_1 \neq 1 \tag{11}$$

For $R_1 = 1$ it follows that

$$NTU_1 = N \cdot \ln \frac{2N(1-P_1) + P_1}{2N(1-P_1) - P_1} \tag{12}$$

3 Accuracy and limitations of the equations

For the cross-counterflow arrangement shown in Fig. 1a, the deviations between the LMTD correction factors F according to eq. (1) and those from exact numerical calculations decrease with increasing number of passes. A comparision for $N = 2$ is shown in Fig. 3. Because of eq. (2), the results from eq. (1) coincide with the exact values for $R_1 \rightarrow 0$ ($NTU_2 \rightarrow 0$) and $R_1 \rightarrow \infty$ ($NTU_1 \rightarrow 0$) as $\xi \rightarrow 0$

and $F \to 1$ in both cases. A very good agreement is also achieved for heat capacity rate ratios of about one. Deviations as large as 25 % can occur for intermediate values of R_1 ($R_1 = 0.3$ and $R_1 = 3$).

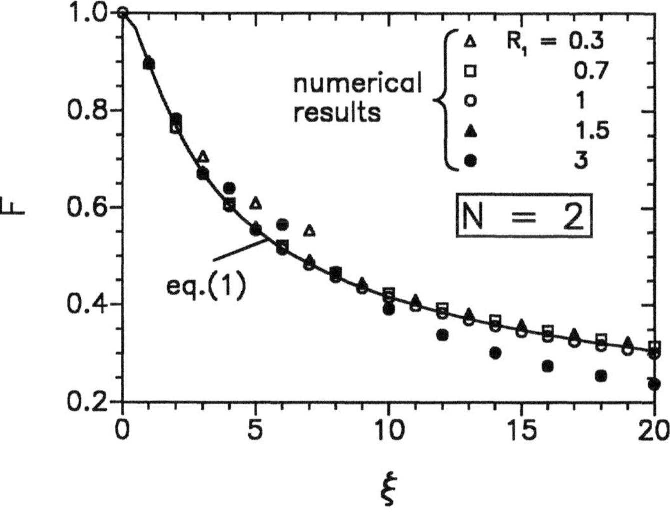

Fig. 3 Comparison of results from eq. (1) with numerical results for cross-counter-flow arrangement shown in Fig. 1a with two passes.

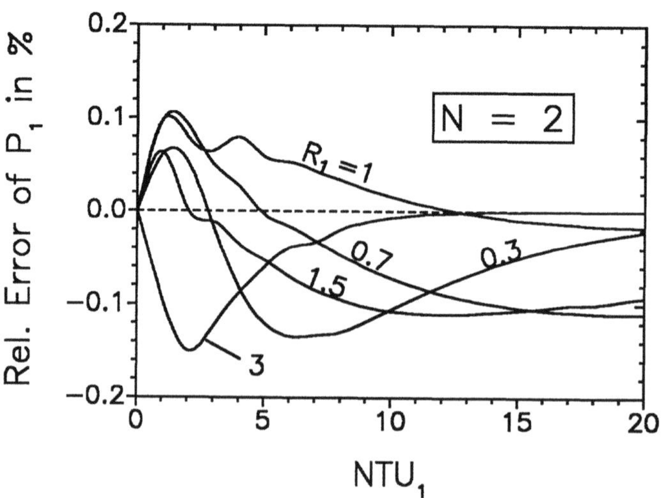

Fig. 4 Relative error of P_1 from eq. (1) for $N = 2$.

However, the correction factor F is only a provisonal result in design calculations. The quantity of primary interest is the thermal effectiveness P_1, which can be calculated from F according to the generally valid counterflow equation

$$P_1 = \frac{1 - \exp\left[(R_1-1)\,NTU_1\,F\right]}{1 - R_1\exp\left[(R_1-1)\,NTU_1\,F\right]} \qquad \text{for } R_1 \neq 1 \tag{13}$$

$$\text{or} \qquad P_1 = \frac{NTU_1\,F}{1 + NTU_1\,F} \qquad \text{for } R_1 = 1 \tag{14}$$

It appears that P_1 is very insensitive to changes in F in the region where the large errors in F occur. Figure 4 shows the relative error of P_1 versus NTU_1 for $N=2$ and several values of R_1.

For $N=1$ (single-pass unmixed-unmixed cross-flow) the proposed equations produce a slightly greater error. Figure 5a shows the $F\text{-}P_1$ chart for this flow arrangement where the continous lines were calculated with the exact equation and the dashed lines arise from eqs. (3), (4) using ξ according to eq. (2). Whereas the accuracy is sufficient for $R_1=1$, the deviations become larger for small values of R_1. Better results for $N=1$ can be obtained if instead of ξ another variable $\tilde{\xi}$ according to eq. (7) is used in eqs. (3), (4) (Fig. 5b).

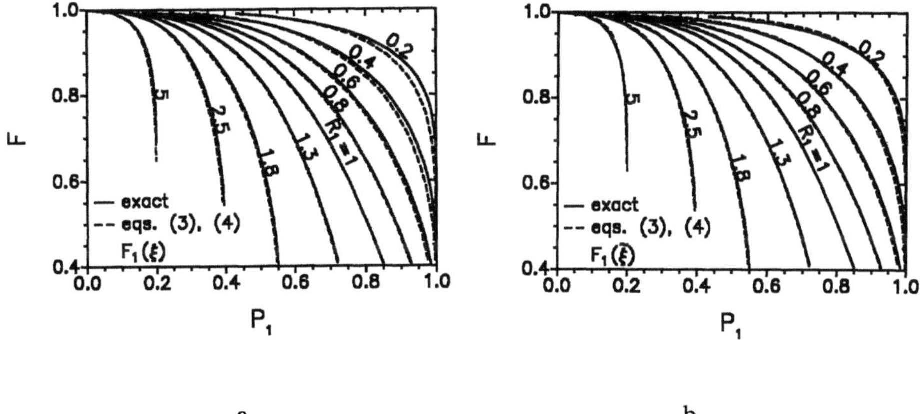

Fig. 5 $F\text{-}P_1$ charts for single-pass unmixed-unmixed cross-flow. Comparison of exact results with eqs. (3), (4) using a) ξ according to eq. (2) or b) $\tilde{\xi}$ according to eq. (7).

For the cross-counterflow arrangements with fluid 1 mixed throughout (Fig. 1b and c), it is expected that the given eqs. (8) and (9) will yield more accurate results if the number of passes is increased. This effect is demonstrated in Figs. 6 and 7. It can be seen that for $R_1 = 1$ the fit is excellent over the entire ξ range. The largest deviations occur for $R_1 \neq 1$ and $F < 0.7$ which are without practical significance. As the exact expressions for two and three passes given by Stevens et al. [3] are very simple, the use of eqs. (8) and (9) is recommended for $N \geq 4$.

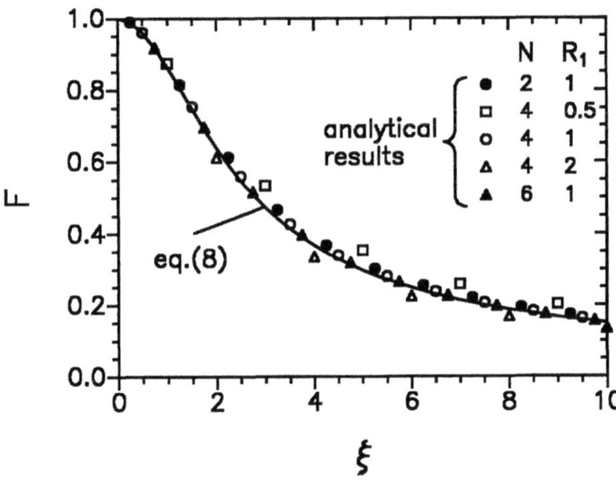

Fig. 6 Comparison of results from eq. (8) with analytical results.

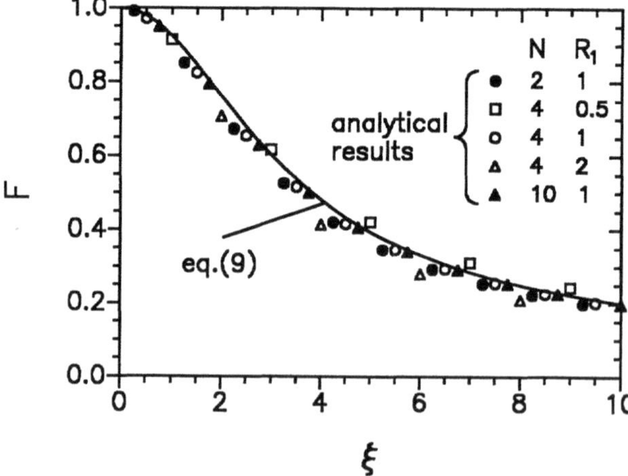

Fig. 7 Comparison of results from eq. (9) with analytical results.

For the equations discussed so far the upper limits of the error of the thermal effectiveness together with the limits of N are summarized in Table 1. The error limits are valid for $0 \leq NTU_1 < \infty$ and $0 \leq R_1 < \infty$.

Table 1 Limits of number of passes and maximum error of thermal effectiveness for approximation equations.

Flow arrangement shown in Fig.	Equations	N	Max. error of P_1 in %
1a	(1) with (2)-(4) and (6)	≥ 2	± 0.2
1a	(3), (4) with (2)	1	± 0.6
1a	(3), (4) with (7)	1	± 0.3
1b	(8)	≥ 4	± 1
1c	(9)	≥ 4	± 0.8

Equations (11), (12) for the flow arrangement with fluid 1 mixed throughout and same flow directions in each pass (Fig. 1c) are useful for sizing of heat exchangers. Figure 8, where NTU_1/N is plotted versus the argument of the logarithm on the right-hand side of eq. (12), shows a comparision with analytical results. The deviations increase with decreasing N and increasing value on the abscissa which means increasing P_1. The use of eqs. (11), (12) is recommended for sizing of exchangers with more than four passes and values of the correction factor F greater than 0.8. In this range the error of NTU_1 is always less than 5 %.

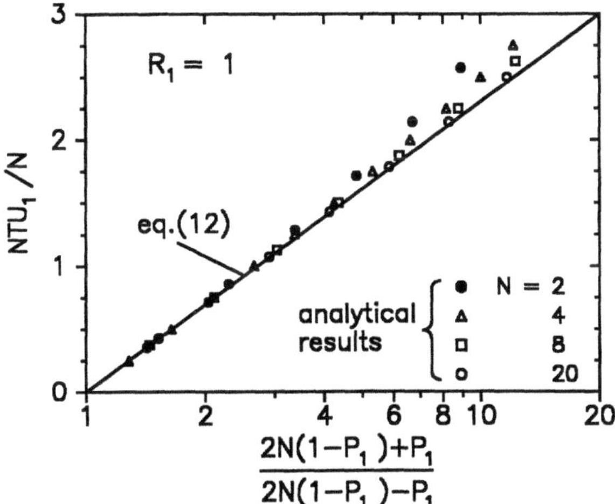

Fig. 8 Comparison of results from eq. (12) with analytical results.

Nomenclature

C	heat capacity rate, $C = \dot{M} c_p$
F	LMTD correction factor, $F = \Delta T_m / \Delta T_{ln}$
N	number of passes
NTU	number of transfer units, $NTU_1 = U A / C_1$, $NTU_2 = U A / C_2$
P	thermal effectiveness, $P_1 = (T_{1,in} - T_{1,out}) / (T_{1,in} - T_{2,in})$
R	heat capacity rate ratio, $R_1 = C_1 / C_2$
ξ	variable defined in eq. (2)
$\tilde{\xi}$	variable defined in eq. (7)
Ψ	variable defined in eq. (10)

Subscripts

1, 2	fluid 1, 2, except F_1 LMTD correction factor for a single-pass unit
∞	for multi-pass unit with number of passes $\rightarrow \infty$
i	pass i

References

1. Nusselt, W.: Eine neue Formel für den Wärmedurchgang im Kreuzstrom. Techn. Mechanik Thermodyn. 1 (1930) 417-422.

2. Mason, J.L.: Heat transfer in cross-flow. Proc. 2nd US Natnl. Congr. Appl. Mech., ASME (1955) 801-803.

3. Stevens, R.A.; Fernandez, J.; Woolf, J.R.: Mean temperature difference in one, two and three-pass crossflow heat exchangers. Trans. ASME 79 (1957) 287-297.

4. Nicole, F.J.L.: Mean temperature difference in cross-flow heat exchange, applied to multipass air-cooled fin-tube units with a finite number of rows. CSIR Special Report CHEM 223 (1972).

5. Braun, B.: Wärmeübergang und Temperaturverlauf in Querstrom-Rohrbündeln bei beliebiger Schaltung der Rohrreihen. Forsch. Ing.-Wes. 41 (1975) 181-191.

6. Shah, R.K.; Mueller, A.C.: Heat exchangers. Chapter 4 of Handbook of Heat Transfer Applications, 2nd ed., New York: McGraw-Hill 1985.

7. Roetzel, W.: Berechnung von Wärmeübertragern. Chapter C of VDI-Wärmeatlas, 5th ed., Düsseldorf: VDI-Verlag 1988.

8. Hausen, H.: Wärmeübertragung im Gegenstrom, Gleichstrom und Kreuzstrom. 2nd ed., Berlin, Heidelberg, New York: Springer-Verlag 1976, p. 209 ff.

9. Bes, Th.; Roetzel, W.: Report on the research project "Schlangenrohrwärmeaustauscher" to Stifterverband für die Deutsche Wissenschaft. Hamburg 1983.

Numerical Analysis of Cross-Flow Heat Exchangers in Order to Establish a New Design Method

O. Marin, S. Petrescu, N. Baran

Technical Thermodynamics Division
Polytechnic Institute of Bucharest

Summary.

Due to the large field of application of the cross flow heat exchangers and to the limits of the classic designing method, there has been created a new one. The analytic solution of the resulting system of differential equations is very difficult; that's why there has been used a numerical method. This one establishes, given the initial conditions and the necessary heat exchange surface, the final medium difference of temperature between the two fluids, as well as the final medium temperatures of them. There have been drawn diagrams that might be used properly. There has been solved a practical problem.

The classical designing method.

The following steps are to be taken:

a) The calculus of the outlet temperatures of the two fluids :

$$T_1{''} = \frac{T_2{'} + (T_1{'} (1 - \phi) - T_2{'}) \exp[-(1 - \phi) U A/(\dot{m}_1 c_1 \eta)]}{1 - \phi \exp[-(1 - \phi) U A)/(\dot{m}_1 c_1 \eta)]}$$

$$T_2{''} = \frac{(1 - \phi)T_2{'} + \phi T_1{'} (1 - \exp[-(1 - \phi) U A/(\dot{m}_1 c_1 \eta)])}{1 - \phi \exp[-(1 - \phi) U A/(\dot{m}_1 c_1 \eta)]}$$

where ϕ is defined by the relation:

$$\phi = (\dot{m}_1 c_1 \eta) / (\dot{m}_2 c_2)$$

b) there are introduced the following dimensionless groups:

$$P=(T_2{''} - T_2{'})/(T_1{'} - T_2{'}) ; \quad R=(T_1{'} - T_1{''})/(T_2{''} - T_2{'}) ; \quad \psi=\Delta\bar{T}_{ln}/\Delta\bar{T}_{ln,c} .$$

c) the heat flow results:

$$\dot{Q} = \dot{m}_1 c_1 (T_2{''} - T_2{'}) R \eta = \dot{m}_2 c_2 (T_2{''} - T_2{'}) = U A \psi \Delta\bar{T}_{ln,c}$$

where $\psi = \psi(P,R)$ results theoretically or experimentally [1, 2].

136

Suggested mathematical method.

The thermal calculus of the plane heat exchangers is based on the following equations:

- thermal survey equation: $\dot{Q}_1 = \dot{Q}_2 / \eta$,

- sizing equation: $\dot{Q} = U A \, \Delta\bar{T}_{ln}$.

Figure 1 presents a cross flow heat exchanger, with the explicit notations.

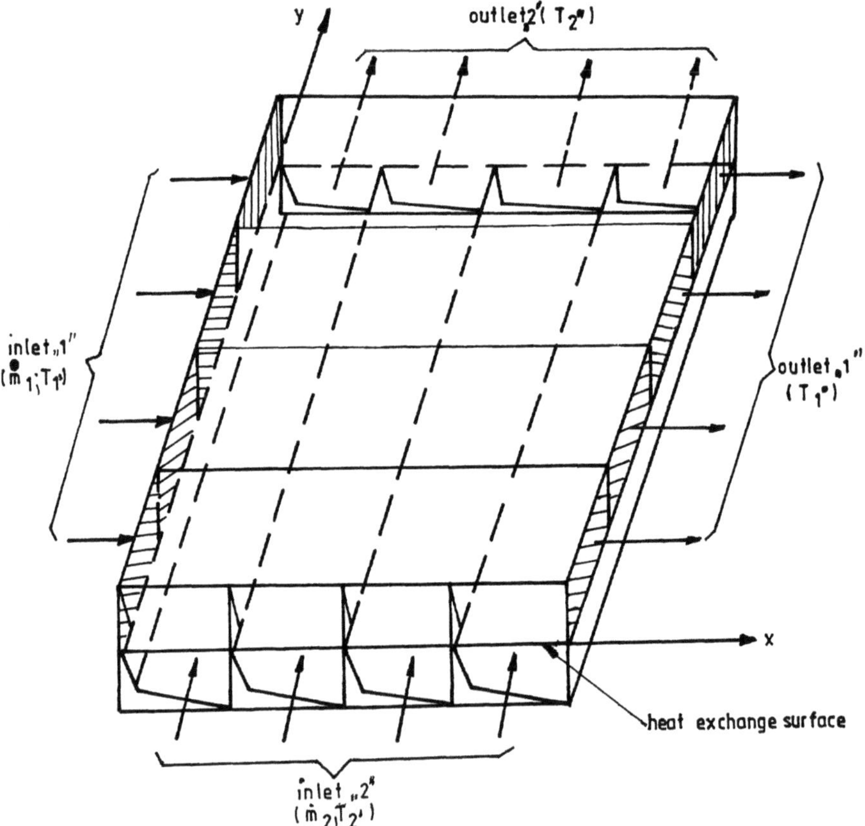

Fig. 1. Cross flow heat exchanger

The previous equation could be rewritten for the given situation:

$$\eta \, \dot{Q}_1 = \dot{m}_1 \, c_1 \, \Delta T_1 \, \eta = \dot{m}_2 \, c_2 \, \Delta T_2 = \dot{Q}_2 \ .$$

The differential equations of the previous ones are:

$$\eta \ \delta\dot{Q}_1 \ = \ \delta\dot{Q}_2 \ = \ U \ \Delta T \ dA \tag{1}$$

The overall heat transfer coefficient is determined with the formula:

$$U \ = \ 1 \ / \ (\ 1/\alpha_1 \ + \ \delta/\lambda \ + \ 1/\alpha_2 \ )$$

The equation (1) could be rewritten:

$$\eta \ \dot{m}_1 \ c_1 \ dT_1 \ = \ \dot{m}_2 \ c_2 \ dT_2 \ = \ U \ (T_1 \ - \ T_2) \ dA \tag{2}$$

In this circumstances, T_1 and T_2 are functions of two coordinates:

$$T_1 \ = \ T_1(x, \ y) \ ; \qquad T_2 \ = \ T_2(x, \ y) \tag{3}$$

Equations (2) form a system of two differential equations:

$$\eta \ \dot{m}_1 \ c_1 \ dT_1 \ = \ U \ (T_1 \ - \ T_2) \ dA$$

$$\dot{m}_2 \ c_2 \ dT_2 \ = \ U \ (T_1 \ - T_2) \ dA \tag{4}$$

The system (4) has been solved numerically ; the heat exchanger has been divided into a certain number of stripes.

There has been elaborated a computer program ; the final objective of it is the establishing of the temperature distributions along the heat exchanger outlet, as well as the medium temperatures at outlet – $\bar{T}_{1";2"}$ and their difference $\Delta T"$ (fig.2).

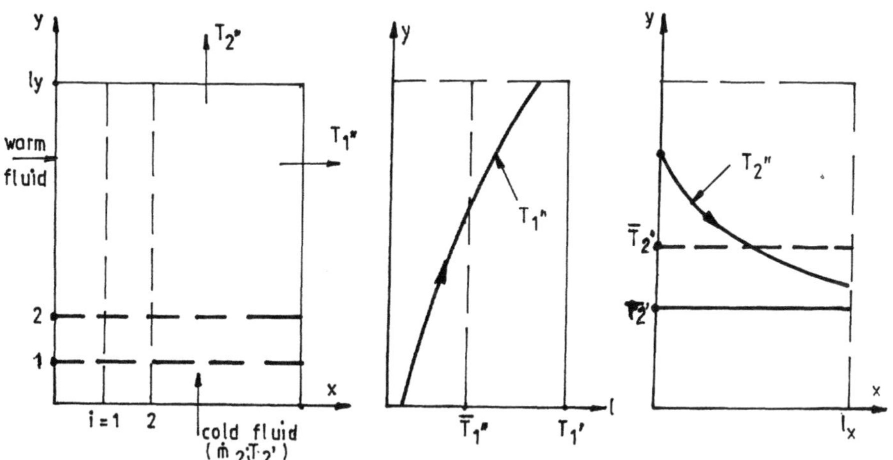

Fig. 2.   The outlet distribution of temperature

There have been used the following dimensionless groups:

$$E = l_y / l_x$$

$$F = (\dot{m}_2 \, c_2) / (\eta \, \dot{m}_1 \, c_1) \, .$$

Given the distribution of temperatures, one could find the medium temperatures:

$$\bar{T}_{1";2"} = \frac{{}_0\!\int^1 T_{1";2"} \, dx}{l_{x,y}}$$

In the end there results the final medium difference:

$$\Delta T \, " = \bar{T}_{1"} - \bar{T}_{2"} \, .$$

Finally there has been drawn a diagram (fig. 3) with some usual values for the length of the heat exchanger – $1 \div 3$ meters and for the terms : $\dot{m}_1 c_1$ - $100 \div 300$ W/K. There have been chosen the following values for the dimensionless groups : $E = 1.5$; $F = 2$. The diagram shows the variation of $\Delta T"_0$ with the overall heat exchange coefficient.

The way of using the diagram is : having chosen the dimensions and the overall heat exchange coefficient, there results the final medium difference of temperatures for an initial difference of them $\Delta T'_0 = 100°C$ ($\Delta T'_0 = T_1' - T_2'$). For any other initial difference of temperatures $\Delta T'$, the final one $\Delta T"$ results:

$$\Delta T" = (\Delta T'/100) \, \Delta T"_0 \tag{5}$$

In the end one could calculate the values of $\bar{T}_{1";2"}$ with the help of the following equations:

$$\bar{T}_{1"} = (\, T_1' + F \, (\Delta T" + T_2') \,) / (1 + F)$$

$$\bar{T}_{2"} = (\, T_1' - \Delta T" + F \, T_2' \,) / (1 + F) \tag{6}$$

As one could see, the presented equations (6) don't have much in common with the first ones from this paper. Diagrams like that shown in fig.3 could be drawn very easily with the help of the computer, for any chosen values.

Entropy generation in the cross flow heat exchanger.

This method permites a facile calculus of the generation of entropy in the heat exchanger. Choosing a sector out of the surface of the heat exchanger (fig. 4), the variation of entropy results:

$$d\dot{S} = - \, \delta\dot{Q}_1/T_1 + \delta\dot{Q}_2/T_2 \, .$$

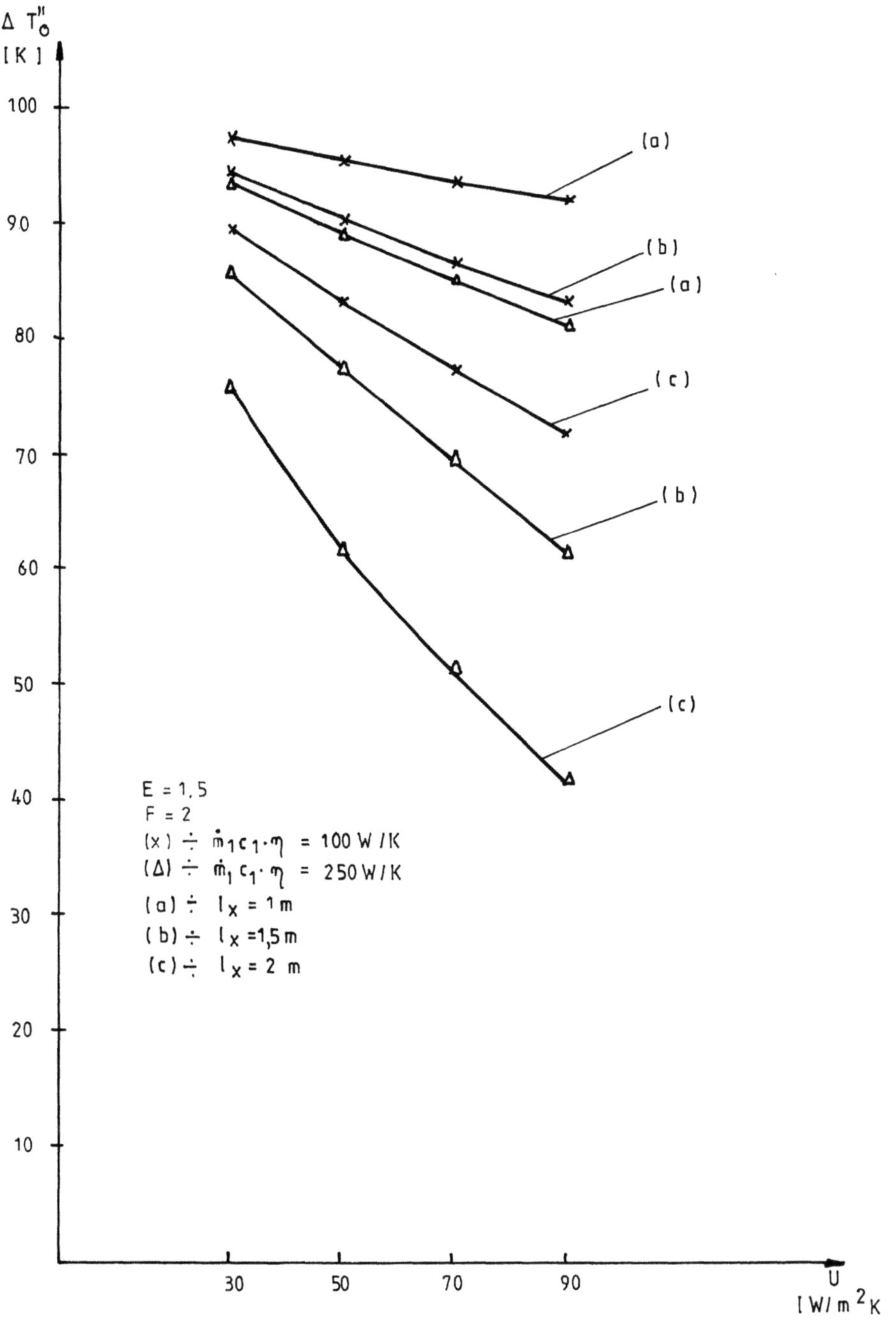

Fig. 3. The final difference of temperatures distribution

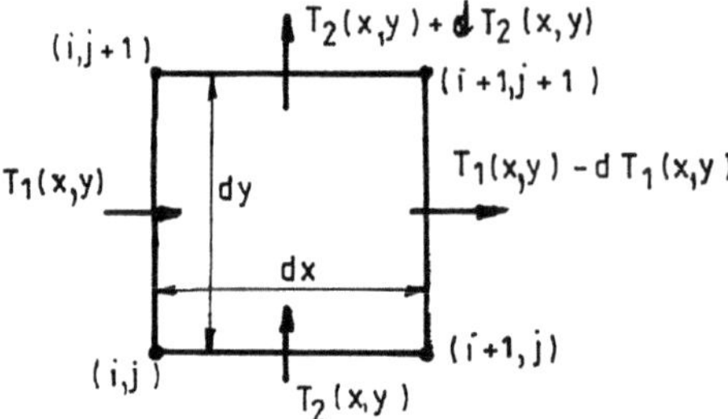

Fig. 4.   Detail of the heat exchanger

From equations (1), (2) results:

$$d\dot{S} = (\,U\,(T_1 - T_2)^2 \,/\, (T_1\,T_2)\,)\;dA\;.$$

There have been drawn two diagrams, but for a more proper plotting, they have been particularized for the application. The first of them (fig. 5) presents the generation of entropy as a function of heat exchange surface and the medium value of it:

$$\bar{\dot{S}} = (\textstyle\int \dot{S}\;dA\,)\,/\,A\;.$$

This diagram could become an efficiency indicator which, for certain situations, could lead to changes in the design of the heat exchanger.

The second diagram (fig. 6) presents a family of curves, indicating the entropy generation as a function of temperature T_1. There have been chosen four stripes, parallel with the cold fluid flow, first at the entrance of the warm fluid (0%), then at 40% of l_x , the next at 70% and the last at the exit of the warm fluid.

Application.

The following example has been chosen:

- $\eta\;\dot{m}_1\;c_1 = 100$ W/K ;

- $l_x = 1.5$ m ;

E = 1.5 ; F = 2 ; U = 50 W/m^{2}K ; $T_1' = 400°$C ; $T_2' = 100°$C .

The final difference of temperatures results (fig. 3):

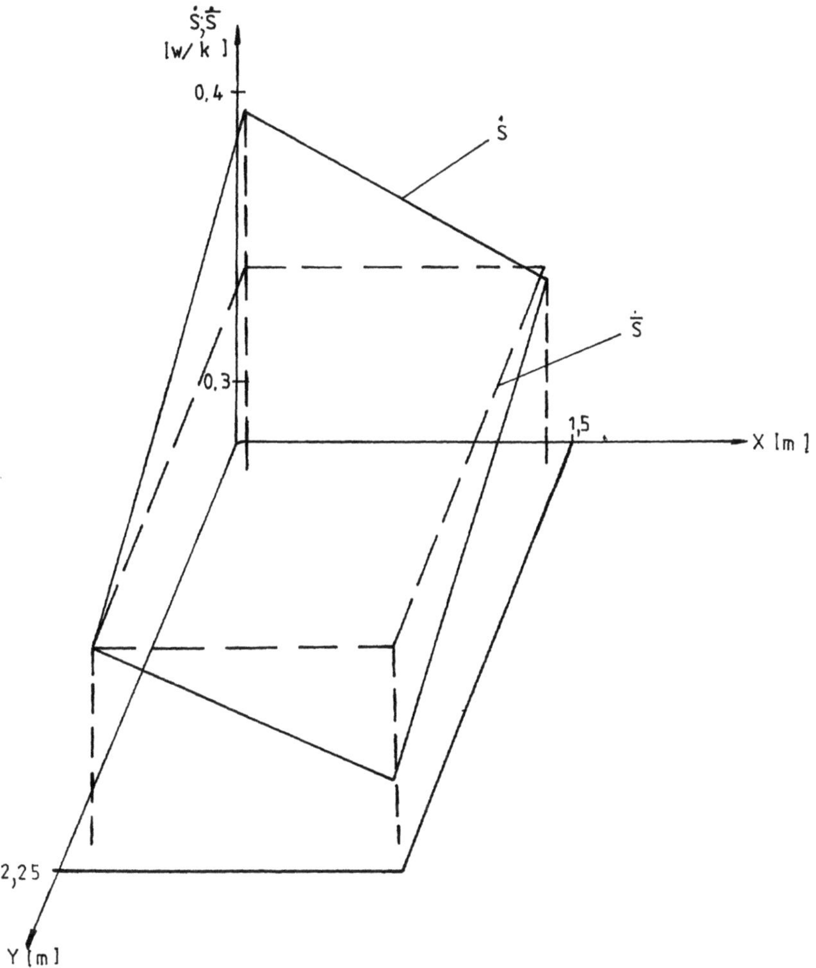

Fig. 5. Entropy generation $\dot{S} = \dot{S}(x, y)$

$$\Delta T''_o = 77.333 \ ^oC \ .$$

The real final difference of temperatures results (5):

$$\Delta T'' = 77.333 \ (400 - 100)/100 = 232^oC \ .$$

The final medium temperatures of the two fluids result (6):

$$T_1{}'' = 354.6°C \quad ; \quad T_2{}'' = 122.66°C \ .$$

These last results corespond with the ones of the computer. As one can see, the mathematical apparatus is quite easy to handle.

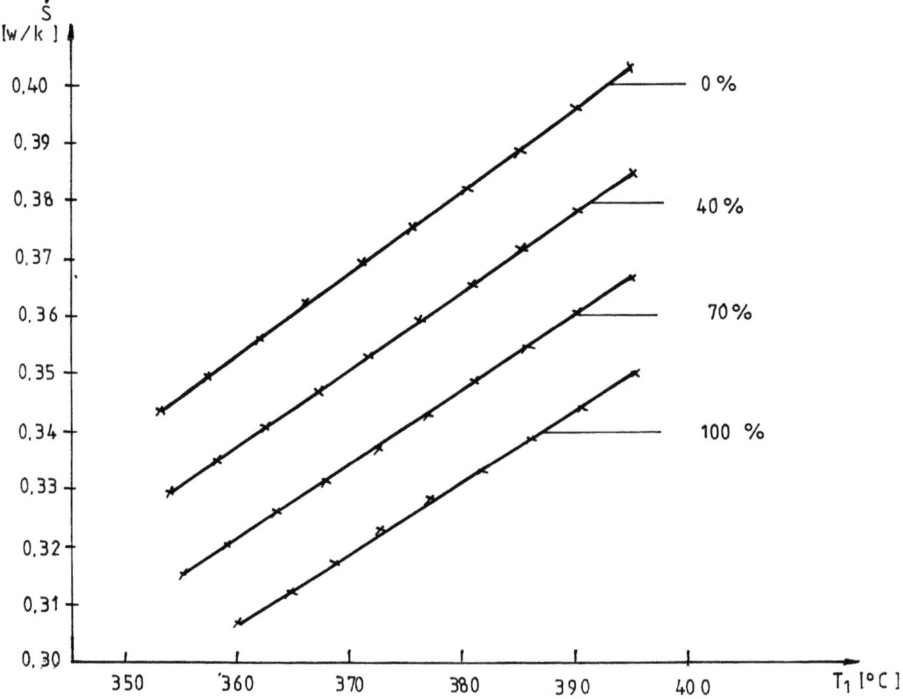

Fig. 6. Entropy generation $\dot{S} = \dot{S}(T_1)$

Conclusions.

a) The method leads easily to the final difference of temperatures at the outlet of the heat exchanger; it is very precise as well.

b) Using the presented diagrams (extrapolated for other initial conditions), it is possible to find out the temperatures at the outlet of the cross flow heat exchanger and, finally, the total heat flow and the generation of entropy.

References.

1. Leonachescu, N. : Technical Thermodynamics. Bucharest 1974.

2. Popa, B. : Technical Thermodynamics and Thermic Machines. Bucharest 1977.

3. Pop, M. G. a.o. : Indrumar, Tabele Nomograme si Formule Matematice. Bucharest 1987.

Nomenclature for physical quantities.

Quantity	Symbol	SI Unit
Area	A	m^2
Efficiency	η	–
Entropy Flow	$\dot{S}$	W/K
Flow rate	$\dot{m}$	Kg/s
Heat flow	$\dot{Q}$	W
Heat transfer		
Overall	U	W/m^2K
Convection	α	W/m^2K
Conduction	λ	W/mK
Length	$l_{x,y}$	m
Temperature		
Cross flow logarithmic difference	$\Delta\bar{T}_{ln}$	K
Counter flow logarithmic difference	$\Delta\bar{T}_{ln,c}$	K
Final medium difference	$\Delta T''$	K
Initial medium difference	$\Delta T'$	K
Instantaneous difference	ΔT	K
Inlet	$T_{1',2'}$	K
Outlet	$T_{1'',2''}$	K
Specific heat capacity	c_v, c_p	J/KgK
Thickness	δ	m
Warm fluid characteristics	$(\ )_1$	–
Cold fluid characteristics	$(\ )_2$	–

Improvement of Fin-Tube Heat Exchangers by Longitudinal Vortex Generators

M. Sanchez, M. Fiebig, N.K.Mitra

Institut für Thermo- und Fluiddynamik,
Ruhr Universität Bochum,
Postfach 102148, 4630 Bochum, Germany

Abstract
Conjugate heat transfer and the flow field between the two fins of a model
fin-tube heat exchanger with and without built-in longitudinal vortex generators in form of winglets on the fin have been numerically investigated
for different fin thickness and Reynolds numbers.

Introduction

Extended surfaces in form of fins are commonly used to increase the heat
transfer rate per unit volume of a heat exchanger. Further increase of the
heat transfer can be achieved by generating longitudinal vortices in the
flow between the fins by suitably punching small triangular or rectangular
pieces of the fin in such a way that they remain attached to the fin at the
base and stick out in the flow with an angle of attack. Depending on their
shape these vortex generators can be named as delta wings, delta winglets,
rectangular wings or restangular winglets.

Experimental investigations of **Dong** [1] showed that vortex generators in
form of delta-winglets placed on the fin in the wake of a tube reduced the
usual deterioration of the heat transfer there, and the maximization of the
heat transfer depended on the location of the vortex generators (Δx and Δy
in *fig.1*).

Because of the large number of parameters ($\Delta x, \Delta y, \beta$, size of the winglet, D,
B, H, δ, λ and Re, see *fig.1*), the optimum design of a fin-tube heat exchanger with longitudinal vortex generators require parameter studies which
can be done from numerical simulations. To this purpose a computational
scheme has been developed that can simulate the fluid flow and heat trans-

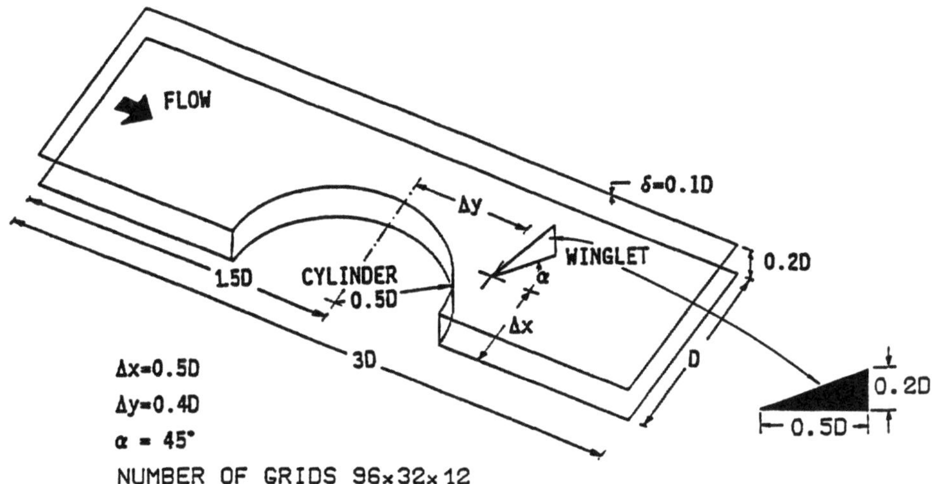

Fig.1 : *Schematic of the computational domain for a one-row fin-tube heat exchanger with a built-in delta winglet.*

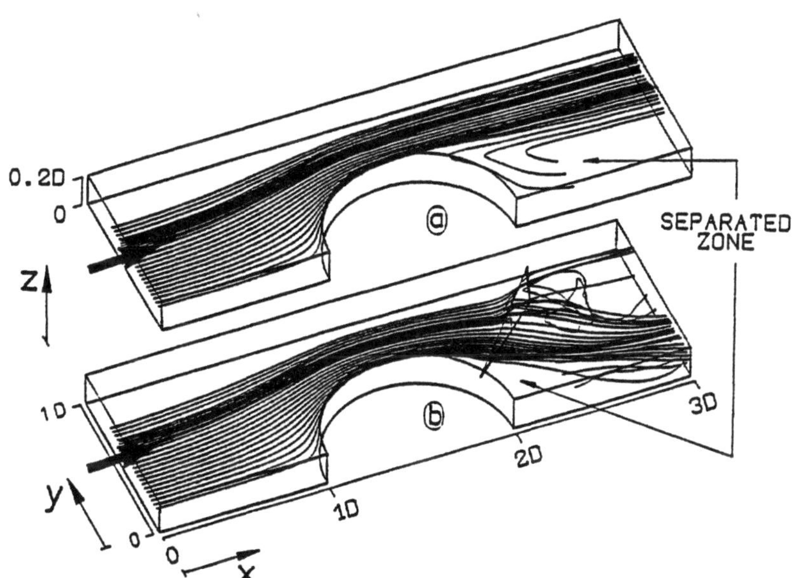

Fig. 2: *Particle tracks for Re=2000 a. Without winglets b. With winglets the separated zone in the tube wake in b. is much smaller than in a.*

146

fer around the tube and vortex generators in the channel formed by two neighboring fins together with the heat conduction in the fin.

The objective of the present paper is an exemplary numerical determination of heat transfer rate, frictional losses, and the fin efficiency of a one-row fin-tube heat exchanger with a pair of punched vortex generators on the fins.

<u>Basic Equations</u>

The model for a one-row fin-tube heat exchanger is shown in Fig.1. The location of the winglet corresponds to the optimum location for heat transfer augmentation found in **ref.[1]**. However, the channel height H (i.e. the fin pitch) and the fin size are different and more realistic here than in the cited reference.

Depending on the ratio of height(H) to diameter(D) and the Reynolds number, it is possible to obtain periodic or steady solutions (i.e. periodic at large Re and small D/H). Previous results **[2]** showed that for H/D=0.4 and Re=2000 only steady solutions were obtained. In the present calculation with H/D=0.2 and Re (defined as Re=V_{av}.Dh/ν , with Dh: hydraulic diameter=2.H, and V_{av}: mean velocity at the inlet) between 400 und 2000 a final steady state is also expected and hence only the half of the channel is modeled.

The flow and temperature field in this channel is calculated by solving the complete unsteady Navier-Stokes and energy equations for incompressible fluid with constant properties. These equations in cartesian coordinates and in dimensionless form are:

$$\frac{\partial u_i^*}{\partial x_i^*} = 0 \tag{1}$$

$$\frac{\partial u_i^*}{\partial t^*} + u_j^* \frac{\partial u_i^*}{\partial x_j^*} + \frac{\partial p^*}{\partial x_i^*} = \frac{2}{Re} \nabla^2 u_i^* \tag{2}$$

$$\frac{\partial T^*}{\partial t^*} + u_j^* \frac{\partial T^*}{\partial x_j} = \frac{2}{Re.Pr} \nabla^2 T^* \tag{3}$$

Here Pr is the Prandtl number and the dimensionless variables are defined as x_i^*=x_i/H; u_i^*=u_i/V_{av}; p_i^*=$(p_i-p\infty)$/$\rho.V_{av}^2$; t^*= t.V_{av}/H and T^*=T/T_∞, with T_∞:temperature at the inlet.

The dissipation term in the energy equation has been neglected.

Uniform velocity and temperature profiles are used as the inlet condition. The tube is assumed to have a constant temperature of T_p ($T_p \neq T_\infty$). No-slip velocity boundary conditions are used on the solid surfaces. At the channel exit the second derivatives of the velocity components and of the temperature in the main flow direction are set equal to zero.

The temperature boundary condition on the fin will couple the heat conduction in the fin to the heat transfer from the gas to the fin. Since the fin is very thin, the temperature in the fin in z-direction is assumed to be constant. The temperature on the fins can then be described by the two dimensional heat conduction equation for the steady case, which in dimensionless form is as follow

$$\frac{\lambda_F}{\lambda}\,\nabla^2 T_F^* = -\frac{1}{\delta^*} - q \tag{4}$$

where

- q : dimensionless heat flux $= \left.\dfrac{\partial T^*}{\partial z^*}\right|_{z^*=0} - \left.\dfrac{\partial T^*}{\partial z^*}\right|_{z^*=1}$

Here the ratio λ_F/λ works as a dimensionless heat conductivity of the fin while δ^* is the dimensionless fin thickness ($\delta^*=\delta/H$). In the following sections the superscript $*$ will be dropped.

The boundary conditions for the equation (4) are adiabatic condition at the fin edges and in the stamping of the winglet and constant temperature T_p at the junction to the tube wall.

Method of Solution

The temperature distribution on the fin is determined by solving the heat conduction equation (4) by a finite element (FE) algorithm. The continuity and momentum equations are decoupled from the energy equation and solved first by a 3-D finite difference (FD) algorithm. This algorithm consists of a time marching procedure where the pressure is determined by a correction scheme like that used by the SOLA algorithm of *Hirt et al*[3] and which is equivalent to the solution of a Poisson equation for the pressure. A standard Jacobi iteration is used to march in time. A central-upwind hybrid scheme is used to discretize the convective terms of eq. (2), while the diffusive terms are discretized by central differences.

Once the velocity field is available, the energy equation (by FD) and the heat conduction equation (by FE) are solved simultaneously.

The solution procedure can be described as follows:

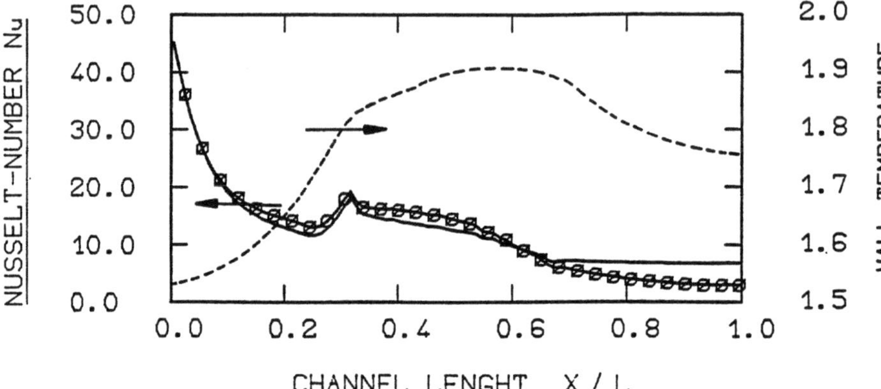

Fig.3 : Cross averaged Nusselt number distribution for a configuration without winglets, Re=2000, (———)Tw=Tp ; (—⊘—)Tw from eq.(4), with δ=0.1. The curves cross at x=0.6 where the wall temperature reaches a maximum. (- - -)Cross averaged wall temperature distribution for a configuration without winglets, Re=2000, δ=0.1.

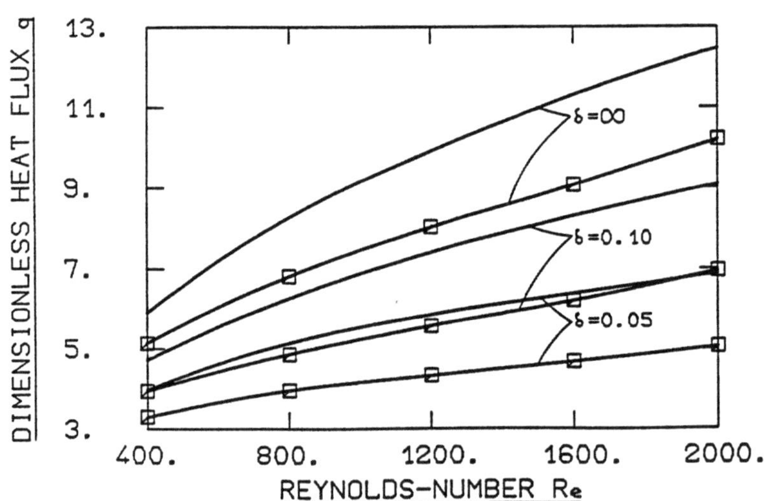

Fig.4 : Mean heat flux q̄ for configurations (—⊟—) without winglets and (———) with winglets in function of the Re and fin thickness(δ). The absolute increase in heat flux does not strongly depend on the fin thickness for a given Re.

(i) The equations (1) and (2) are first solved and the computational results give the velocity field in the channel. This is stored.

(ii) The temperature of the fin is assumed to be constant equal to T_p.

(iii) Using the stored velocity field, the energy equation (3) is solved.

(iv) Equation (4) is solved and a new wall temperature distribution is obtained, which of course differs from that assumed in step (ii).

(v) Steps (iii) and (iv) are repeated using each time the new wall temperature and the procedure continues until that two consecutive wall temperature fields do not differ by more than 10^{-5}.

This procedure has typically taken about 20 iterations to converge. The interpolations between the two different grids (FE and FD) was found not to be a draw-back for the computational scheme either in the use CPU time (much less than 1%) or in the accuracy of the results (we could observe no deviations between the contour lines of wall temperature in both grids).

The coupled solution of eqs. (3) and (4) for a case took 40 CPU hours in a 0.33 MFLOPS computer (SUN 3/60) with 46000 cells in FD and 2000 elements in FE. Of this time only 10% was needed for the solution of eq.(4).

The calculation of the flow field took in average 400 CPU hours per Re-number in SUN 3/60. Because of the huge time requirement, the programm was stored in the CRAY-YMP of the HLRZ at KFA-Julich, where part of the calculations together with the studies of the grid dependency was performed.

<u>Results</u>

Numerical results have been obtained for Pr=0.7 (air), λ_F/λ=7500 (aluminium to air) and Re between 400 and 2000 with a grid of 12 x 96 x 32 cells. Grid independence studies using Richardson extrapolation (see *ref.*[6]) and three different grids (of 12, 18, and 27 cells in height while the number of cells in the other two directions was not changed) were performed for the configuration without winglets and Re=2000. The mean Nusselt-number (Nusselt number defined with the hydraulic diameter and the bulk temperature) and the dissipation coefficient c_φ (c_φ=fapp $-\frac{\Delta\alpha}{2L}$, where the apparent friction factor fapp=2Δp$\frac{H}{L}$ and $\Delta\alpha$ is the kinetic energy correction factor which accounts for the pressure regain after the heat exchanger due to uniformization of the velocity profile) were extrapolated. The grid independent values (GI) and the deviations for the three grids are displayed in table I.

Table I: *Grid independent value (GI) and error in percent with respect to the grid independent mean value for Nu and c_φ for Re=2000 in the channel without winglets.*

	(GI)	G_I (12x98x34)	G_{II} (18x96x32)	G_{III} (27x98x34)
Nu	13.82	8.9%	6.4%	4.1%
$c\varphi$	0.052	2.9%	1.0%	0.3%

Figs 2a and **2b** compare particle tracks in channels without **(2a)** and with **(2b)** delta winglets. We notice that with vortex generators, the particles are guided closer to the tube and the separated zone in the wake is much reduced in size. The mixing effect of the winglets is also evident.

Figs.3a and **3b** show the effect of the fin temperature distribution on the Nusselt-number. The cross-averaged Nu for a reference case with Re=2000, and constant wall temperature (=> $\delta=\infty$) is compared with equivalent results for a fin with $\delta=0.1$ **(fig.3a)**. In the same figure **(fig.3b)**, the distribution of the cross averaged wall temperature is shown. The location of the maximum of the wall temperature corresponds to the crossing point of the two Nusselt number curves. The growing wall temperature in flow direction on the first part of the channel (0< x/L< 0.6) increases the Nu number, on the other side, the decreasing wall temperature in the rear part of the channel (0.6 <x/L< 1.0) strongly reduces the Nu number (up to 100% near the exit).

Fig.4 displays the mean heat flux q as a function of Re and fin thickness δ for the one row fin tube heat exchanger without and with winglets.

The relative increase in the heat flux due to the inclusion of the winglets is evident in **fig.5** which shows the ratio of the mean heat fluxes (q/q_o) with and without vortex generators. The curves show a maximum for a Re which depends on the fin thickness. The highest improvement of 37% occurs for the thinnest fin ($\delta=0.05$) at Re=800 The cause of this maximum is that at low Re numbers, the winglets have a low mixing effect and at high Re numbers the increase of the heat transfer due to the inclusion of vortex generators reduces the fin efficiency (an increase in the heat transfer produces higher temperature gradient in the fin reducing the fin efficiency).

Fig.6 shows the fin efficiency as a function of the fin thickness and the Re for flows without and with winglets. The fin efficiency is higher with winglets although here the opposite would be expected due to the higher heat transfer for this case.

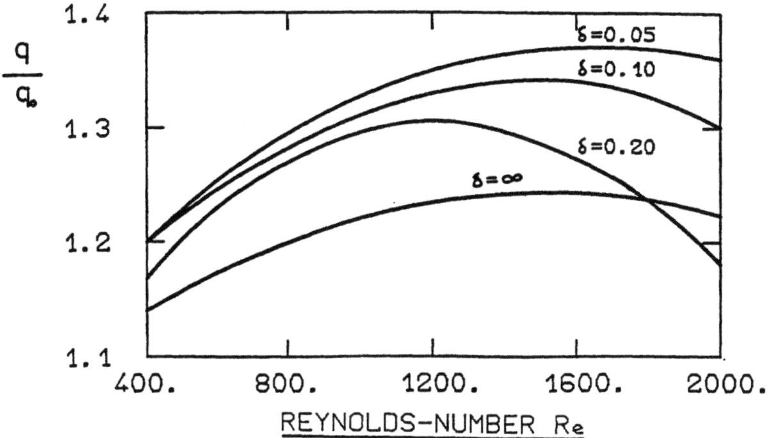

Fig.5 : *Relative heat flux $\frac{q}{q_0}$ in function of Re and fin thickness(δ). The curves show a maximum for a Re which depends on the fin thickness. For thinner fins grows the increase in the heat flux and the maximum will be displaced to the region of higher Re.*

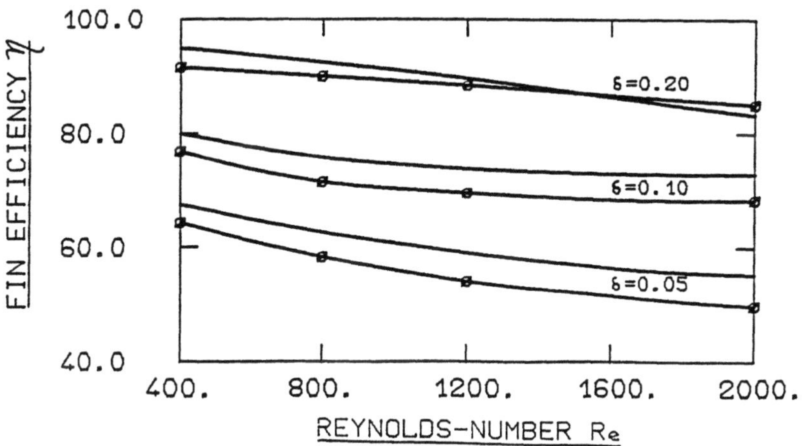

Fig.6 : *Fin efficiency versus Reynolds number for three different fin thickness(δ) and for configurations without(—⊘—) and with (———) winglets. The fin efficiency increases with an increase in the fin thickness and decreases with increasing the Re. The fin efficiency is higher for the case with winglets exept for the thickness $\delta=0.2$ and Re>1500.*

Calculations for Re=2000 and δ=0.1 with winglets where the stamping was not considered in eq. (4) showed only less than 2% deviation in efficiency from those where the stamping was considered. Hence it can be concluded that the stamping has little effect on the fin efficiency.

Fig.7 shows the apparent friction factor fapp together with the dissipation factor $c\varphi$ against Re for configurations with and without winglets. The product $c\varphi.Re^3$ is directly proportional to the pumping power [4], while the mean heat flux is proportional to the heat exchange power. *Figs. 4 and 7* can be used to compare different cases.

As an example we consider a case with a constant pumping power and fin thickness and observe the behavior of the heat flux. For Re=2000 and δ=0.1 without winglets $c\varphi.Re^3=4.16\times10^{+8}$ is obtained. The mean heat flux $\bar{q}$ is equal to 6.95 (see *fig.4*). The case with winglets with Re=1460 (obtained by interpolation) and δ=0.1 has the same $c\varphi.Re^3$, but an average heat flux of 8.0.

In the same way it is possible to maintain a constant heat flux and fin thickness and observe the dissipative losses or investigate the changes in fin thickness keeping heat flux and dissipation constant.

Conclusions

An algorithm which is able to simulate the conjugated heat transfer in very complex 3-D domains (i.e. a one-row fin-tube heat exchanger model) has been developed.

The determination of the wall temperature distribution is accurate and consumes less CPU-time than the flow field calculations.

The deviations on the computed mean values of Nu are under 10% of the computed grid-independent mean values.

The temperature gradients on the fin strongly influence the local Nu-number. The winglets increase the heat transfer along with an increase in the dissipative losses. The fin efficiency also increases by winglets due to a better distribution of the heat flux over the fin, and the stamping on the fin has little influence on the fin efficiency.

Aknowledgement: the authors are grateful for the support provided by HLRZ at KFA-Julich where many calculations in the CRAY-YMP where performed.

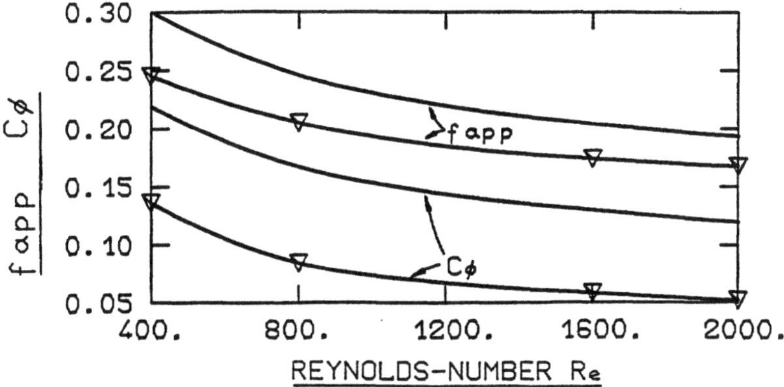

Fig.7 : *Apparent friction factor (fapp) and dissipation coefficient (cφ) versus Re for (—▽—) configuration without winglets ; (——) configuration with winglets*

References

[1] Dong Y. "Experimentelle Untersuchung der Wechselwirkung von Längswirbe lerzeuger und Kreiszylindern in Kanalströmungen in Bezug auf Wärmeübergang und Strömungsverlust", *Doktorarbeit, Ruhr-Universität Bochm, BRD*, (1988).

[2] Sanchez,M., N.K.Mitra, M.Fiebig, "Numerical Investigation of Three Di mensional Laminar Flows in a Channel with a Built-in Cylinder and Wing-Type Vortex Generators," *Proc. of the* 8th *GAMM Conference on Numerical Methods in Fluids, pp 484-492, Delft*, (1989).

[3] C.W.Hirt, B.D.Nichols, N.C.Romero,"SOLA a Numerical Solution Algorithm for Transient Fluid Flow", *Los Alamos Scientific Laboratory, Report LA-5652,New Mexico*, (1975)

[4] P. Kiehm, N.K.Mitra, M.Fiebig, *in Proc. 6 GAMM Conf. on Numerical Methods in Fluids , pp.153-160, Göttingen*, (1985)

[5] U. Brockmeier, N.K.Mitra, M.Fiebig, *in Proc. 7 GAMM Conf. on Numerical Methods in Fluids , pp.48-55, Louvain-La-Neuve*,(1987)

[6] Davies, G. de Val, "Natural Convection of Air in a Square cavity: a benchmark solution", *I.J. for N.M. in Fluids, vol.3, pp 249-264*, (1983).

Numerical Studies of a Compact Fin-Tube Heat Exchanger

A. Bastani, M. Fiebig, N. K. Mitra
Institut für Thermo- und Fluiddynamik
Ruhr-Universität Bochum, Postfach 10 21 48
4630 Bochum, Germany

ABSTRACT

A numerical scheme has been developed to compute the flow field between neighboring fins of a compact fin-tube heat exchanger. Exemplary computations show that at low Reynolds number (~ 400) the Nusselt number in the neighborhood of the second tube of a two-tube in-line configuration is close to the Nu given by the periodically fully developed flow.

INTRODUCTION

In a compact fin-tube cross flow heat exchanger a bank of tubes share common fins, see fig. 1. Generally a liquid flows through the tubes and a gas flows through the channels formed by the neighboring fins, around the tube bank. The heat transfer between the liquid and the gas is primarily governed by the large gas side thermal resistance caused by the poor thermal conductivity of the gas. By increasing the transfer area, the fins serve to reduce this resistance.

The heat transfer between the gas and the fins and the tube surfaces is determined by the flow structure which is three dimensional and may contain horse-shoe vortices, vortex streets and separated zones.

For compact heat exchangers, the flow can be treated as laminar since the characteristic Reynolds number based on the average velocity and the hydraulic diameter of the channel will be less than 2000.

The flow field will also depend on such geometrical parameters as the tube arrangement and tube spacing, diameter of the tubes and the channel height (i. e. the distance between two neighboring fins).

The distribution of the heat transfer coefficients on the fin and on the tube surfaces is not homogeneous. Areas of fin surface with poor heat

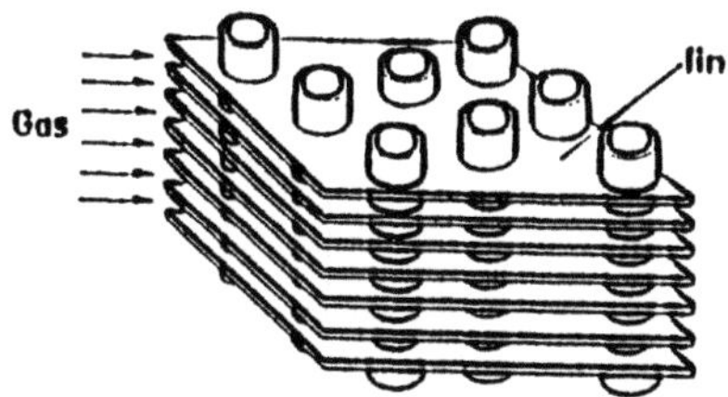

Fig. 1: Schematic of a compact fin-tube cross flow heat exchanger.

transfer coefficient can be manipulated with turbulators when their locations are known. The goal of manipulation is the maximization of heat transfer enhancement with the least increase in flow losses.

Because of the large number of geometrical parameters with or without surface manipulation, the detailed information about the flow field will be of great help for the optimum design of a heat exchanger.

One way to obtain the detailed flow field is through the numerical simulation. This will require a numerical scheme for the solution of unsteady three dimensional Navier-Stokes and energy equation in a rectangular channel with a built-in bank of cylindrical obstacles. This is the purpose of the present work.

Two dimensional simulations of flow past a tube bundle has been reported by Launder and Massey /1/, Chen et.al/2/, Fuji et. al /3/ and Wung and Chen /4/. In all these works, the simulation of a tube bank has been achieved by selecting a small rectangular element with periodically fully developed flow. This has been achived by imposing periodic boundary conditions at the inlet and exit of the element /5/. The implicit assumption here is that the flow in the complete heat exchanger can be modeled by the flows in the repeating elements. In interrupted plate passages, periodically fully developed flow will appear after 5 and before 10 repeating elements from the inlet /5/. With tube bank it is not known where the periodically fully developed flow will appear. The appearance of periodically fully developed flow will also depend on Reynolds number. Two-dimensional flow fields can not predict heat transfer between the fluid and the fin, hence their simulations have limited application. In the present work a computational scheme has been developed to calculate three-dimensional laminar flows in a rectangular channel with built-in cylindrical obstacles. By solving complete Navier-Stokes and energy equations this scheme can generate a data base with parameer studies for an optimum design of a fin-tube heat exchanger. In this paper exemplary computational results comparing flow and heat transfer in a rectangular channel with one tube and periodic boundary conditions and in a channel with two in-line tubes without periodic boundary conditions are presented.

MATHEMATICAL FORMUALTION

The computational domains consists of an element of the heat exchcanger. Assuming symmetry conditions on the mid-plane between two fins, the computational domain is then represented by a rectangular channel whose bottom boundary simulates the fin, the top boundary the mid-plane and the side boundaries the symmetry planes. Fig. 2a shows the channel with two tubes in an in-line arrangement. Fig. 2b shows this geometry with a built-in circular tube of diameter D in the middle. This will be the computational domain for the periodic inlet and exit conditions.The height of the domain is $H/2$ where H is the distance between two fins. Other dimensions are shown in fig. 2.

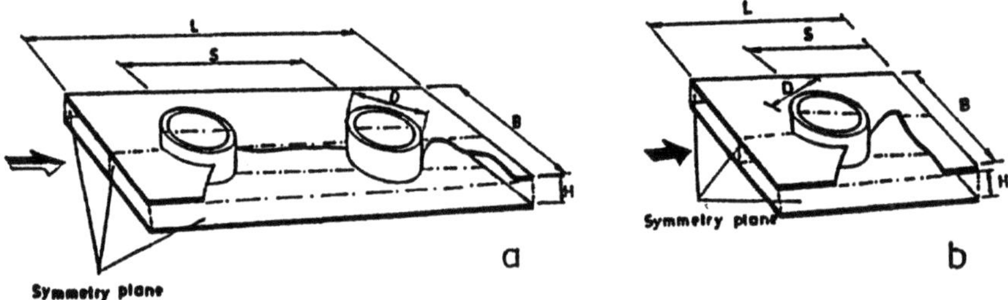

Fig. 2 : Schematic of the computional domain. a) with two in-line tubes in a rectangular channel, L/H=18. b) perodically fully developed flow, L/H=9. Bottom boundary ist the fin, top boundray is symmetrical mid-plane. Computations are preformed in half of the channel width. B/H=4.5 , s/H=9 D/H=3.6 .

The computational domain with symmetry on the tube axial plane precludes periodic vortex street in the wake. This is justified because previous computations without this assumed symmetry gave steady solutions without a vortex street up to Re_H = 2000, where Re_H is the Reynolds number based on H /6/. The flow in the computational domain are given by incompressible Navier-Stokes and energy equations which in nondimensional cartesian form for a medium with constant properties read as

$$\frac{\partial u_i}{\partial x_i} = 0 \tag{1}$$

$$\frac{\partial u_i}{\partial t} + \frac{\partial(u_i u_j)}{\partial x_j} + \frac{\partial p}{\partial x_i} = \frac{\nabla^2 u_i}{Re_H} \tag{2}$$

$$\frac{\partial T}{\partial t} + \frac{\partial(u_i T)}{\partial x_i} + \frac{\nabla^2 T}{Re_H Pr} \tag{3}$$

The unsteady form of the equations are written since a time-dependent numerical procedure is used to solve them. All length coordinates have been nondimesionalized with the channel height H, the velocity components with the average velocity at the inlet u_{av}, the pressure with ρu^2_{av} and the temperature with the inlet temperature T_∞. The solid surface temperature T_w is taken to be $2T_\infty$. The Reynolds number Re_H is u_{av} H/ν and the Prandtl number is ν/a where these symbols have usual meanings. The dissipation terms in the energy equation have been neglected. No-slip conditions and symmetry conditions are used on surfaces and symmetry planes respectively. The periodic conditions for fig. 2b require that the velocity profiles (u, v, w) at the exit and inlet are identical. Following Patankar /5/ the periodic boundary condition for the temperature has been satisfied by imposing

$$(T_{in} - T_w) / (Tb_{in} - T_w) = (T_{ex} - T_w) / (Tb_{ex} - T_w) \tag{4}$$

where subscripts in and ex stand for inlet and exit conditions and Tb stands for the bulk temperature.
For fig. 2a homogeneous velocities (u = u_{av}, v = w = 0)
and temperature are used at the inlet. At the exit, $\partial/\partial x$ (u, v, w, T) are set equal to zero.

COMPUTATIONAL SCHEME

The basic equations have been solved by a modified version of the marker and cell (MAC) technique /7/. To this purpose the computational domain is first discretized into grids in form of rectangular parallelopipeds. The cylinder is simulated by cartesian grids. When a converged solution in these grids is obtained, a second step of computations with polar grids in the neighborhood of the tube follows.
The scheme uses staggered grids i.e. the dependent variables are not defined at the same point. The pressure and the temperature are defined in the center of the cell, the velocity components on the midpoint of the cell faces on which they are normal. The computation proceeds in two steps. In the first step the momentum equations are solved explicitly from known velocity and pressure fields in order to obtain these field values are the time $(t + \delta t)$. In the second step a pressure-velocity updating is performed in a way that corresponds to the solution of the Poissen equation for pressure. This updating is continued until the continuity equation is locally satisfied on each cell. Then the solution for the next time step is performed. The procedure is continued until a steady or periodic solution is obtained. Since the momentum equations are uncoupled from the energy equation, the latter can be solved whenever required. The details of the computational scheme along with a flow chart is given by Kiehm /7/.

RESULTS AND DISCUSSION

Computations have been carried out on cartesian girds of 123x32x12 ($\Delta x = \Delta y = 0.15$; $\Delta z = 0.1$) on a SUN 3/60 Workstation. Typical CPU times are 40 hours for periodically fully developed flow and 200 hours for two-tube configuration. Grid independence studies have not been performed in this work. However, a grid independence study of a similar problem with the same computational scheme as used in this work shows that typical flow parameters (eg. mean Nusselt number on the fin) computed with $\Delta x = \Delta y = 0.155$, $\Delta z = 0.085$ deviates less than 10% from the grid independent value /8/.

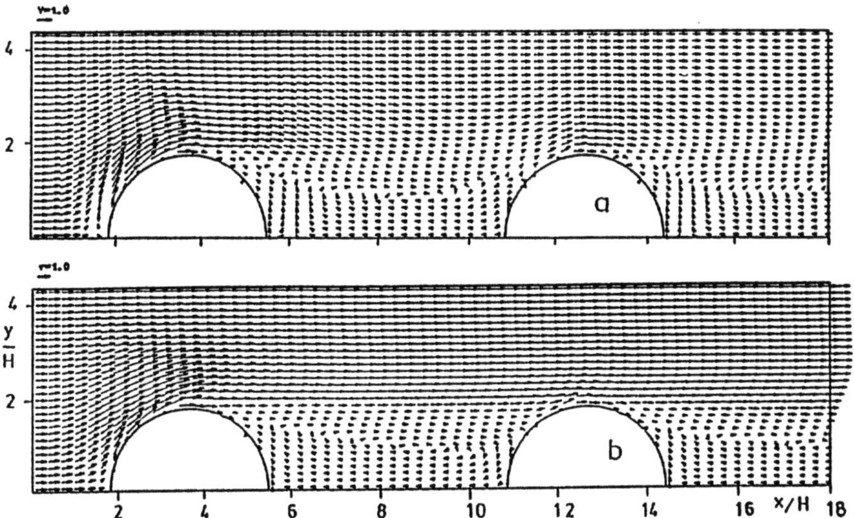

Fig. 3: Velocity vectors in x-y plane for Re=1200 for geometry of fig.2a (a) at z=0.1 (b) on the mid plane z =1/2 . Backflow at geometrical stagnation point of the second tube. Secondary vortices are present.

Figures 3a and 3b show velocity vectors near the bottom fin (z = 0.1) and the midplane (z = 1/2) respectively in the channel with two tubes for Re based on average velocity and hydraulic diameter of 1200. Similar pictures in channels with one tube and periodic boundary conditions are shown in fig. 4a and 4b. The space between the two tubes contains a

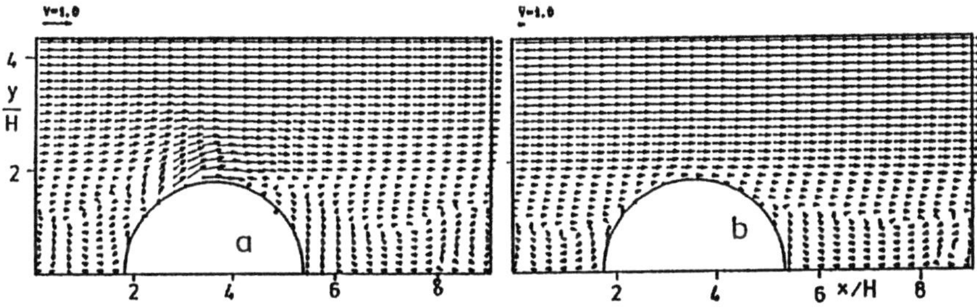

Fig. 4: Velocity vectors on x-y plane. for Re=1200 for the geometry of 2b. (a) at z=0.1, (b) on the midplane. Secondary vortices are present.

160

recirculating bubble (fig. 3). The wake of the second tube also consists of a recirculating bubble. The stagnation areas before the second tube in two-tube configuration and in front of the tube of with periodic b.c. have some structural similarity. Back flow appears at the geometrical stagnation points. Flow separation appears at approximately 30° from the geometrical stagnation point. Secondary vortices appear in the stagnation areas.

Figures 5a and 5b compare the Nusselt number contours on the fin of the channel with two tubes and the channel with one tube and periodic boundary conditions for Re = 1200. The Nusselt number Nu is defined with the difference of the bulk and the wall temperatures and the hydraulic diameter of the channel. The Nu with periodic b.c. are smaller than Nu in two tube configuration. The stagnation area and the wake of the second tube show small Nu ($\leq$ 4) as those area with periodic b.c. ($\leq$ 2). Figs. 4a and 4b show the need and location of surface manipulation on the fin in order to guide the flow in the separated zones.

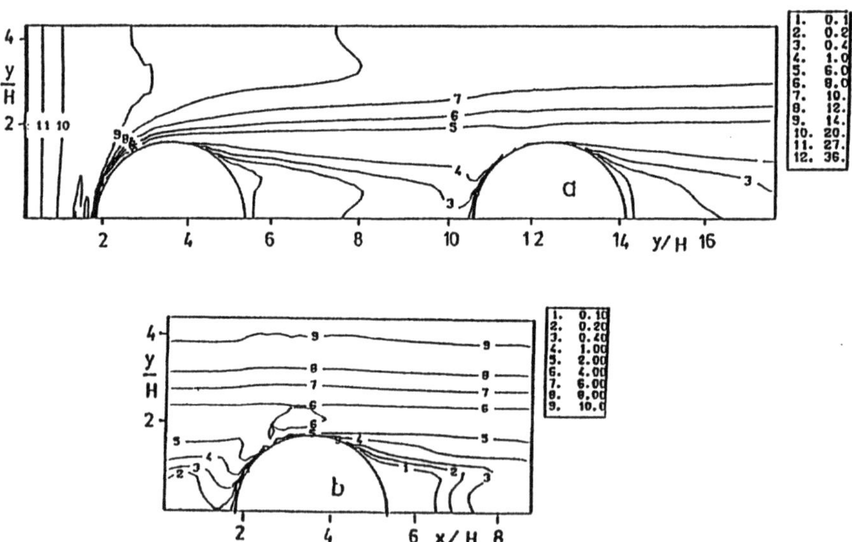

Fig.5: Contours of Nusselt numbers on the fin for Re=1200 (a) for fig.2a Large Nu in the stagnation area of the first tube is due to the horse shoe vortex, (b) for fig.2b.

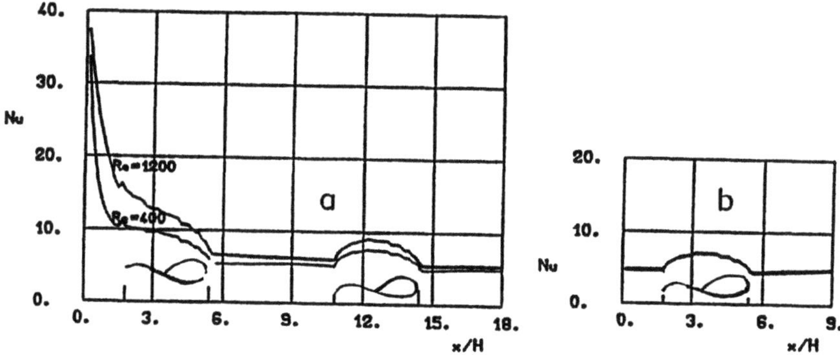

Fig. 6: Cross averaged Nu on the fin for Re=1200 and Re=400 (a) for fig.2a, (b) for fig.2b.

Figures 6a and 6b compare spanwise averaged Nu on the fin for Re of 400 and 1200 with two tubes (fig. 5a) and one tube and periodic boundary conditions. The behaviour of Nu from the rear of the first tube downstream is qualitatively similar to that in the channel with one tube and periodic boundary conditions The increase in Nu on the locations where the tube is placed, is caused by the increase in the velocity due to the blockage of the tube. The quantitative difference in Nu between fig. 5a (around the second tube) and fig. 6b is less than 20% in peak value and less than 5% in the stagnation and wake regions.

For Re = 400 and 1200 the maximum difference in Nu in fig. 5a in the neighborhood of the second tube (ie. $9 \leq x/H \leq 18$) amounts to less than 10 %. For periodic b.c. (fig. 5b) this difference becomes less than 2 %. The average Nu on the fin for periodic b.c. is practically the same for Re = 400 and 1200 and equals to 5·24. The average Nu in the neighborhood of the second tube ($9 \leq x /H \leq 18$) in two-tube configuration is 5·76 and 6·59 for Re = 400 and 1200 respectively. For Re = 4oo, the flow around the second tube closely resembles the flow with periodic b.c. and hence the Nu distributions are nearly identical.

The flow losses in the channel is given by the dissipation coefficient c_ϕ where

$$c_\phi = f_{app} + \Delta\alpha \, \frac{H}{2L} \tag{5}$$

where f_{app} is the apparent friction factor and $\Delta\alpha$ is the difference in the kinetic energy at the inlet and exit. With increasing Reynolds number c_ϕ decreases. With Re = 400, c_ϕ = 0.1225 and 0.094 for two-tube configuration and one tube with periodic b.c. respectively. The corresponding values at Re = 1200 are 0.06 and 0.036 respectively. For the region with the second tube only ($9 \leq x/H \leq 18$) in two tube configuration, c_ϕ is 0.099 and 0.039 for Re = 400 and 1200 respectively. Again these values are comparable to those for one tube with periodic b.c..

CONCLUSION

For Re = 400, the heat transfer and flow losses for the second tube in an in-line arrangement of a fin-tube heat exchanger deviate less than 9 % and 5 % from the corresponding values for periodically fully developed flow. At Re = 1200, the deviation in heat transfer becomes 20 % and the deviation in flow loss remains less than 5%. Detailed flow field computations show the areas of poor heat transfer on the fin where surface manipulation of the fin should be considered.

REFERENCES

1. Launder, B.E. and Massey, T.H., "The Numerical Prediction of Viscous Flow and Heat Transfer in Tube Banks", I. Heat Transfer (ASME), V.100, pp. 565-571, (1978).

2. Chen, C.J. and Chen, H.C., "Development of Finite Analytic Numerical Methods for Unsteady Two Dimensional Navier-Stokes Equations", I. Comp. Phys., V.53, pp. 209-226, (1984).

3. Fujii, M., Fujii, T. and Nagata, T., "A Numerical Analysis of Laminar Flow and Heat Transfer of Air in an In-Line Tube Bank", Numerical Heat Transfer, V. 7, pp. 89-110, (1984).

4. Wung, T.S. and Chen, C.J., "Finite Analytic Solution of Convective Heat Transfer for Tube Arrays in Cross Flow", Part I and Part II, J. Heat Transfer (ASME), V. 111, pp. 633-648, (1989).

5. Patankar, S.V. and Prakash, C., "An Analysis of the Effect of Plate Thickness on Laminar Flow and Heat Transfer in Interrupted-Plate Passages", Int. J. Heat Mass Transfer, V. 24, pp. 1801-1810, (1981).

6. Bastani, A. Mitra, N.K. and Fiebig, M. "Numerical Simulation of Flow Field in a Fin Tube Heat Exchanger in Fluid Machinery Components, FED-Vol. 101, ed. D.L. Rhode, J. Tuzson, ASME WAM, pp. 91-96, (1990).

7. Kiehm, P., "Numerische Untersuchung der laminaren Strömungsfelder und des Wärmeübergangs in einem Kanal mit quereingebautem Kreiszylinder", Dissertation, Ruhr-Universität Bochum, FRG (1986).

8. Sanchez, M., Fiebig, M. and Mitra, N.K. "Improvement of Fin-Tube Heat Exchangers by Longitudinal Vortex Generators", in this Proceeding.

Heat Transfer and Pressure Drop in Single Rows

M. Beziel and K. Stephan

Institut für Technische Thermodynamik und Thermische
Verfahrenstechnik der Universität Stuttgart
Pfaffenwaldring 9, 7000 Stuttgart 80, F. R. Germany

ABSTRACT

Results of experimental investigations on the influence of turbulence intensity and pitch-to-diameter ratio on heat transfer and pressure drop in single rows of plain tubes are presented. The vertically arranged tubes with pitch-to-diameter ratios between 1.26 and 3.44 were heated by saturated steam, condensing inside, and cooled outside by air in cross flow. The turbulence intensity in the entrance cross-section was enhanced by means of different biplanar grids placed in the test section of the wind tunnel upstream before the single rows. The mean streamwise turbulence intensity behind the grids, measured with a constant temperature hot-wire anemometer, varied between 0.8% and 38.8%. Reynolds-numbers ranged from $3 \cdot 10^3$ to $3 \cdot 10^5$. The measurements show for single rows an increase of the Nusselt-number for a given turbulence intensity with increasing Reynolds-number. For constant Reynolds-number and constant turbulence intensity the Nusselt-numbers increase until a pitch-to-diameter of $a = 2.69$. If the pitch-to-diameter ratio was enlarged the Nusselt-numbers decrease. In the investigated range of Reynolds-numbers the drag coefficient is only a weak function of the inlet turbulence intensity. The use of turbulence grids thus leads to higher efficiencies of heat exchangers.

1. INTRODUCTION

Compact heat exchangers consisting of tube bundles with a few rows of plain tubes in cross flow are often used in industry. Heat transfer coefficients in the first row of these heat exchangers are much lower than further downstream. Hence, the heat transfer rate of tube bundles with a few rows is mainly determined by the heat transfer of the first row. In order to increase the heat transfer some investigators installed turbulence generators in the oncoming fluid. Bressler [1] reported that the heat transfer coefficient thus could be increased from the first to the fifth row of tube bundles with ten rows. Schellerich [2] correlated the measured Nußelt-numbers for a single row and for the first row of a tube bundle consisting of four rows with the turbulence intensity. Stephan [3] determined the effect of a high turbulence on heat transfer and flow resistance experimentally found that the effect of turbulence on heat transfer is very pronounced for the first row and becomes smaller further downstream. Other investigations [4] showed an increase of the heat transfer with increasing pitch-to-diameter ratio.

This paper presents the results of experimental investigations on heat transfer and pressure drop in single rows consisting of plain tubes with pitch-to-diameter ratios between a = 1.26 and a = 3.44 at high turbulence intensities.

2. EXPERIMENTAL PROCEDURE

2.1. Apparatus

The tubes were tested in an open wind tunnel. The wind tunnel was described in detail in [5]. The experimental set-up is shown schematically in Figure 1.

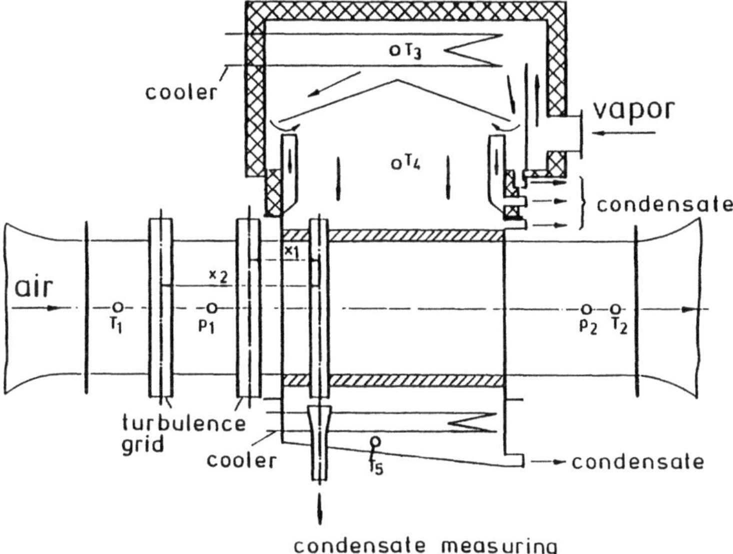

Fig. 1   Experimental set-up.

The cross-section of the channel was 258 mm x 258 mm. These dimensions allowed tube arrangements presented in Table 1.

Table 1. Tube arrangements.

s_q [mm]	d_a [mm]	$a=s_q/d_a$ [—]	number of tubes
43.0	34	1.26	6
43.0	28	1.54	6
25.8	15	1.72	10
43.0	24	1.79	6
43.0	34	2.53	3
43.0	16	2.69	6
86.0	28	3.07	3
51.6	15	3.44	5

The tubes were heated by saturated steam, condensing inside, and cooled by air in cross flow. The turbulence intensity of the air in the entrance cross-section could be varied by means of different grids, placed at two different distances before the row.

2.2 Instrumentation

The mass flow rate of the air was measured by a nozzle at the inlet of the wind tunnel. All temperatures were measured by calibrated resistance thermometers. The air-side pressure drop across the single row was obtained from the difference of the average static pressures before and behind the turbulence grid and the single row (see p_1 and p_2 in Fig. 1). For heat transfer studies the mass flow rate of the condensate was determined. Four different biplanar grids were used, with the dimensions given in Table 2.

Table 2. Dimensions of turbulence generating grids.

grid	M [mm]	D [mm]	M/D [--]
grid G 4/10	40	10	4
grid G 2/20	40	20	2
grid G 5/4	20	4	5
grid G 10/4	40	4	10

The mean streamwise turbulence intensity was measured as a function of the distance x behind the grids by means of a constant temperature hot wire anemometer. For the distance $x_1 = 100$ mm turbulence intensities from 8.0% to 38.8% were obtained. Minimum streamwise turbulence intensity (no grid) was 0.8% for air velocities from 5 m/s to

30 m/s. In Figure 2 the measured turbulence intensities are compared with the turbulence intensities for one biplanar grid [6], with results from an equation found for latti-

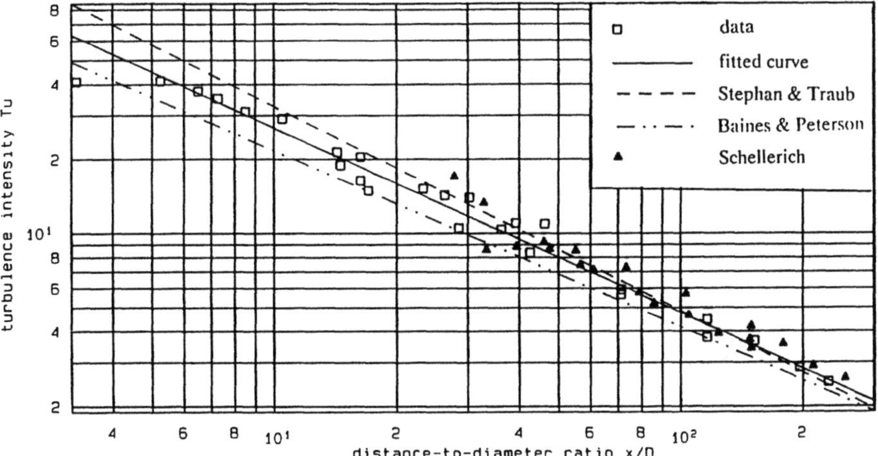

Fig. 2 Turbulence intensity behind grids.

ces [7] and with measurements made by Schellerich [2]. It is obvious that for smaller ratios of distance-to-diameter the turbulence intensity with biplanar grids is greater than that with lattices. High shear stresses exist at the sharp edges of bars or strips and produce big eddies with high turbulence.

2.3 Calculations

The heat transfer rate of tube bundles is given by

$$\dot{Q} = k \, A \, \Delta T_m \quad . \tag{1}$$

The overall heat transfer coefficient k is calculated from the energy balance for the condensate

$$\dot{Q} = \dot{M}_K \, \Delta h_v \tag{2}$$

and for the air

$$\dot{Q} = \dot{M}_L \, c_{pL} \, (T_{L2} - T_{L1}) \tag{3}$$

with the specific heat of condensation Δh_v and the mean temperature difference $(T_{L2} - T_{L1})$ of the air flow. The heat transfer coefficient α_a on the outside of the tubes is given by

$$\alpha_a = \left(\frac{1}{k} - \frac{d_a}{\alpha_i \, d_i} - \frac{d_a}{2\lambda_R} \ln \frac{d_a}{d_i} \right)^{-1} . \tag{4}$$

λ_R is the thermal conductivity of the tube material and α_i the heat transfer coefficient inside the vertical tubes obtained from Nusselt's film theory [8]. With the heat transfer coefficient α_a the Nusselt-number is given by

$$Nu = \alpha_a \, d_a / \lambda_L . \tag{5}$$

The Reynolds-number is defined by

$$Re_e = u_e \, d_a \, \rho_L / \eta_L \tag{6}$$

where u_e is the velocity of the minimum flow area in the single row. The drag coefficient for the pressure drop Δp across the single row is obtained from

$$\varsigma = 2\Delta p \, / (\rho_L \, u_e^2) . \tag{7}$$

3. RESULTS

3.1 Heat Transfer

In Figure 3 the Nusselt-number Nu is shown as a function of the Reynolds-number Re_e for a single row with $a = 2.53$. The data for constant turbulence intensity can be approximated by straight lines. In Figure 3 only the curves for the minimum and the maximum turbulence intensity represented by the dashed lines are shown. The full line represents values calculated from an empirical equation reported by Gnielinski [9], which does not account for the turbulence. It represents well the values for $Tu = 8\%$ for low Reynolds-numbers and for $Tu = 0.8\%$ for high Reynolds-numbers. The measurements show for single rows an increase of the Nusselt-number for a given turbulence intensity with increasing Reynolds-number. In Figure 4 the influence of pitch-to-diameter ratio on the heat transfer of single rows at constant turbulence intensity of $Tu = 0.8\%$ is shown. For constant Reynolds-number the Nusselt-number increases with the pitch-

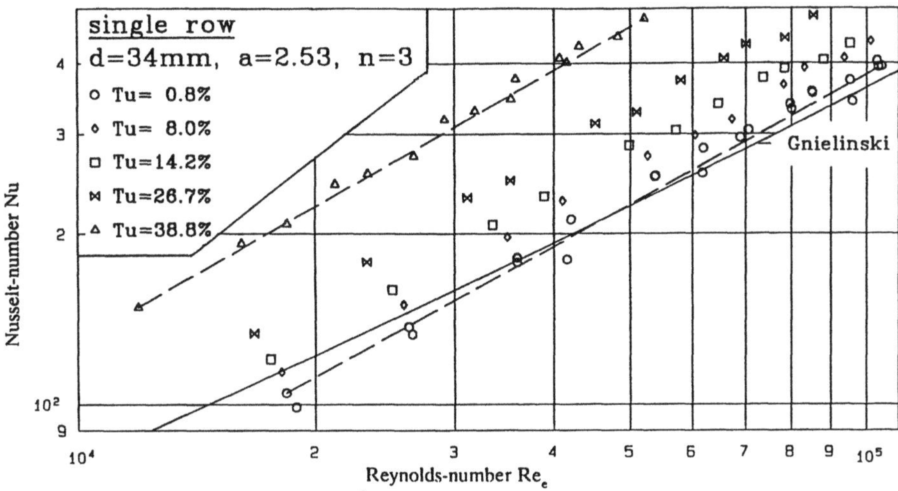

Fig. 3 Heat transfer for a single row with a = 2.53.

to-diameter ratio. If the pitch-to-diameter ratio exceeds the value of a = 2.69 the Nusselt-number and hence the heat transfer rate decreases. To obtain constant Reynolds-numbers in the minimum cross-section of the single row the air velocity at the stagnation point of the tubes has to be higher for high pitch-to-diameter ratios and thus the mean Nusselt-numbers were enhanced. The higher the pitch-to-diameter ratio, easier cold air streamlines can build up which do not participate in the heat exchange. Therefore the heat transfer rate decreases at pitch-to-diameter ratios above a = 2.69.

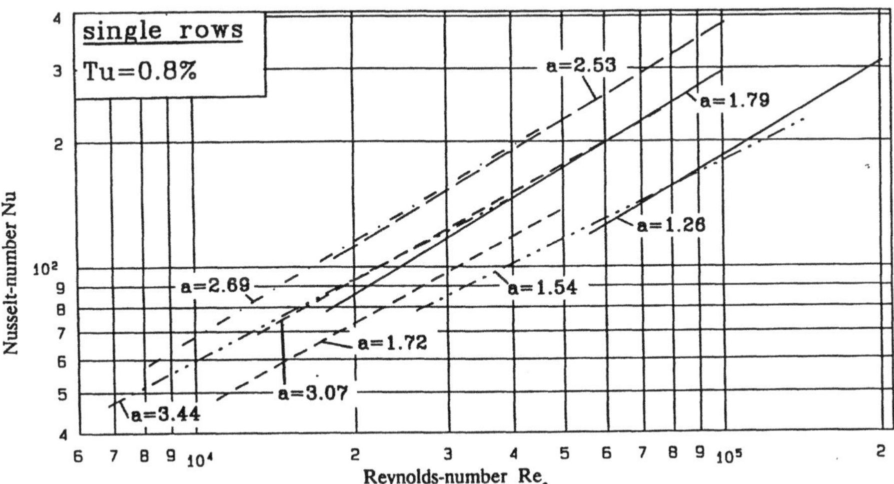

Fig. 4 Heat transfer for single rows at Tu = 0.8%.

3.2 Pressure drop across the single rows

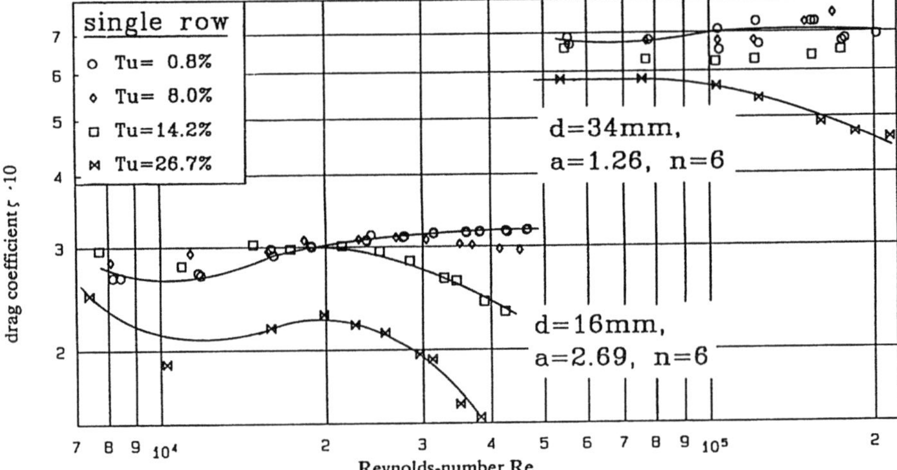

Fig. 5 Pressure drop across the single row.

Figure 5 shows the drag coefficient, defined by eq. (7) as a function of the Reynolds number for single rows with pitch-to-diameter ratios of a = 1.26 and a = 2.69. The drag coefficient is only a weak function of the inlet turbulence and almost independent from the Reynolds number. For high turbulence levels and high Reynolds numbers the drag coefficient decreases rapidly, which indicates a change of the flow pattern around the tubes. The point of boundary layer separation shifts further downstream and thus the recirculation zone becomes smaller. Figure 5 shows, that the drag coefficient strongly depends on the pitch-to-diameter ratio.

4. DISCUSSION

As the experimental results demonstrate, a turbulence grid can considerably intensify the heat transfer of a single row with plain tubes. Simultaneously the drag coefficient of turbulence grid and single row also increases and it is still an open question whether the enhanced heat transfer goes on the expense of the higher pressure drop across the system. As Stephan and Mitrovic [10] reported the dimensionless number St^3/ζ_{tot} is a criterion for the enhancement of heat transfer of a given heat exchanger with different turbulence intensities but the same ventilation power. In Figure 6 this dimensionless

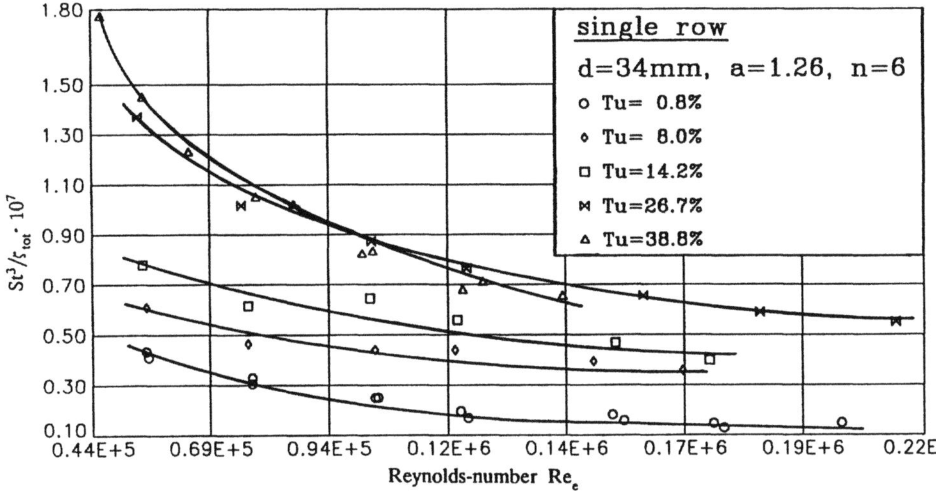

Fig. 6 Criterion for a single row with a=1.26.

number is plotted as a function of the Reynolds-number for a pitch-to-diameter ratio of a=1.26 and different turbulence intensities. The curve with the open circles represents the data for the minimum turbulence intensity. All data above this reference curve represent heat exchangers with grids. The ventilation power is the same in all cases. A considerable enhancement is obtained with the grids. For the grid with Tu=38.8% the higher pressure drop of the system dominates for Reynolds-numbers above $Re_e = 10^5$ and the efficiency of this exchanger is lower than that for the exchanger with Tu=26.7%. The results showed no improvement of the efficiency for heat exchangers with a>2.53. This is due to the additional pressure drop of the grids. The pressure drop of single rows with large pitch-to-diameter ratios is low compared to that of single rows with small pitch-to-diameter ratios. Thus the additional pressure drop of the grids has considerable influence on the efficiency of single rows with large pitch-to-diameter ratios.

5. CONCLUSIONS

1. The free stream turbulence intensity considerably influences the heat transfer of a single row tube bundle. Additionally the results show a significant influence of the pitch-to-diameter ratio of about a=2.69 on the enhancement of the heat transfer.

2. The drag coefficient turned out to be a weak function of the inlet turbulence intensity provided that the pitch-to-diameter ratio is not too large.

3. As a criterion for the enhancement of the heat transfer of a given heat exchanger with different turbulence grids the dimensionless number St^3/ς_{tot} seems to be useful. It is shown that for single rows with small pitch-to-diameter ratios the use of turbulence grids leads to higher efficiencies of heat exchangers. For a pitch-to-diameter ratio of $a=2.53$ the additional pressure drop caused by the grid is too large compared to that of tubes, whereas for smaller ratios the heat transfer is considerably increased and the additional pressure drop of the grids is low.

ACKNOWLEDGEMENT

We gratefully acknowledge the financial support by the Deutsche Forschungsgemeinschaft (DFG).

NOMENCLATURE

$a=s_q/d_a$	pitch-to-diameter ratio
A	area
c_p	specific heat capacity at constant pressure
d	diameter
D	grid bar width
k	overall heat transfer coefficient
M	mesh size
$\dot{M}$	mass flow rate
$\dot{Q}$	heat transfer rate
Δp	pressure drop
s_q	distance between the tubes
u_e	velocity of the minimum flow area
α	heat transfer coefficient
λ	thermal conductivity
η	kinematic viscosity
ρ	density
ς	drag coefficient

Subscripts

a	outside
e	minimum cross-section
i	inside
K	condensate
L	air
tot	total system

Dimensionless Groups

Nu	Nusselt-number
Re	Reynolds-number
St	Stanton-number

REFERENCES

1. Bressler, R.: Wärmeübergang und Druckverlust bei Rohrbündel-Wärmeübertragern. Dissertation, TH München, 1956.

2. Schellerich, W.: Wärmeübergang und Druckabfall an querdurchströmten Glattrohrbündel. Forsch.bericht des Deutschen Kältetechn. Ver. Nr.8, 1983.

3. Stephan, K.: Wärmeleistung und Strömungwiderstand von Spezialrohren für Wärmeaustauscher. Verfahrenstechnik 4 (1968), 158-160.

4. Beziel, M.; Stephan, K.: Heat transfer and pressure drop in heat exchangers with a single row at high turbulence intensity. Proc. 9th Int. Heat Transfer Conf., Jerusalem, Israel, 1990, Vol. 5, Hemisphere, Washington D.C., pp. 67-71.

5. Beziel, M.; Stephan, K.: Einfluß der Turbulenz auf den Wärmeübergang und Druckabfall an quer angeströmten Rohrbündeln. Chem.-Ing.-Tech. (in press).

6. Stephan, K.; Traub, D.: Einfluß von Rohrreihenzahl und Anströmturbulenz auf die Wärmeleistung von quer angeströmten Rohrbündeln. Wärme- und Stoffübertragung, 21 (1987), 103-113.

7. Baines, W. D.; Peterson, E. G.: An investigation of flow through screens. Trans. ASME 73 (1951), 467-480.

8. Nusselt, W.: Die Oberflächenkondensation des Wasserdampfes. Z. VDI 60 (1916), 541-546 und 569-575.

9. Gnielinski, V.: Gleichungen zur Berechnung des Wärmeübergangs in quer durchströmten einzelnen Rohrreihen und Rohrbündeln. Forsch. Ing. Wes. 44 (1978), 15-25.

10. Stephan, K., & Mitrovic, J.: Maßnahmen zur Intensivierung des Wärmeübergangs. Chem.-Ing.-Tech. 56 (1984), 427-431.

Pressure Losses in Tube Bundles of Close Spacings

J. Buxmann
Technical University of Hamburg-Harburg
Institute of Energetics

Summary

For a special plastic heat exchanger module the pressure drop outside
the tubes at 17 staggered rows has been measured in a hot air test
rig. The results can be expressed in form of a pressure loss coefficient
dependent on the Reynolds-number. The closely packed tubes, caused by
the manufacturing process, have transverse spacing ratios of 1.05.
By a simple relation between loss coefficient, Reynolds-number and
transverse spacing ratio the experimentell results can be extended to
higher spacing ratios up to 1.50. The calculated results are in good
correspondence with those of other authors obtained by experiments.

Heat Exchanger Modules

Though the use of technical thermoplastics in heat exchanger design

does not allow temperature ranges above 200 °C, there are many applica-

tions at temperatures up to 150 °C, where the plastic material can be

used profitable for tubes, plates and ducts. The advantages are corrosion

and chemical resistance to acids, caustic solutions, water, detergents,

oil and others. The non porous, smooth surface tends to small fouling and

the cleaning of the surface is simple. The low weight is profitable to

construction, transport and installation. A manifold geometric form and

simple manufacturing methods compared to metallic materials have economical

consequences. In Figure 1 you see different modules of plastic heat exchan-

gers. The single tubes are enlarged to a hexagon at their ends and can be

welded together to a tube bundle. The widening of the tubes is limited, so

that the transverse and longitudinal spacing ratios are near to one. With-

in this range of small spacings in the literature you find relatively few

investigations of the pressure drop and heat transfer. Therefore we tested

a special module to get some information about the pressure drop of the

staggered tube bank at a transverse spacing ratio of 1.05 and at a tube

outer diameter of 20 mm. The longitudinal spacing was 0.89.

Fig. 1: Different modules of plastic heat exchangers
(Werkbild Röhm GmbH, Darmstadt)

Hot Air Testing Plant

Figure 2 shows a scheme of the hot air testing plant. The air flow within the duct can be heated up to 180 °C. The flow outside the tubes coming from the hall is produced by a fan. The freestream velocity of the air to the tube-bank can vary between 0.6 and 1.6 m/s at 17 rows of staggered tubes and at a transverse spacing of 21 mm. The module is assembled by 255 tubes with a tube length of 500 mm.

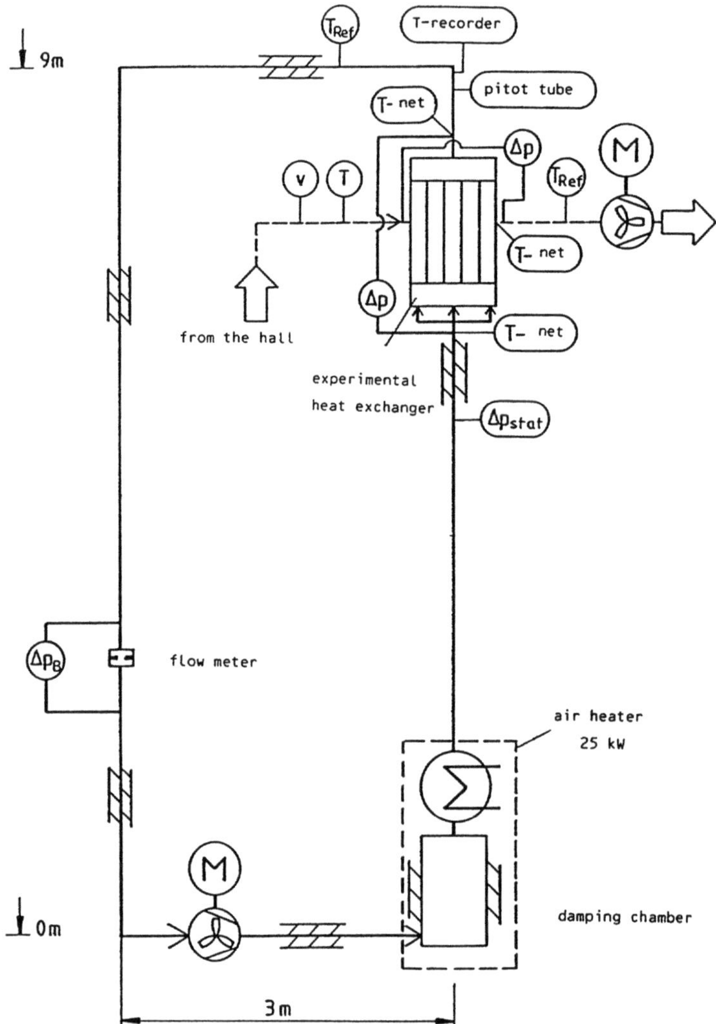

<u>Fig. 2:</u> Hot air test rig for heat exchangers

Experimental Results

Figure 3 shows the measured pressure drop Δp dependency on the
free-stream velocity of the air w_o to the tube bank. Δp can be expressed
as a function of w_o by

$$\Delta p = 9{,}1 \cdot (w_o)^{1{,}9}$$

for the existing isothermal air flow through the 17 rows of staggered

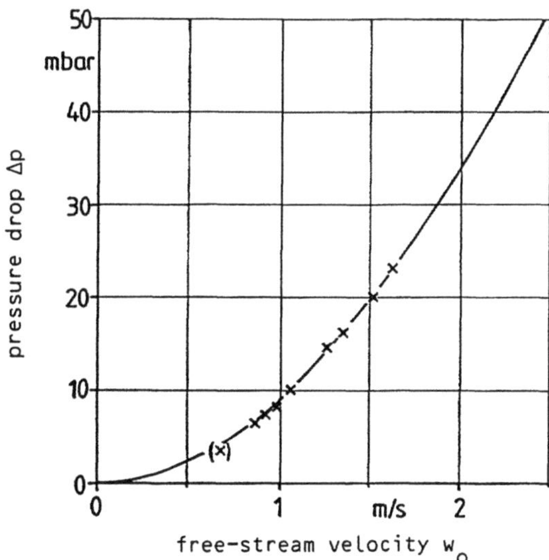

Fig. 3: Measured pressure drop Δp at a tube bundle of 17 staggered rows

$$\Delta p = 9{,}1 \cdot (w_o)^{1{,}9}$$

tubes. The pressure drop coefficient ζ with reference to one row will be

$$\zeta = \frac{\Delta p}{n \cdot \rho \frac{w^2}{2}} \qquad (1)$$

at n number of rows, ρ density and w a reference velocity.

Pressure Drop Coefficients

At cross-current through the tube bank the pressure drop coefficient ζ
for one row, see equation (1), can be determined from the test results
plotted in Figure 3. It will be dependent on the Reynolds-number and the
transverse spacing s_q related to the outer diameter of the tubes d. In
the actual case you get $a = s_q/d = 1{,}05$. In Figure 4 the coefficient ζ
is plotted against the Reynolds-number $Re = w_m \cdot d'/\nu$, from literature
sources /1/. In the Re-number, the velocity w_m is defined as

178

$$w_m = w_o \frac{1}{(1 - \frac{\pi}{4 \cdot a})} \qquad (2)$$

and the diameter

$$d' = d \cdot (\frac{4 \cdot a}{\pi} - 1) \quad . \qquad (3)$$

The pressure drop for a tube bundle becomes

$$\Delta p = \zeta \cdot \frac{d}{d'} \cdot n \cdot \rho \cdot w_m^2/2.$$

From the tests of the plastic module the following relationship is obtained

$$\zeta = 2,177 \cdot \left[\frac{64}{Re} + \frac{2,2}{Re^{0.13}}\right] \qquad (4)$$

which is represented in the upper curve of Figure 4.
Extending the equation (4) by

$$\zeta = \frac{2,4}{a^2} \left[\frac{64}{Re} + \frac{2,2}{Re^{0,13}}\right] \qquad (5)$$

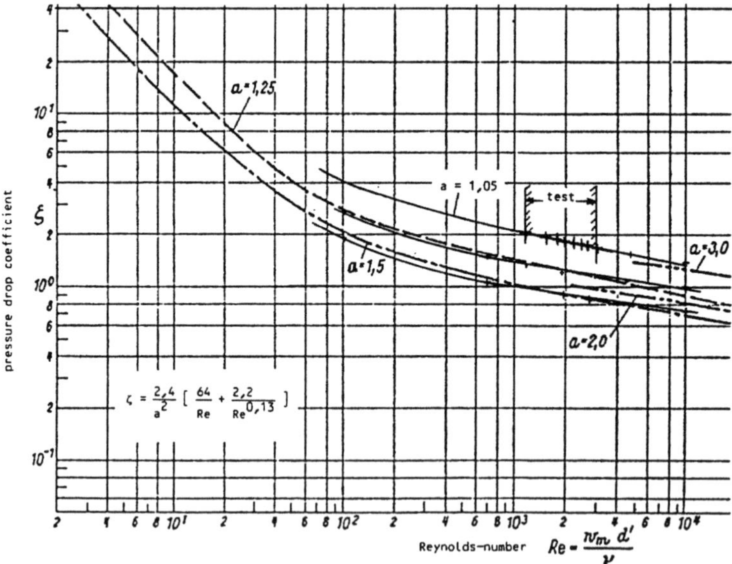

Fig. 4: Pressure drop coefficient ζ at transverse flow for a staggered tube bundle /1/.

the test results can be converted to other transverse spacing ratios a. This has been done for a = 1,25 and a = 1,5. The calculated values of ζ from equation (5) are also plotted in Figure 4. They correspond very well to the curves of other authors /1/.

References
/1/ VDI-Wärmeatlas 1977; Ld1, Ld2.

Improvement of Existing Fuel-Air Heat Exchangers of Modern Air-Breathing Engines

V. M. EPIFANOV, D. V. EFREMOV

Moscow State Technical University named after N. E. Bauman, Moscow, USSR

L. S. YANOVSKIY

Central Institute of Aviation Motors named after P. I. Baranov, Moscow, USSR

Introduction

Results of initial investigations of heat and mass transfer processes in fuel-air heat exchangers (FAHE) of modern air-breathing engines are discussed in this work for the case of thermo-chemical break-down of aviation kerosene. This investigation was initiated by the interest in the fact that the heat sink capability of hydrocarbons increases when they are heated up to the temperature at which the processes of decomposition (pyrolysis) starts. The increase of the heat sink capability is caused by disrupting intermolecular links.

The aim of the work is to create methods of computation and design of the heat exchangers to be installed in the following flying vehicles (FV).

1a. Subsonic FV with gas turbine engine (GTE) having high gas temperature (stoichiometric) at the turbine inlet. It is used for increasing the heat sink capability of the air coolant in the conventional air cooling systems of GTE (Figure 1a);

1b. Small-size, hypersonic FV in which hydrogen fuel can not be used for cooling the engine and the airframe (Figure 1b).

In both cases the fuel from the FAHE passes to the combustion chamber.

Preliminary estimates [1] show that heating standard kerosines up to 1300 K corresponds to doubling the total heat sink capability (up to $4 \cdot 10^6$, J/kg). The total heat sink capability is determined by the physical heat sink capability and by the thermal effect of endothermic reactions of pyrolysis (chemical heat sink capability). The above value of the total heat sink of the fuel makes it possible to increase the gas temperature at the turbine inlet up to **1800 K** without using ceramic elements in its design or to create conditions for flight of hypersonic FV at Mach number **Ma = 10**.

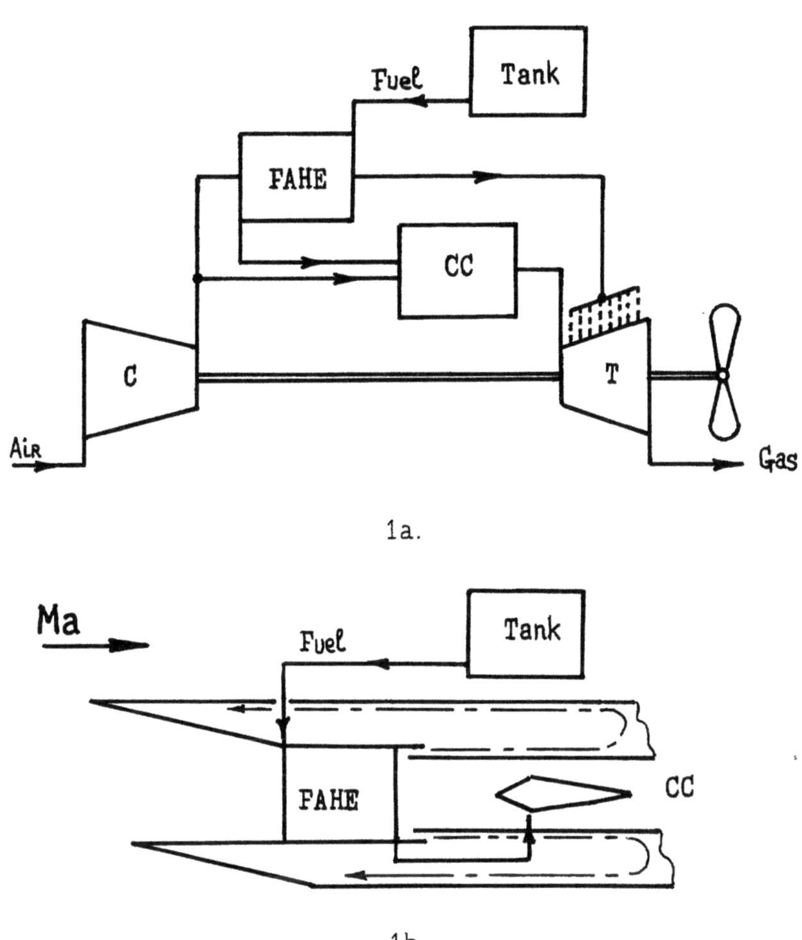

Fig.1. Principal layout scheme of a flying vehicles FAHE: C - compressor; CC - combustion chamber; T - gas turbine.

However some problems should be solved to make this idea practic-

able. The first problem is probably the deposition of coke on
the heat exchange surfaces when the fuel is heated and the re-
moving the deposits in the process of repair and maintenance.
The authors formulated some recomendations in order to decrease
this phenomenon in aviation FAHE. The second difficulty: there
are no reliable methods of computing and designing the cooling
system of such type.

Coke Deposition

When hydrocarbons are heated two temperature zones of increased
coke deposits occur (see Figure 2).

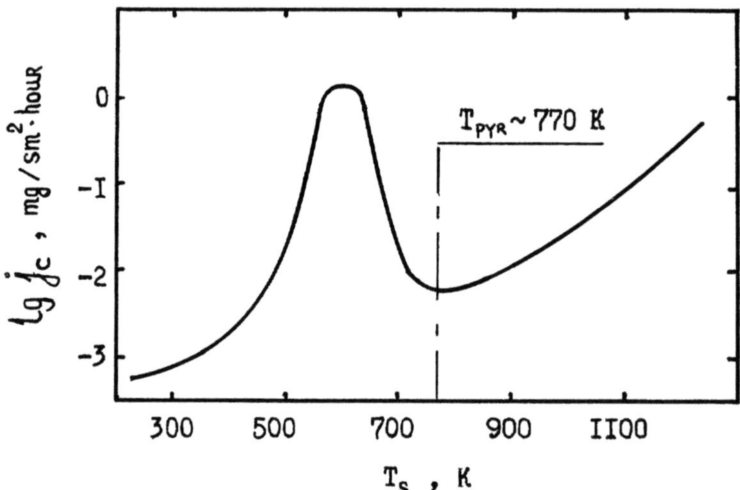

Fig.2. The dependence of the time averaged rate of coke
deposit formation upon the kerosine temperature

The first zone corresponds to the temperature range of 500-650 K
(lower than temperature at which the pyrolysis starts T_{pyr}, which
is approximately 700-800 K) and is due to the reactions of hydrocarbon oxygenation reactions. The second zone is created by formation of atomic carbon during the pyrolysis.
There are some possibilities to reduce the coke deposit process
by taking the following measures:

1. Application of the fuel which consists of paraffine and na-
phthene hydrocarbons having 8-12 carbon atoms in a molecule;
2. Preliminary cleaning of the fuels from admixtures of dissolved oxygen, nitrogen and halogen containing compounds, mechanical
and biological particles;
3. Greater uniformity of heating the heat exchange surfaces with
simultaneous decrease of their thermal stress;
4. Decreasing roughness of the fuel channel surface, rounding
their leading and trailing edges, elimination of stagnation
zones;
5. Application of such structural materials which inhibit coke
deposition (for example **Au, W, Cr, Mo, Va, Cu**);
6. Creation of protective films on the fuel channels surface;
7. Use of two-layer tubes for fuel channels – the outer layer
being made of structural steel and is the structural layer while
the inner layer is the passivator of the coke. It is also suggested to spray the surface of the channels creating a protective layer of Al_2O_3, TiO_2, ZrO_2;
8. Dilution of the fuel with hydrogen;
9. Reduction of the time when fuel remains in the zone of heating (increasing the pumping speed).
However all these measures only decrease the rate of deposition
but do not stop the processes of coke formation. Therefore it
is necessary to choose an optimum method of cleaning the surface
of heat exchange channels from the deposits. The following
methods may be used:
1. Mechanical cleaning;
2. Burning out the coke deposit using oxygen or air;
3. Pumping strong oxydizers (ozone, atomic oxygen) which is
followed by cleaning with alcohol;
4. Using the magnetostriction phenomenon;
5. Cavitation method;
6. Ultrasonic cleaning.
When the above requirements are met it is possible to create
FAHE in which the aviation kerosene is heated up to the temperature when the process of pyrolysis starts (approximately 770 K)
and even higher.

Calculation Method

The second (from the point of view of its importance) problem is development of reliable methods of calculating and designing FAHE. The authors developed a package of applied programs for personal computers intended for designing cooling systems. This package is used in particular for creating a prototype heat exchanger with a bank of aligned smooth-walled tubes in cross-flow.

The calculations were carried out separately for the section where the fuel is heated to the temperature at which pyrolysis begins and for the section where chemical decomposition of the fuel occurs. The correlation (agreement) between the two calculation results is obtained using the successive approximations method on the basis of the air temperature at the boundary of the two sections under consideration in the air circuit. Parameters of the heat-mass transfer processes and hydrodynamic characteristics in the air and fuel (at the section of fuel heating up to T_{pyr}) circuits are determined in accordance with well-known standard criterial relationships for a multipass cross-flow heat exchanger (See "Heat Exchanger Design Handbook" by Spalding and Taborek [2]).

In the fuel channels

$$Nu = Nu \ (d, \ Re, \ Pr, \ T_w/T_s)$$
$$Eu = Eu \ (Re, \ \eta_w/\eta_s) \ . \tag{1}$$

In the air circuit

$$Nu = Nu \ (D, \ S_1, \ S_2, \ Re, \ Pr, \ T_w/T_s)$$
$$Eu = Eu \ (Re, \ S_1, \ S_2) \ . \tag{2}$$

In accordance with calculations carried out by the authors the number of the fuel circuit passes at the section where the chemical heat sink capability of the fuel is fully manifested comprises **30 %** of the total volume of the tube bank. Determination of the laws governing the variation of heat- and mass transfer and hydraulic resistance coefficients under the conditions of hydrocarbon thermochemical decomposition in smooth-walled tubes are made using the following assumptions:

1. Multi-component chemically reacting mixture is considered in its binary approximation: A (fuel) $\rightarrow$ B (products of decomposition);

2. Fuel decomposition in the flow is neglected, it is assumed that the reaction takes place entirely at the tube wall;

3. The heat flux into the wall is considered steady along the tube length;

4. By analogy to the majority of real reacting mixtures the Lewis number is taken close to unity.

For the turbulent flow in a tube the mean equations may be written in the one-dimensional form

$$\frac{d\,p}{d\,x} = \frac{\dot{M}^2}{\rho} \cdot \left[\frac{1}{\rho} \cdot \frac{d\,\rho}{d\,x} + \frac{\xi}{2\,d} \right] \tag{3}$$

$$\frac{d\,H}{d\,x} = \frac{4\,\dot{q}_w}{\dot{M}\,d} \tag{4}$$

$$\frac{d\,X_B}{d\,x} = \frac{4\,\dot{j}_w}{\dot{M}\,d} \;. \tag{5}$$

Pressure drop (friction coefficient) is calculated from the relationship for turbulent flow condition (see Equation 6)

$$\xi = \frac{1}{[1.82 \cdot \lg Re - 1.64]^2} \;. \tag{6}$$

In accordance with the theory of Kutateladze and Leontiev [3] the effect of non-isothermicity and thermochemical transformations on flow characteristics can be taken into account by introducing relative laws of heat- and mass transfer. In the case under consideration ($Le = 1$) these corrections will be structurally uniform for these laws

$$St = \Psi_T \cdot \Psi_B \cdot St_O \qquad \text{and} \qquad St_D = \Psi_{TD} \cdot \Psi_{BD} \cdot St_{DO} \;. \tag{7}$$

For calculating the values of thermal and diffusion Stanton number at steady properties of isothermal flow the relationship of Petukhov and Kirillov [4] is used

$$St_O = \frac{\xi}{8} \cdot \left[1 + \frac{900}{Re} + 12.7 \sqrt{\frac{\xi}{8}} \left(Pr^{2/3} - 1 \right) \right]^{-1} \qquad . \qquad (8)$$

The influence of the endothermic reaction of the flow decomposition upon heat and mass transfer is taken into account in accordance with the model of "effective" injection. Then in accordance with [5] we can write

$$\Psi_B = b_{T_1} \cdot \frac{\exp b_{T_1}}{\left[\exp b_{T_1} - 1 \right]} \quad \text{and} \quad b_{T_1} = b_T \cdot \frac{Pr}{\left[Pr + 0.25 \right]} \qquad . \qquad (9)$$

where the injection parameter is determind from the relationships

$$b_T = \frac{v_B \, \rho}{\dot{M} \, St_O} \qquad \text{and} \qquad v_B = - \frac{d}{4} \cdot \frac{d \, \bar{u}}{d \, x} \quad .$$

Here v_B takes into account the effect the chemical reaction which takes place entirely at the channel wall.

In Equation (5) j_w - mass injection rate, taking into account the reaction kinetic relationship written in the Arrenius form

$$\dot{j}_w = \frac{\dot{\xi}_R}{A_w} = B_k \cdot \exp \left[- \frac{\Delta E}{R \, T_w} \right] \cdot \left(1 - X_{Bw} \right) \sqrt{P_s \, \rho_w} \qquad . \qquad (10)$$

Correction for non-isothermal character of the flow is introduced in the form suggested by Valyuzhinich, Eroshenko and Kuznetchov [6]

$$\Psi_T = \left[\frac{\theta_w - 1}{\theta_w^{-m_1}} \left[\frac{m_2}{\theta_w - 1} + \frac{(1 - m_2) \, c_p T_s}{H_w - H_s} + m_3 \, F \, T_s Pr^{-0.6} \theta_w^{-m_1} \right] \right]^{-1} \qquad . \qquad (11)$$

In Equation (11) the coefficients m_1, m_2, m_3 take into account the effect of the initial thermal region of the tube and the heat flow direction. Parameter F takes into account variation of the flow density with temperature and the composition of the reacting mixture.

It should be pointed out that the laws of mass transfer are determind from similar relationships only if the effective Schmidt number is used instead of the Prandtl number.

Thus the system of equations (3...11) with the additional equation of state for a perfect gas may be solved when there are data concerning the dependence of thermophysical properties of the reacting mixture upon pressure, temperature and the initial fuel decomposition ratio.

Calculations are carried out using the "step-by-step" technique along the whole channel length.

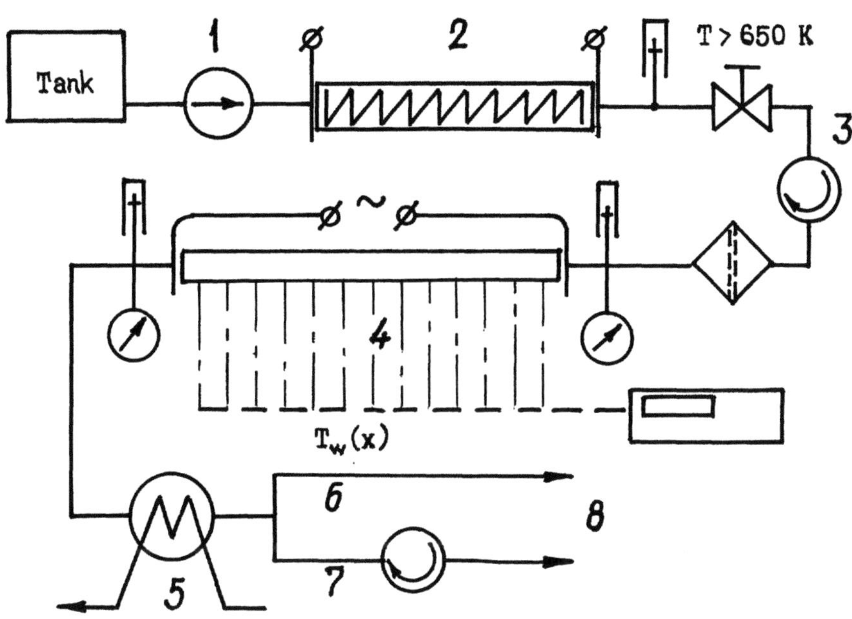

Fig.3. Scheme of the experimental unit for determining reacting kerosene flow parameters: 1 - pump, 2 - preliminary heater, 3 - meter of mass flow rate in liquid phase, 4 - measuring section (the tube with $d = 1 \cdot 10^{-3}$ m, $L = 1$ m), 5 - water cooler, 6 - gas line, 7 - liquid line, 8 - chemical analysis of decomposition products.

Constants of the kinetic equation of the chemical reaction of the fuel decomposition (B_k and $(\Delta E /R)$ in Equation (10)) were obtained by the authors in the course of processing the results of special experiments: they heated hydrocarbons within stainless steel smooth-walled tubes. The experimental units is schematically presented in Figure 3.

The calculations of FAHE were made using the suggested methods. These calculations allowed the selection of an optimum geometry of the tube bank with dense aligned arrangement.

From these calculations it is clear that utilization of chemical heat-sink capability of the fuel requires a multi-pass heat exchanger in the fuel circuit (side). In the prototype version of the heat exchanger designed for the propulsion system with the temperature of the gas at the turbine inlet about 2100 K the passes ratio (fuel/air) is **24:1**.

The calculations results showed that permissible value of the hydraulic losses in the air circuit determines the geometry of the heat exchanger (see Figure 4). It should be pointed out that in order to provide the required level of the air and fuel pressure losses in the FAHE, the pumping speeds are not high and correspond to $Re = (3-6) \cdot 10^4$.

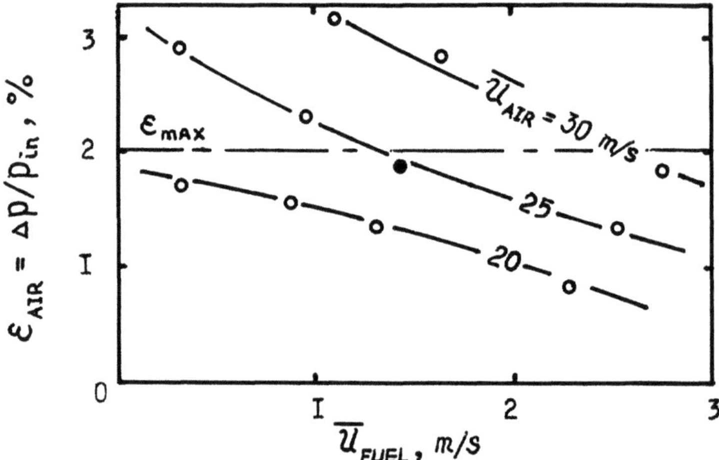

Fig. 4. The dependance of relative pressure drop in air circuits upon pumping speeds of air ($\overline{u}_{AIR}$) and fuel ($\overline{u}_{FUEL}$)

Preliminary Experimental Work

The authors carried out experimental investigation using the prototype version of the FAHE under operating conditions of the aviation gas turbine engine being developed now.

The results showed that even incomplete decomposition of the
fuel (up to **10 %**) when it is heated up to **870 K** increases its
total heat sink by **15 %** in comparison with a non-reacting fuel
heated up to the same temperature. The experimental data obtained are in general in agreement with the calculated values.
Unfortunately the authors could not test the FAHE at higher
temperatures. The authors hope that the investigations will be
finished this year.

<u>Nomenclature</u>

A_w – area of the inside tube wall surface, m^2;

B_k – kinetic constant, preexponential factor in Arrenius relationship which is depended on catalytic activity of tube wall material, $1/s$;

$b_T = j_w/(\rho\, \bar{u}\, St_0)$ – non-dimensional parameter of the injection;

c_p – specific heat capasity of nonreacting flow at constant pressure, $J/kg\cdot K$;

$D,\ d$ – outside and inside diameter of smooth tube, m;

$Eu = \Delta p/(0.5\cdot\rho\bar{u}^2)$ – Euler number, –;

$(\Delta E/R)$ – kinetic constant, temperature coefficient, –;

$\dot{j}_w$ – relative mass flux on inside tube surface, mass injection rate, $kg/m^2 s$;

H – total enthalpy of reacting mixture, J/kg;

$\dot{M}$ – mass flow rate, kg/s;

$Nu = \alpha\cdot\ell/\lambda$ – Nusselt number, –;

p – static pressure, Pa;

$Pr = c_p\cdot\eta/\lambda$ – Prandtl number, –;

$\dot{q}_w = \alpha\cdot(T_w - T_s) + j_w\cdot\Delta H^0$ –
      – heat flux on inside tube surface, W/m^2;

R – universal gas constant, $J/mol\cdot K$;

$Re = u\cdot\ell/\upsilon$ – Reynolds number, which is calculated by mass average parameter of the flow, –;

$S_i = s_i/D$ – transversal and longitudinal relative steps of the aligned tube bank, m/m;

$St_0 = Nu_0/(Re\cdot Pr)$ – Stanton number at "standard" conditions (without influence of the disturbance factors on the boundary layer in the tube: non-iso-termicity, injection, pressure gradient etc.);

T	– total temperature, K;
$\bar{u}$	– mass average velocity of reacting flow, m/s;
$v_B = \dot{j}_w/\rho_B$	– "effective" injection velocity, m/s;
x	– axial tube coordinate, m;
$X_B = 1 - X_A$	– decomposition ratio of the initial fuel, relative concentration of decomposition products, kg/kg;
$\dot{\xi}_R$	– mass rate of reaction, kg/s;
ρ	– mass average density of the flow, kg/m^3;
$\theta_w = T_w/T_s$	– factor of non-isotermicity, –.

Subscripts:

"w" – parameter on the wall;   "s" – parameter of the main flow;

"в" – injection;   "т" – non-isotermicity;

"d" – diffusion parameter.

References

1. Lander, H.; Nixon, A.C.: Endothermic fuels for hypersonic vehicles/ AIAA 5th Annual Meeting and Technical Display, Philadelphia, USA. AIAA Paper No. 68-997 (October 1968) 1-12.
2. Heat exchanger design handbook/ Vol. 1. Heat exchanger theory Contributers: D. Brian Spalding, J. Taborek. New York, Philadelphia, Washington, London: Hemisphere Publishing Corporation 1983.
3. Kutateladze, S.S.; Leontiev, A.I. Heat transfer, mass transfer and friction in turbulent boundary layers. New York, Washington, Philadelphia, London: Hemisphere Publishing Corporation 1990.
4. Petukhov, V.S.; Kirillov V.V.: On the problem of heat transfer in the turbulent fluid flow in tubes. Teploenergetica No. 4 (1958) 63-68.
5. Eroshenko, V.M.; Ershov, A.V.; Zaichik, L.I.: An influence of variable physical gas properties on turbulent flow and heat transfer in a tube with permeable walls. Journal of Engineering Physics 60 (1986) 195-200.
6. Valyuzhinich, M.A.; Eroshenko, V.M.; Kuznetsov, E.V.: Experimental investigation of heat transfer by turbulent convection in helium at a supercritical pressure under the condition of high non-isothermicity. Teplofizika vysokikh temperatur 24 (1986) 89-94.

A Model for Predicting the Performance of Domestic Gas-Fired Water Heaters

K K Yau* and J W Rose

Department of Mechanical Engineering
Queen Mary and Westfield College
University of London
London U.K.

* GEC Alsthom Turbine Generators Ltd., Manchester, U.K.

Abstract

The paper outlines a model for predicting the performance of downward-firing, gas-fired, domestic water heaters employing horizontal high-finned tubes. The model includes radiative heat transfer in the combustion zone, convective heat transfer for flow over side channels of the combustion zone and the tube bank, and condensation where the surface temperature is lower than the dew point of the combustion gas mixture. The inputs to the model are the inlet gas temperature, pressure, flow rate and composition (mainly air and methane), the water flow rate, inlet temperature and flow path arrangement, as well as details of the geometry. The model calculates the heat-transfer rates:- (a) by radiation to the combustion zone side channels and top tube row, (b) by convection to the combustion zone side channels and to each tube row and (c) by condensation on those surfaces for which the wall temperature is below the dew point of the combustion gases. Predictions of the model are compared with experimental data obtained from measurements on an instrumented commercially-available heater. Predictions for alternative designs are given.

Introduction

Significant improvements in the efficiency of gas-fired domestic water heaters have been achieved in recent years by arranging that water vapour in the combustion products is condensed in the heater. The design of these compact heat-exchangers, where radiation, convection and condensation all contribute to the heat transfer to the water, is complex. This paper describes a newly-developed, comprehensive design code for a particular configuration, namely a downward-firing, multi-pass, cross-flow, finned tube arrangement. For given air, gas and water inlet conditions and flow rates, the model computes the heat transfer by all three mechanisms to the various elements (tube rows or side channels) of the heat exchanger.

<u>General description</u>

A typical water heater geometry is illustrated in Fig. 1. The gas-air mixture enters the combustion zone via the horizontal burner plate at the top of the heater. Radiative heat transfer takes place from the gas to the side walls of the combustion zone and to the top row of tubes. The combustion gases flow vertically downwards over the rows, leaving the heat-exchanger at the base, together with condensate formed on surfaces below the dew-point temperature. All surfaces receive heat transfer by convection and, depending on their location and temperature, may also have radiative and condensation components.

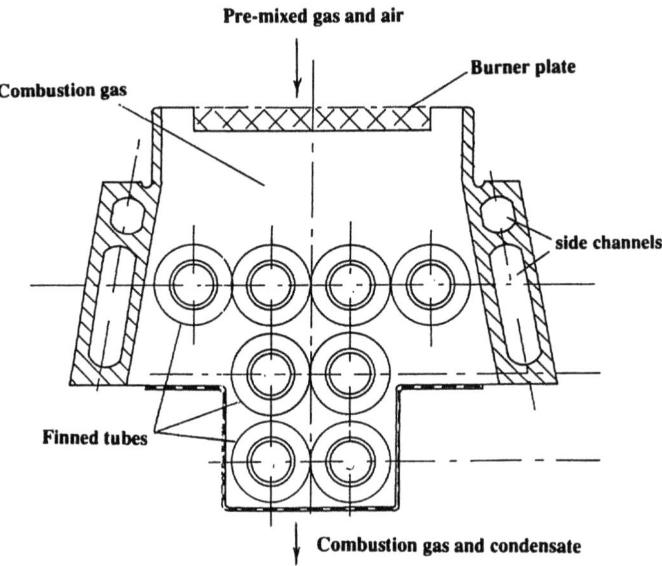

Fig.1 Downward firing, finned-tube water heater

The inputs to the model are the relevant parameters of the gas, air and water at their respective inlets to the heat-exchanger:-

1.   air temperature, pressure, humidity and flow rate.
2.   gas temperature, pressure, flow rate and composition.
3.   water temperature and flow rate.

The geometrical details, surface emissivities, and water flow path arrangement are also specified.

The programme calculates the heat-transfer rates to each tube row and side channel, taking account of convective, radiative and condensation heat transfer as appropriate.

Combustion gas composition

The composition of the combustion gas mixture is needed for the determination of relevant properties for calculation of radiation to the top tube row and side channels and for convective heat and mass transfer calculations throughout the heat-exchanger.   This is found from the gas and air flow rates assuming complete combustion of all hydrocarbons.

Radiative heat transfer

Radiation contributes typically less than 15% of the total heat transfer and a simple treatment was deemed adequate.   The model used is that of Hottel, described in Hottel and Sarofim [1].   The combustion zone is modelled as a four-surface enclosure: (1) burner plate, (2) end walls, (3) cooled side walls and (4) the top tube row.   The burner plate is treated as an adiabatic surface and top tube row is modelled as a plane surface with emissivity unity. All surfaces are assumed to have uniform radiosities with emissivities assigned to the side and end walls.   The gas is assumed to have a uniform temperature and carbon dioxide and water to be the only participating constituents.   The relevant view factors were found by approximating the combustion zone as an orthogonal region having the same volume, height and length.

The approximate method of Hottel and Sarofim [1] requires the "total emissivity" of the gas.   This quantity is determined by the pressure and temperature of the gas mixture and the partial pressures of the participating gases, together with a characteristic dimension known as the "mean beam length", which depends on the geometry. Also required are gas absorptivities for radiation between the gas and the individual surfaces. These latter depend on the same parameters as the gas emissivity and also on the temperatures of the relevant surfaces. With the exception of the gas and wall temperatures, all of the variables required to calculate the radiative heat-transfer rates to each of the surfaces can be obtained. In the overall model, initial guesses are used for these temperatures which are subsequently redetermined and used in an iterative process as described below.

<u>Convective heat transfer</u>

The heat-transfer coefficient for flow over the side channels is obtained using the equation for laminar flow over a flat plate:-

$$Nu_L = 0.664 \ Re_L^{1/2} \ Pr^{1/3} \qquad\qquad (1)$$

where Nu_L is the mean Nusselt number,

               Re_L is the Reynolds number, using the mean downward gas velocity,

               Pr is the Prandtl number of the combustion gas.

For flow over finned tubes several correlations are available. All indicate that the heat-transfer coefficient varies as Reynolds number raised to a power near 0.65 and as Prandtl number to a power near to 0.33. The heat-transfer coefficient, h, is obtained from an appropriate correlation and the convective heat-transfer rate is then given by:

$$Q_c = \eta \ h \ A \ (\overline{T}_g - T_w) \qquad\qquad (2)$$

where η is the "surface effectiveness", which incorporates the fin efficiency and can be found when the heat-transfer coefficient is known and the geometry specified,

            A   is the total external area of the finned tube,

            $\overline{T}_g$ is the mean of the upstream and downstream temperatures of the gas,

            T_w is the temperature of the tube wall.

Thus, when the upstream and downstream temperatures and the wall temperature are known for a side channel or tube row, the convective heat-transfer rate can be calculated.

Condensation

When the wall temperature (side channel or tube row) is less than the dewpoint temperature of the combustion gas, condensation occurs. The condensation rate, M, is governed by the rate at which water molecules in the gas are transported to the surface and can be written:

$$M = GA\,(\overline{W} - W_i)/(1 - W_i) \tag{3}$$

where G is the mass-transfer coefficient,

            $\overline{W}$ is the mean of the upstream and downstream mass fractions of the vapour.

            W_i is the mass fraction of the vapour at the gas-condensate interface.

The composition and temperature, T_i, at the gas-condensate interface are related by the equilibrium condition which, for ideal-gas mixtures, gives:

$$W_i = 1 - [P - P_{sat}(T_i)]/\left\{P - (1 - M_v/M_g)P_{sat}(T_i)\right\} \tag{4}$$

where M_v and M_g are the molar masses of vapour and gas respectively.

The mass-transfer coefficient is obtained using the Colburn-Hougen "analogy" between convective heat and mass transfer and making use of the above-mentioned convective heat-transfer correlations for flat plate and row of finned tubes. To obtain the corresponding mass-transfer correlation, Nusselt and Prandtl numbers in the heat-transfer correlation are replaced by Sherwood and Schmidt numbers respectively. The validity of this method is discussed by Rose [2] and Lee and Rose [3]. When the condensation rate, or more precisely the "suction parameter", is small, as in this case, the method gives good results. Thus, when the bulk mass fraction of H_2O and the value adjacent to the surface (related to the temperature as indicated in equation (4)) are known, the condensation rate can be calculated.

Condensate film, wall and water-side resistances

The thermal resistances of the condensate film (where present) on a tube or side channel wall, and of the water side, are small. They are, however, included in the model, using the Nusselt approximations for condensation on the vertical and cylindrical condensing surfaces, one-dimensional conduction in the walls and the Petukhov [4] equation for the water-side.

Outline of programme structure

The composition of the combustion products is first determined. A first estimate is made for the temperature of the combustion zone gas (e.g. 0.8 times the adiabatic flame temperature). The relevant thermophysical and radiation properties are calculated and the following steps executed:-

1. A first estimate for the water exit temperature is made.

2. Radiative and convective heat-transfer rates to the side walls and top tube row are calculated using a first estimate for the wall temperature of the top tube row and side channels. (Radiative heat-transfer rates were insensitive to wall temperatures over moderate ranges.) A coolant energy balance gives the temperature at inlet to the top tube row and side channels.

3. Convective heat-transfer rates are calculated for each "element" (tube row or side channel) in turn, following the path of the gas. In each case water- and gas-side energy balances are used iteratively to determine, for the element, the inlet water temperature and exit gas temperature. This gives the heat-transfer rate for the element together with the exit water temperature and the gas approach temperature for the next element. When, for any element, the wall temperature is less than the dew point of the combustion gas at approach to an element, the condensation rate, and consequent additional heat-transfer rate, are determined, and appropriate adjustment to the gas composition at exit from the element is made.

4. Completion of step 3 gives a calculated value of the water temperature at inlet to the heat-exchanger. This is compared with the known value and a better estimate for the exit water temperature is made. Step 3 is repeated until the calculated and specified water inlet temperatures converge.

5. To achieve a closed solution the total heat-transfer rate (i.e. sum of heat-transfer rates for all elements) is compared with that given by an overall energy balance for the gas stream:

$$\Sigma Q = \Sigma(n_i \, \Delta h_{oi})_{fuel} + m_c \, h_{fg_2} - \Sigma\{m_k(h_{k_2} - h_{ko})\}_{gas \; products}$$
$$+ \; \Sigma\{m_k(h_{k_1} - h_{ko})\}_{reactants} - m_c(h_{f_2} - h_{fo}) \tag{5}$$

where ΣQ is the total heat-transfer rate,

           Δh_o is the molar enthalpy of combustion (water in gas phase),

           n is mole flow rate

           m_c is mass condensation rate,

           m is mass flow rate,

           h is specific enthalpy,

           h_f is specific enthalpy of water,

h_{fg} is specific enthalpy of phase change of H_2O,

i denotes hydrocarbons species,

k denotes gas species,

o denotes reference state, 25 ^{O}C,

$_{1,2}$ denote inlet and outlet respectively.

A better estimate of the combustion zone gas temperature is made by comparing the total heat-transfer rate given by the energy balance, equation (5), with that found by summing the contributions of the individual elements found by the heat-transfer calculations.

6. Steps 2 to 5 are repeated until convergence of both water inlet temperature and gas exit temperature are achieved.

<u>Case studies</u>

Measurements have been made by Rose et al. [5] for a water heater with the configuration illustrated in Fig. 1. The heater was instrumented to measure the heat-transfer rates to the separate elements (tube rows and side channels) and tests made for various water inlet temperatures in the range 22 ^{O}C to 70 ^{O}C (i.e. above and below the dew point temperature of the combustion products). Calculations have also been made for the same geometry and conditions using the present model. A convective heat-transfer correlation based on accurate measurements, using a specially designed apparatus for high temperature downward gas flow over high-finned tube banks [5] was used in the model. This gives convective heat-transfer coefficients substantially higher than those indicated by handbook correlations for close-packed, high-finned tube arrangements, at low Reynolds numbers. The Reynolds number (based on the minimum flow area and tube diameter) in the present application is typically in the range 150-800.

The experimental data for the instrumented heater are compared with the predictions of the model in Fig. 2. The prediction of the overall performance is seen to be excellent. Discrepancies between the calculated and measured values of the contributions of the individual elements may be attributable in part to imperfections in the model and also to experimental error. (The overall heat-transfer rate was measured with significantly higher accuracy than were the heat-transfer rates for the separate elements, as indicated by Rose et al. [5]). Both experimental data and model show an increase in heat-transfer rate when the water inlet temperature falls below the dew point temperature of around 52 $^{\circ}$C.   It is also seen that the model predicts the measured condensation rate satisfactorily.

To demonstrate the flexibility and utility of the programme, performance predictions have been made for alternative heater configurations.   Fig. 3 shows the design discussed above (B1), together with arrangements with one (B2) and two (B3) fewer tubes in the top row.

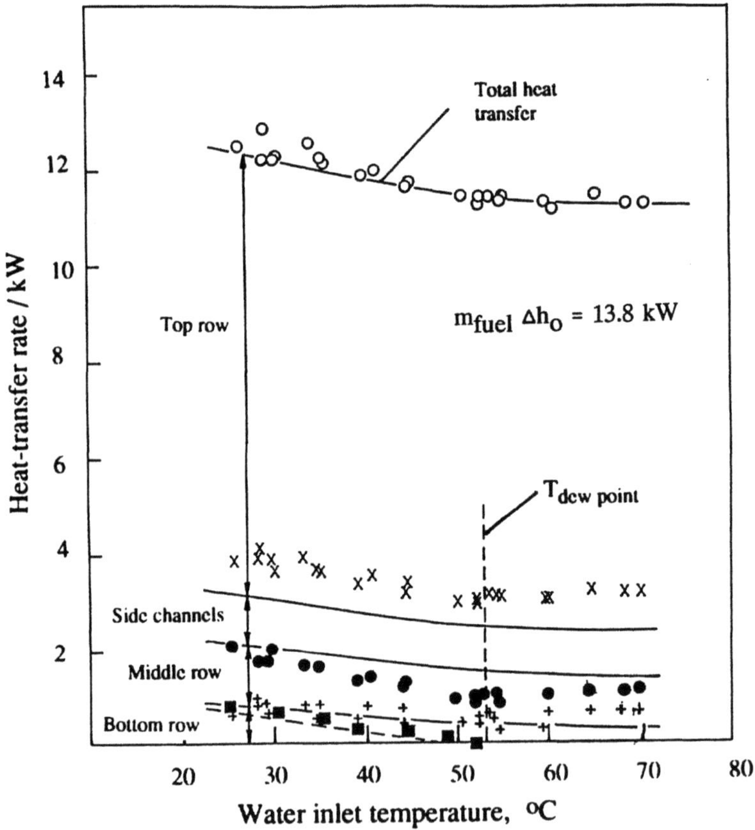

Fig.2 Comparison of prediction with data from instrumented water heater
    ■ condensation, + bottom row,   o bottom and middle rows,
    x bottom and middle rows and side channels, o total,
    - - - - - predicted condensation heat transfer.

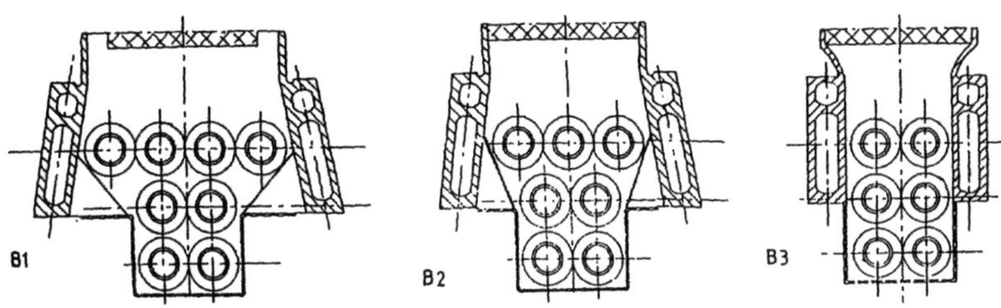

Fig.3   Three water-heater designs

Fig.   4 compares the three designs for the same air, gas and water flow rates and inlet temperatures.   The performance of B2, with one fewer tube, is seen to be slightly better than B1.   This indicates that the effect of the increase in gas velocity, caused by the reduction in flow area, outweighs the decrease in heat-transfer area. Although B3 has around 20% less heat-transfer surface, its performance is only slightly worse than B1.

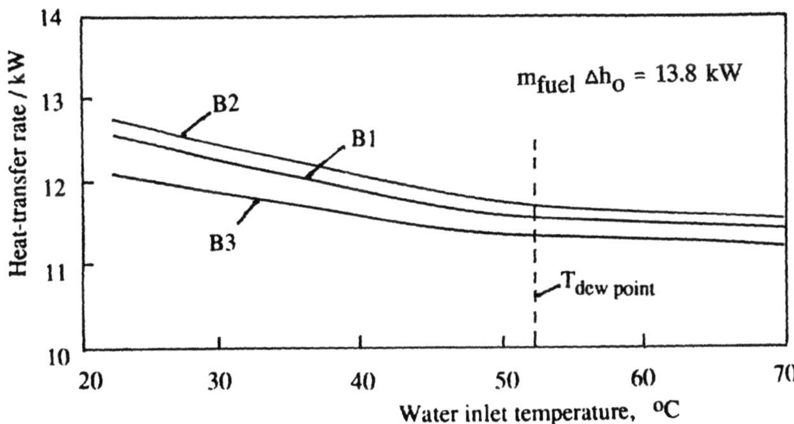

Fig.4   Comparison of performance for three heater designs

Conclusion

An apparently successful model has been developed for prediction of the performance of downward-firing, gas, water-heaters employing horizontal finned tubes.   The model can readily be adjusted to accommodate modifications in design including:- number of tube rows, number of tubes per row, water flow path, as well as fin, tube and duct geometry.   It should be possible to use the same general approach for heaters with different gas-flow and heat-transfer surface arrangements.

References

1. Hottel, H.C.; Sarofim, A.F.; Radiative Transfer, McGraw Hill, New York, 1967.

2. Rose, J.W.; Condensation in the presence of non-condensing gases, Power Condenser Heat Transfer Technology, Hemisphere, 151-161, 1981.

3. Lee, W.C.; Rose, J.W.; Comparison of calculation methods for non-condensing gas effects in condensation on a horizontal tube, IChemE Symp. Ser. no. 75, 342-355, 1983.

4. Petukhov, B.S.; Heat transfer and friction in turbulent pipe flow with variable physical properties, Advanced heat transfer, Vol. 6, Academic Press, New York, 503-564, 1970.

5. Rose, J.W.; Cooper, J.R.; Crookes, R.J.; Yau, K.K.; Yau, W.K.; Experimental studies of heat transfer between hot gases flowing downwards over water-cooled, horizontal, close-packed, banks of finned tubes.   In preparation, 1991.

Acknowledgement

This work was supported by a grant, under the U.K. Science and Engineering Research Council (SERC) co-funding scheme, by SERC and the British Gas Corporation.

Plastic Heat Exchangers

Philippe BANDELIER, Jean Claude DERONZIER, Fernand LAURO

Groupement pour la Recherche sur les Echangeurs Thermiques
CEA-CENG
85 X
38041 GRENOBLE Cedex
FRANCE

<u>Summary</u>

Plastic properties may be used in the manufacturing of heat exchangers. Considering their advantages, their field of application is quite large : low temperature heat recovery, concentration of solutions. Three equipment are described : a liquid-liquid shell and tube exchanger, a gas-gas shell exchanger and a falling flow evaporator.

1) INTRODUCTION :

A recent French survey [1] shows that most thermal energy consumed in industry is at low temperature levels, between 50 and 200°C (fig. .1.).

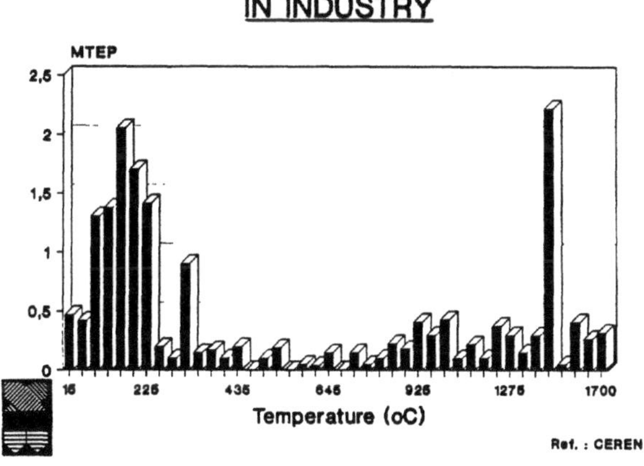

Fig. .1. : The energy consumption in industry (CEREN [1])

Heat recovery is advantageous inside this temperature range. The main difficulties encountered in this task are corrosion and fouling of the exchange surfaces, combined with the investment cost and the upkeep of the exchangers. This often prevents the recovering of heat.

Recent advances in the development of plastics and a better understanding of their properties allow the manufacturing of plastic heat exchangers. This was begun in 1965 by Dupont de Nemours who is still a leader in this field.

In Europe, the first equipment appeared only during the 80's. The European market is still very small (0.4 %) but growing steadily (30 % per year) (fig. .2.).

THE HEAT EXCHANGER MARKET
By materials

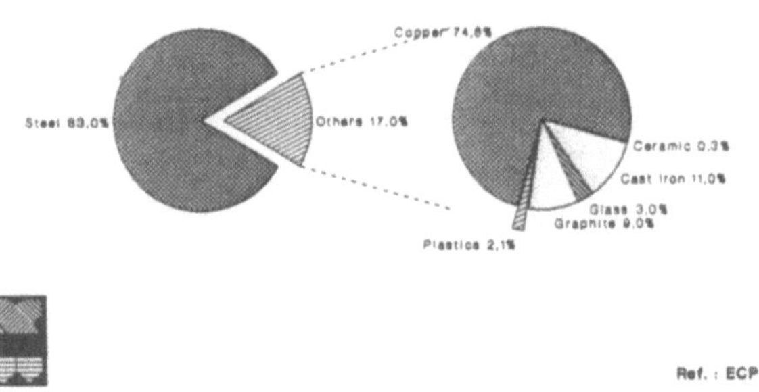

Fig. .2. : The European heat exchanger market (classified by materials) (ECP [1])

<u>2) THE USE OF PLASTICS IN HEAT EXCHANGERS</u> :

The main characteristics of plastics are reviewed in terms of their use in the manufacturing of heat exchangers. Plastics have very attractive properties for this application, but some disadvantages must be carefully dealt with when designing an heat exchanger.

2.1 Lightness :
Plastics density varies from 900 to 2200 kg/m^3. This is 4 to 5 times less than the densities of metals. For a same given volume, plastic exchangers will be appreciably lighter and the casing cheaper.

2.2 Surface aspect :
Plastics have a very smooth surface. The consequences are friction factors and pressure drops which are lower than for metallic surfaces. Moreover, plastic wettability is very low, so steam leads to dropwise condensation instead of filmwise condensation ; this phenomenon may be significant in heat transfer enhancement.
Organic and mineral deposits are less adherent on the plastic surface, which is less sensitive to fouling. Lastly, exchangers are easier to clean.

2.3 Chemical resistance :
Most plastics have good or excellent behavior in the presence of corrosive fluids such as mineral or organic acids, oxidizing agents and many solvents. The most resistant thermoplastics are fluored polymers (i.e. PVDF or PTFE). Each plastic's resistance has been tabulated for a wide range of chemical substances. These are always completed by test carried out under real conditions.

2.4 Mechanical resistance :
The specific resistance of plastics (ratio of mechanical resistance to density) is the highest of all materials. However, the mechanical resistance drops rapidly when the temperature rises. Even at ambiant temperature, plastic resistance is ten times less than for metal.
For pure thermoplastics, tensile strength ranges from 10 to 100 MPa and from 100 to 200 MPa for filled thermoplastics (fig. .3.).
Finally, plastic resistance to erosion is often better than that of metal.

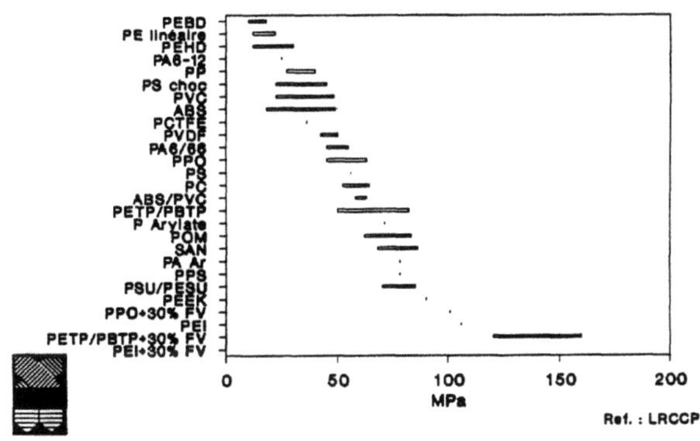

Fig. .3. : Tensile yield strength for some resins (filled or not)

2.5 Thermal expansion and creeping :
Non-filled plastic thermal expansion is about ten times greater than for metals. Suitable clearances have to be specified when designing a plastic-metal assembly.
Creeping is a slow irreversible deformation. It increases with temperature and load.

2.6 Ageing and maximum temperature of use :
For a plastic, ageing means that its properties are not constant in time, a sign of degradation appears more or less quickly, depending on the temperature level. This phenomenon has to be taken into account when choosing the material used in the design of a heat exchanger.
Plastic properties are greatly affected by temperature. Maximal temperature use is chosen based on mechanical load and the equipment's lifetime.
Figure .4. shows examples of lifetimes for some materials.

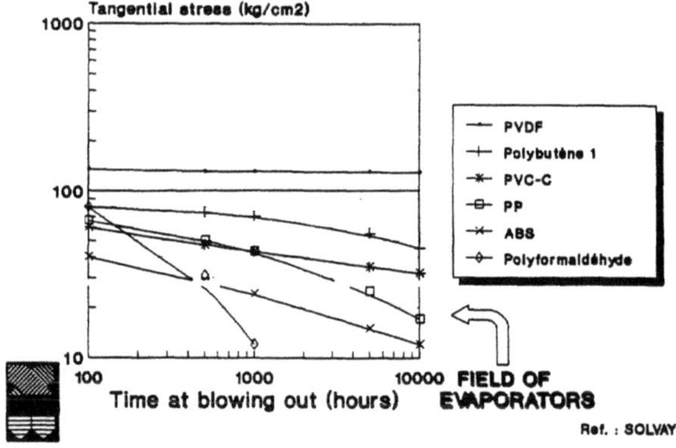

Fig. .4. : Ageing characteristics of some polymers

Basic resin suppliers specify the following maximum operating temperature, without mechanical stress :

PTFE (polytetrafluoroethylene)	250°C
PS (polysulfone)	160°C
PVDF (polyvynilidene fluoride)	140°C
HDPE (high density polyethylene)	110°C
PP (polypropylene)	80°C
PVC (polyvinyl chloride)	60°C

2.7 Thermal conductivity :
Plastics have a very low level of thermal conductivity : 100 to 300 less than metals, between 0.1 to 0.4 W/m.°K (fig. .5.).
Overall heat transfer coefficients may appear very low. In fact, relative value of fluid side coefficients must be analyzed.
The overall heat transfer coefficient is given by the relation :

$$\frac{1}{U} = \frac{1}{k_1} + \frac{1}{k_2} + R_w + R_f$$

With :

U : overall heat transfer coefficient $(W/m^2.°K)$
k_1 : local heat transfer coefficient side 1 $(W/m^2.°K)$
k_2 : local heat transfer coefficient side 2 $(W/m^2.°K)$
R_w : wall thermal resistance $(W/m^2.°K)^{-1}$
R_f : fouling thermal resistance $(W/m^2.°K)^{-1}$

Neglecting R_f, U may be drawn, in function of U_0, which is the overall heat transfer coefficient for R=0 (fig. .6.).
It clearly appears that for gas-gas exchangers, the wall resistance has no or a very low effect on the overall heat transfer coefficient.

Plastic Heat Exchangers

Philippe BANDELIER, Jean Claude DERONZIER, Fernand LAURO

Groupement pour la Recherche sur les Echangeurs Thermiques
CEA-CENG
85 X
38041 GRENOBLE Cedex
FRANCE

Summary

   Plastic properties may be used in the manufacturing of heat exchangers. Considering their advantages, their field of application is quite large : low temperature heat recovery, concentration of solutions. Three equipment are described : a liquid-liquid shell and tube exchanger, a gas-gas shell exchanger and a falling flow evaporator.

1) INTRODUCTION :

A recent French survey [1] shows that most thermal energy consumed in industry is at low temperature levels, between 50 and 200°C (fig. .1.).

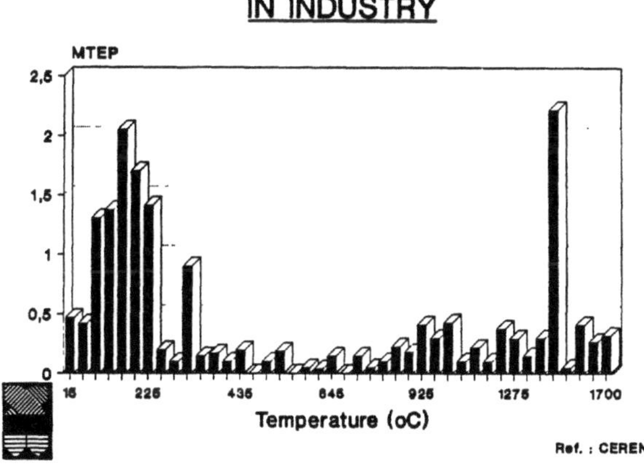

Fig. .1. : The energy consumption in industry (CEREN [1])

Heat recovery is advantageous inside this temperature range. The main difficulties encountered in this task are corrosion and fouling of the exchange surfaces, combined with the investment cost and the upkeep of the exchangers. This often prevents the recovering of heat.

Recent advances in the development of plastics and a better understanding of their properties allow the manufacturing of plastic heat exchangers. This was begun in 1965 by Dupont de Nemours who is still a leader in this field.

In Europe, the first equipment appeared only during the 80's. The European market is still very small (0.4 %) but growing steadily (30 % per year) (fig. .2.).

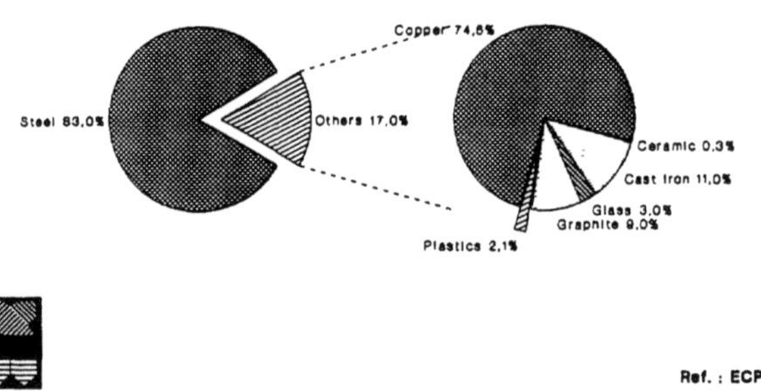

Fig. .2. : The European heat exchanger market (classified by materials) (ECP [1])

2) THE USE OF PLASTICS IN HEAT EXCHANGERS :

The main characteristics of plastics are reviewed in terms of their use in the manufacturing of heat exchangers. Plastics have very attractive properties for this application, but some disadvantages must be carefully dealt with when designing an heat exchanger.

2.1 Lightness :
Plastics density varies from 900 to 2200 kg/m^3. This is 4 to 5 times less than the densities of metals. For a same given volume, plastic exchangers will be appreciably lighter and the casing cheaper.

2.2 Surface aspect :
Plastics have a very smooth surface. The consequences are friction factors and pressure drops which are lower than for metallic surfaces. Moreover, plastic wettability is very low, so steam leads to dropwise condensation instead of filmwise condensation ; this phenomenon may be significant in heat transfer enhancement.
Organic and mineral deposits are less adherent on the plastic surface, which is less sensitive to fouling. Lastly, exchangers are easier to clean.

3.2 Gas-gas heat exchanger :

For this type of exchanger, the wall resistance has no effect on overall performances (fig. .5.). Today, for economical reasons, the wall thickness is within 60 and 100 μm.

This exchanger is called a flexible sheath exchanger.

The technology uses vertical tubes, with a parallel or cross flow. The thermal expansion is compensated by a lower floating tubular sheet.

The applied stress is very low (0.3 MPa). This is caused by the differential pressure between circuits (no more than 100 mmCE) and by the weight of the lower tubular sheet.

A sample flow configuration is shown in figure .7..

If the material used is PVDF, the hot air inlet working temperature may reach 140°C, leading to a reasonable temperature in the wall.

The main advantages of this exchanger are :
    . excellent behavior in corrosive working conditions,
    . lightness,
    . good behavior in fouling conditions,
    . low cost.

This equipment may be used in all cases of low temperature heat recovery on gaseous wastes, especially when steam is able to condense, thus washing soluble gases.

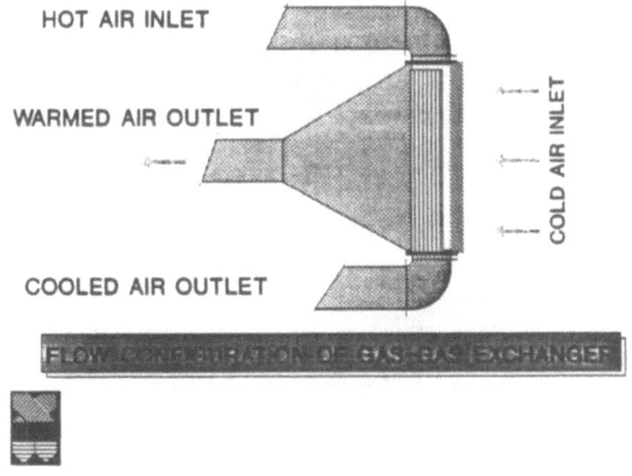

Fig. .7. : Gas-gas plastic heat exchanger

3.3 Falling film evaporator :

In metallic tube evaporators, the heating fluid (generally steam) condenses outside the tubes, yielding latent heat to boiling liquid film flowing inside the tube. The outside pressure is higher than the inside pressure.

To maintain a proper heat transfer level, the plastic wall thickness has to be very low (about 50 μm) (fig. .6.). To prevent the tubes from flattening, the pressure difference has to be reversed : the steam flows inside the tubes and the liquid outside as drawn on figure .8..

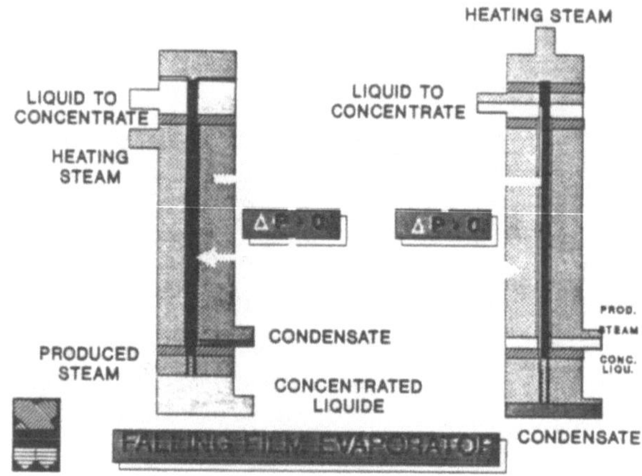

Fig. .8. : Falling film evaporator

Figure .9. gives the pressure difference and the stress in a 30 mm diameter and 100 μm sheath. This is drawn according to the temperature, with the temperature difference as the parameter. The stress reported on figure .4. gives the lifetime of the selected material.

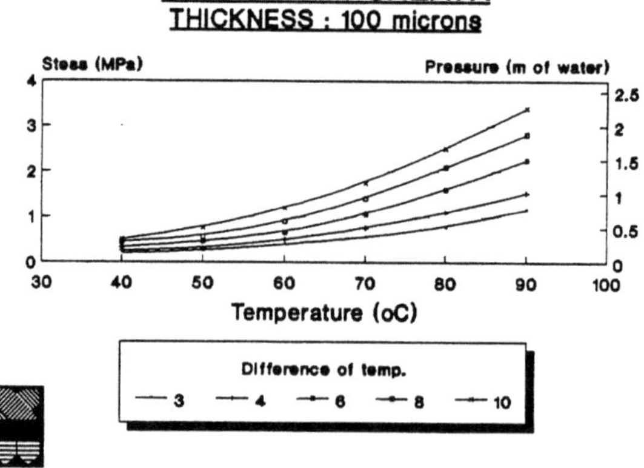

Fig. 9. Pressure difference and stress in sheaths

Some particular patented devices have been studied to obtain a regular steady film flow around the sheath and to allow free individual expansion.

An experimental program has shown that the heat transfer coefficient level may be the same as that of metallic equipment (fig. 10.).

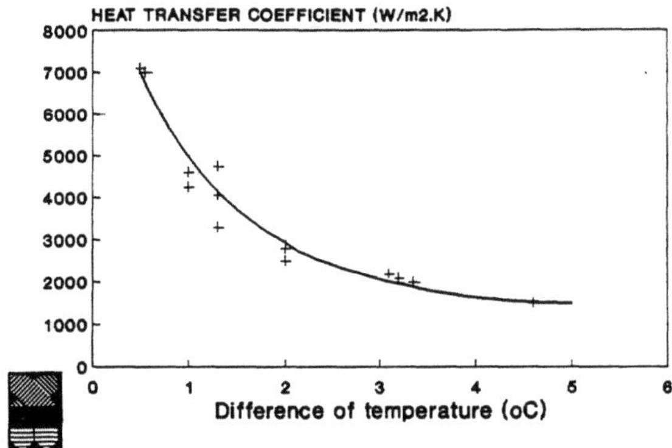

Fig. 10. : Sample of heat transfer results

Applications : This kind of equipment may be used under the same conditions as a metallic evaporator providing that the plastic selected is compatible with the temperature and stress levels.

The good behavior of plastics in corrosive conditions and the low cost of the surface (some ECU/m^2 for plastics and some 10 ECUs/m^2 for metals) may lead to the replacement of high cost metals by cheap plastics .

Since the surface is cheap, the exchanger surface may be increased while decreasing the temperature difference. This leads to a significant saving of energy.

As shown by the GRETh's experiments on the use of the ocean thermal gradient (OTEC) for the desalination of sea water, this equipment is able to work at low temperatures (from 15 to 50°C) ; the effect of the low temperature is compensated by a narrow temperature difference and a satisfactory heat transfer coefficient level is still reached.

REFERENCES :

[1] : Ecole Centrale de Paris ; Groupe de Recherche Stratégie et Technologie ; AFME ; "Les échangeurs de chaleur - Analyse Stratégique" Rapport publique. Juin 1987

[2] : J.C. DERONZIER, F. LAURO ; Les échangeurs de chaleur en matière plastique. Lavoisier. 1989

[3] : M. BIRON ; Les thermoplastiques - Eléments de technologie pour l'utilisateur de pièces plastiques ; LRCCP ; 1986

Condition Monitoring of Air Cooled Heat Exchangers

R J BERRYMAN

Heat Exchanger Advisory Service
AEA Technology
Harwell Laboratory
Oxfordshire
United Kingdom

Summary

The efficient and optimised operation of Air Cooled Heat Exchangers is vital in all Petro-chemical and Refinery operations where low grade heat is released to the environment. The need for continuous operation of these exchangers has often discouraged routine condition monitoring with the inevitable fall-off in performance with time. This paper explains how these exchangers can be performance tested and describes instruments, test methods, data analysis and diagnostic procedures used in maintaining and improving performance. The use of condition monitoring and good design and installation practices can lead to financial benefits with very short pay-back periods.

Background

Air Cooled Heat Exchangers (ACHEs) are used extensively on Oil Refineries and Petro-chemical complexes for condensing and cooling numerous process streams. The efficiency of such units for the rejection of low grade heat to the atmosphere depend a very great deal upon the thermal and mechanical performance of the tube bundle and fans. Optimisation is carried out in the initial design stage to minimise the heat transfer surface and provide an exchanger which can cope with the original process stream.

Since ACHE performance is subject to climatic conditions, they can become the cause of "bottle-necks" in overall plant throughput particularly during times of high ambient air tem-perature. Such events have become commonplace on many process plants in Europe dur-ing recent years and have led to production departments looking more closely at the parameters which effect ACHE performance.

The Heat Exchanger Advisory Service (HEAS), which is a trouble-shooting and consultancy organisation, has carried out many evaluations on existing ACHE units in recent times. Set up some six years ago with technical support from the Heat Transfer and Fluid Flow Service

(HTFS), the Advisory Service uses methods and techniques developed from many years research into ACHE thermal and mechanical problems. Berryman and Russell (1985) describe various aspects of trouble-shooting and include specific case histories.

This paper describes the techniques and instrumentation used to evaluate the performance of an ACHE both initially and after any remedial steps have been taken to improve the situation. The reasons for possible loss in performance are discussed as are methods which can be used to improve or uprate an existing unit.

Also covered are the important basic criteria which should be considered when designing new units. This includes thermal, mechanical and aerodynamic aspects for the optimisation of ACHE performance.

By the application of on-line condition monitoring it is possible to increase the performance of most ACHEs by as much as 25%. This applies especially to older units which have not been regularly cleaned and/or where less efficient fans have been used. The use of modern computer software for the design and performance simulation of existing units has proved valuable in getting the best out of air cooled heat exchangers.

The financial benefit to the user of having ACHEs working at peak performance can be very large with the pay-back period on the return of money invested in performance testing being a matter of just a few days.

Performance Testing

The essential operating data required in order to assess the performance of ACHEs is usually obtained from the airside. By far the most important parameter is airflow and its temperature at inlet and outlet. Most exchangers are of the forced draught design with axial fans driving cold air into plenum (box) chambers which support extended surface, finned tube bundles. Measurement of air inlet temperature is easily made at 5 locations below the fan. Any noticeable difference between this temperature and the surrounding ambient value usually indicates hot air recirculation or radiant heating from adjacent pipework or vessels.

At exchanger outlet, i.e. above the tube bundle in forced draught designs, air velocity and temperature is measured simultaneously at multiple locations. The bundle is divided into known flow areas and measurements are made at 150 mm above the finned tube surface. At this distance any effects of high velocity "jetting" between the fins is minimised. About 30 determinations are usually made over a tube bundle measuring, typically, 2 metres x 9 metres. Two such bundles are normally served by two fans (3.66 m diameter) and the data on airflow and temperature rise are all that is required to determine the airside heat load. Berryman and Russell (1986) give further details on how to measure airflow in ACHEs including comparisons using different instruments.

Other data which can be of use are the static pressure developed by the fans within the plenum chambers and the electrical power consumed by the fan/motor drive system. These quantities, together with the airflow, can be compared with the fan performance characteristics supplied by the fan manufacturer. Such curves are the reference material which is used to determine whether the existing fan performance can be improved, generally by increased airflow.

Information on fan speed can be obtained using an infra-red tachometer but checks on blade angle are difficult unless the fans are shut down for a reasonable period of time. Usually such information has to be assumed from the original specification unless measured at a recent routine shutdown.

Process side parameters are normally only available from "on-line" instrumentation which is read out to the Control Room computerised data acquisition system. Process flowrate, temperatures, pressures and fluid composition are required to determine the process heat load.

To complete the performance testing procedure, observations are made of the general mechanical condition of the exchanger. In this case it is important to note the extent of fouling on the bottom tube row, physical damage to the top of the bundle (see Figure 1), air leakage through holes in the plenum or gaps between plenum and bundle. With regard to the fans, any possibility of speed reduction from belt drive slippage must be checked. For direct drive this does not normally occur but a check on the fan speed is still required. If the fans can be stopped for a short period then the tip clearance between blade and fan ring can be measured. The normal gap is normally 1% of the fan diameter. Any greater clearance or significant variation around the fan ring circumference, reduces the air throughout.

Instrumentation

Since gathering data on petro-chemical plant and refineries is a somewhat hazardous activity, it is important that all instrumentation is both portable and electrically safe. In this latter respect, battery operated equipment is permissible if used in conjunction with personal gas detectors.

For simultaneous measurement of air velocity and temperature, a miniature vane type anemometer is used which incorporates a thermocouple (see Figure 2). This instrument gives a digital readout and can be used to provide an integrated time-averaged value of both air velocity and temperature. This very small anemometer can be as accurate as +/- 3% and is better than the conventional vane type also shown in Figure 3.

For fan and plenum static pressure, a Pitot tube may be traversed through tappings in the plenum wall. This, when coupled to a micro-manometer, can give the static pressure accurate to +/- 10 Pa (1 mm water gauge).

As noted above, fan speed is readily determined with an infra-red tachometer. The fan must be shut down for a short period so that a reflective marker may be attached to the main shaft or pulley. Blade angles are determined at shutdown using an inclinometer and straight edge near the blade tips. At the same time the tip clearance may be easily checked.

Electrical power consumed by the fan/motor/drive system is usually obtained from "on-line" ammeters and voltmeters. Knowing the local distribution power factor and efficiencies of motor and drive, the actual fan power can be calculated.

<u>Data Analysis</u>

The data obtained from performance testing are used primarily to determine the existing thermal performance. With adequate airside and process side information the heat loads on both hot and cold sides of the exchanger can be determined and then compared. The resulting heat balance should be less than 10% if the data have been correctly determined.

The total volumetric airflow determined from velocity measurements is compared with the expected performance of the fan. Using other data such as fan/plenum static pressure, fan tip speed and blade angle and consumed fan power, the available additional capability of the exchanger, if any, can be determined.

By using up-to-date computer software for the simulation of ACHE performance, it is possible to determine the maximum thermal capacity of the exchanger in its "as new" condition. Comparing these results with the data obtained from actual performance tests gives an indication of any short-fall in thermal and aerodynamic capability. Such computer codes can also be used to determine the degree of fouling which may be present assuming that airflow is satisfactory and evenly distributed.

In the final stage of analysis computer software can be used to establish the conditions required to achieve a different or increased thermal performance. This is essential in any exercise where the nature of the process stream is changed or a "re-vamp" exercise is being considered.

In work of this nature the Heat Exchanger Advisory Service has found the HTFS Computer Code ACOL4 to be invaluable in simulating the performance of air cooled heat exchangers.

<u>Diagnostics - Reasons for Poor Performance</u>

Diagnosing what may be the cause of poor performance on an ACHE is not straightforward and can only be achieved from sufficient previous experience. Whist it is commonplace for loss in performance to be due to a short-fall in airflow, this is not always the only reason. Often reduced performance is due to a number of factors each contributing a few percentage points to an overall loss in efficiency which can be in excess of 25%.

Looking firstly at reduced airflow, this can be due to incorrectly optimised fan settings or, more probably, a layer of dirt on the first and second tube rows. Dirt in the atmosphere is filtered by the successive layers in the tube bundle and is encouraged to stick onto the finned surface, particularly in the presence of lubricating oil and grease applied to the motor/drive system. Such an occurrence is more commonplace with forced draught units. The layer of dirt not only reduces the total airflow but adds an extra thermal resistance to heat transfer. On occasions the air quantity supplied by the fan may be adequate, but losses through holes in the plenum and gaps between plenum and bundle can lead to by-passing. Such losses are seldom greater than 2-3% of total airflow.

Sometimes the fault can lie mainly with the fan, its settings and general mechanical condition. With older, less efficient fans with few blades and narrow chord width, small changes in blade angle tip clearance and speed can have significant effects. Often, exchangers designed some years ago were not given sufficient allowance for inefficiencies when installed. Software was then less developed and design and rating procedures were not so rigorous. Such units are often found to be inadequate when called upon to provide a larger duty and sometimes the only solution is to replace the fans with more efficient ones.

Troubles can occur with the process stream. This can be due to a build-up of fouling inside the tubes although, in general, the airside resistance is more dominant. The process fluid can be maldistributed to the exchanger. A large diameter inlet feed pipe often supplies a manifold which splits into 4, 8 or even 16 small diameter pipes before reaching the individual bundles of an exchanger. Unequal flow to the sections of the exchanger can give an overall reduction in thermal performance.

A common process side problem, particularly with single pass condensing ACHEs, is the build-up of non-condensable gas. In steam condensers small quantities of air can accumulate in the upper rows of the bundle unless adequate venting takes place. This gas blanketing reduces the available heat transfer area and condensing performance drops off. In rare cases, where there is a marked increase in fluid viscosity as a process stream is cooled, the phenomenon of "freezing" can occur. This happens with some waxy hydrocarbons and requires careful control of the air temperature by the use of inlet steam coils and/or louvres.

Despite the above comments on process side problems, it is invariably the airside which is the focus of diagnosis and subsequent problem solving. Ambient conditions cannot be controlled and interactions between fan and bundle are critical to the performance of air cooled heat exchangers. Table 1 conveniently summarises some of the symptoms, possible faults and solutions to ACHE problems.

How to Improve Performance

Having diagnosed the problem or problems we must now seek to overcome them as quickly and economically as possible. The complete replacement of an existing ACHE by a new larger unit is something which should only be considered as a last resort. Such an action is both costly and time consuming.

If the expected airflow is not achieved and the bundle appears dirty, then cleaning is the obvious first step. The use of high pressure steam or water/detergent jet spraying is normally recommended although consideration should be given to dry rotary brushing if the deposit is well attached to the finned surfaces. Such a procedure has led to increases in airflow of 20-25% in many cases studied by the Heat Exchanger Advisory Service and, on occasions, as much as 50% extra airflow has been achieved.

Often the removal of airside fouling allows minor adjustments to be made to the fan settings (increased blade angle) so that the fan then operates at a more efficient point on its characteristic curve. Sometimes the speed may be increased to provide even more airflow but this will depend upon the available electrical power of the motor. This is another reason why it is important to determine the current and voltage of the fan/motor/drive system during performance testing.

During periods of high ambient temperature on an exchanger where the temperature difference between process side and airside is reduced, it can be beneficial to spray water into the incoming air. Such a practice has often been carried out on existing exchangers with disastrous results. Water cascading onto the top of the bundle from coarse sprinklers has flowed down counter-current to the air passing through from below. This has usually led to a reduction in airflow, corrosion in the bundle and the plenum walls and floor, and a deluge of wasted water falling onto pipe racks and equipment below the exchanger. As much as 95% of this water is wasted with only marginal benefit to the thermal performance. A better way is to use very fine water/air atomising sprays at fan or bundle inlet thus using only a minimum amount of water and ensuring the maximum benefit from the evaporative cooling effect of the most air. Data on this technique is not yet readily available but a study is very soon to be carried out by HEAS.

Where it can be shown that there is sufficient fouling on the process side, then there is justification for cleaning the tubes in the bundle. This requires a major shutdown but gives the opportunity for the tubes to be examined for corrosion. Tube cleaning is by manual rods or abrasive spheres and examination is carried out using an intra-scope.

Maldistribution effects on the process side have already been described. The effect on airside performance of any maldistribution in airflow through the bundle is relatively small. Berryman and Russell (1987) carried out tests which showed that a maldistribution in terms of a standard deviation of 23% on a single phase cooler gave a reduction in performance of only 1%. Even at 50% standard deviation a loss of only 4% was found in the resulting thermal performance. For a condenser the effects were greater but still no more than 8% at 50% standard deviation in airflow. A shortfall of only 10% in total airflow through an ACHE however can have a much greater effect than any major maldistribution.

If, following all the above potential methods of improvement, the resulting performance is still below that required, then the only possibility is to change the fans or fit larger tube bundles. This is an economic decision and changing fans can be cost effective if the resulting improvement provides a short pay-back period. Alterations involving new and larger tube bundles with more complex pass arrangements are costly and are a last resort.

<u>Good Design and Installation</u>

Looking now at new air cooled heat exchangers, many steps can be taken to ensure that these units will work satisfactorily.

Firstly, it is important in the initial planning stage to use modern, well-tested, computer software now available for heat exchanger design. Codes such as the HTFS program ACOL4, incorporate the latest correlations for heat transfer and pressure drop based on a comprehensive research programme. Recently much work has been carried out on the effects of fan inlet geometry on airside pressure drop. Additional effects of pressure recovery through the fan, flow through the plenum and resistance through the bundle have been considered so that a realistic total pressure drop can be specified to the fan supplier.

A realistic airside specification is necessary to prevent some of the problems described previously. Some degree of over-design can be advantageous since most process streams will change over the 20 year life of the exchanger. Having spare capacity can allow flexibility to switch such process streams as sources of crude feed change. Care needs to be given to the choice of both summer and winter maximum and minimum ambient air temperatures in the design specification. This is particularly so where temperature differences are 50 K or less.

Having determined the size of exchanger for a given duty, allowances must be made for an overall loss in thermal efficiency for the equipment "as installed". Mechanical and thermal tolerances exist which will prevent the unit performing at its absolute best. Fouling is one such element and realistic values should be used at the design stage. Good practice suggests a 10% safety margin should be applied to both heat transfer area and air throughput.

The correct siting of new exchangers is important to achieve good performance and reduce potential problems. However, general constraints on space and the need to fit auxiliary and ancillary heat exchangers, vessels etc. close to ACHEs, invariably leads to some problems. A fan when fitted to a bundle will only give its best performance if air can freely flow into the inlet from all directions. When installed in long, multiple banks, fans will compete for air with adjacent units and it is then that air starvation can occur if the fans are at the limit of their aerodynamic capability. The presence of pipe racks alongside and underneath and tall adjacent buildings can all lead to airflow problems (see Figure 3). Included with the air starvation effect can be thermal performance losses, caused by hot air recirculation. This can be minimised by the use of induced draught exchangers where the exit velocity from the fan is 2 to 3 times that above a forced draught unit.

Finally, and most importantly in the context of this paper, a regular maintenance schedule should be established which incorporates the condition monitoring aspects described above. Frequent condition monitoring provides information on how the unit is performing and the rate at which this performance decreases. From this data it is possible to predict the time at which such reduced performance will have a noticeable effect on the rest of the process stream.

Financial Benefits

The financial benefit of condition monitoring will be mainly dependent upon the value of the process stream. In addition to this, however, the value of "on-line" maintenance of the fan and tube bundle will also be of significance. Fewer mechanical breakdowns mean fewer losses in product. Optimised operation results in a high process throughput.

Frequent examples of condition monitoring carried out by HEAS over the last 2 years have shown airflows to be typically 25% below the optimum. Following bundle cleaning and re-setting of the fans a 20% increase in throughput has been achieved.

In one such case an extra 1000 barrels per day of product was recovered after cleaning and optimisation. With a modest value of $5 per barrel, the total return on an investment of $20 K was $300 K over a period of two summer months. Even at lower ambient air temperatures a worthwhile daily financial benefit was achieved. On this plant, condition monitoring is now carried out every six months and regular maintenance checks have been instigated with a major overhaul of the fan/drive system and cleaning of the airside at each annual shut-down.

Conclusions

For existing air cooled heat exchangers regular condition monitoring, maintenance and cleaning are necessary to keep the thermal performance at its peak. With older plant, which has not been regularly cleaned, a build-up of dirt on the fans in the bundle can reduce airflow by 25% or more. This, together with the use of less efficient, poorly optimised fans is the usual cause of lost performance.

Performance testing techniques using portable equipment have been established to evaluate ACHE performance before and after cleaning and maintenance. Such techniques can also be used to validate the performance of new units as part of any acceptance testing requirements.

Modern computer software is a valuable tool in comparing measured performance with that to be expected under ideal conditions. ACHE software of this type is used to generate good designs for new units when used in conjunction with realistic specifications.

Allowances for installed efficiency, correct siting and steps taken to minimise air starvation and hot air recirculation, lead to trouble-free operation of ACHEs.

Worthwhile financial benefits can be achieved by regular maintenance, condition monitoring and airside cleaning.

Condition monitoring addresses problems which can be experienced in the operation and maintenance of all heat exchangers. Berryman (1988) describes how performance testing techniques and trouble-shooting can be applied to all large process heat exchangers.

<u>References</u>

Berryman, R.J. and Russell, C.M.B. Trouble-Shooting on Air Cooled Heat Exchangers. Process Engineering, Vol.66, No.4, April 1985, pp.25;29.

Berryman, R.J. and Russell, C.M.B. Airflow in Air cooled Heat Exchangers. 4th Symposium on Multiphase Transport and Particulate Phenomena, Miami Beach Florida, 15-17 December 1986.

Berryman, R.J. and Russell, C.M.B. The Effect of Maldistribution of Airflow on Air Cooled Heat Exchanger Performance. Proceedings of 24th ASME National Heat Transfer Conference, Pittsburg, Aug 9-12, 1987.

Berryman, R.J. Performance Testing and Trouble-Shooting on Large Process Heat Exchangers. Proceedings of 1st International Conference on Experimental Heat Transfer, Fluid Mechanics and Thermodynamics, Dubrovnik, Sept. 4-9, 1988.

Table 1:   Air Cooled Heat Exchanger Fault Finding Chart

Figure 1:   Close up of Typical Bundle Damage

Figure 2:   Examples of Vane Anemometers used for Airflow Measurement

Figure 3:   Example of a Restricted Air Inlet to an Air Cooled Heat Exchanger

TABLE 1

Air cooled heat exchanger fault finding chart		
Some of the symptoms, faults and solutions. (Note: This list is by no means exhaustive!)		
Symptom	Possible Fault	Solution
High process outlet temperature	Exchanger undersized Excessive tubeside fouling Tubeside flow maldistribution Low air flow Air flow maldistribution Hot air recirculation Excessive airside fouling	Check using rating computer code Clean - if possible Redesign heater or fit inserts Check fan settings and adjust Fit more efficient fan Difficult - try wind fences/deflectors Clean - stream/water spray
High process pressure drop	Excessive tubeside fouling Overcooling/increased fluid viscosity Failure to condense vapour	See above Reduce air flow Increase air flow
High inlet/outlet pressure/temperature of condenser	Low condensation rate	Increase air flow, reduce fouling etc (see above)
High air outlet temperature	Low air flowrate Tubeside flow maldistribution Air recirculation	See above
Low air outlet temperature	High air flowrate Tubeside flow maldistribution Low ambient temperature	Check fan setting See above Check fan, adjust louvres, if fitted
Excessive noise	Poor fan/drive/motor selection Fan near stall conditions Worn, slipping, misaligned belt drives Worn gear box Worn fan/motor bearings Resonating plenum/structure/pipework	Check noise levels, compare with design data Check fan curve Replace worn items Fit cross-bracing, support brackets, etc.
Excessive vibration	Badly balanced fan/blades Worn fan/motor bearingts Worn or misaligned drive belts Loose mounting bolts Resonating plenum/structure	Check/balance/adjust Replace worn items Tighten See above

Figure 1: Close up of Typical Bundle Damage

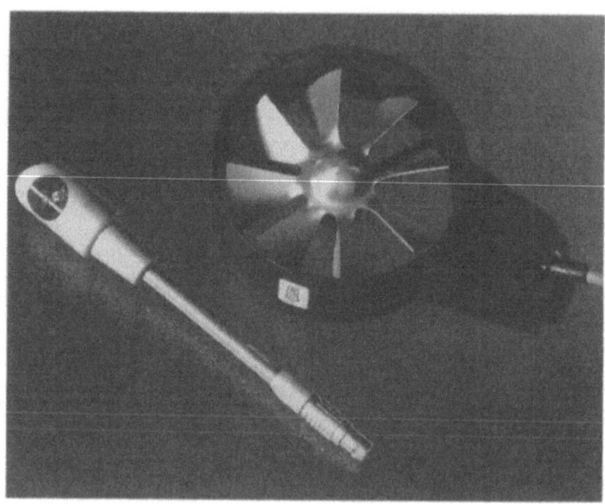

Figure 2: Examples of Vane Anemometers used for Airflow Measurement

Figure 3: Example of a Restricted Air Inlet to an Air Cooled Heat Exchanger

Plate Heat Exchangers

Approximate Theory of Spiral Heat Exchanger

Th. BES and W. ROETZEL
Institute of Thermodynamics
University of the Federal Armed Forces Hamburg

Summary
Based on the spiral of Archimedes, an analytical solution is developed which describes the thermal behaviour of the countercurrent Spiral Heat Exchanger (SHE) including the characteristic maximum of effectiveness occurring with increasing values of NTU. For the analysis, the overall heat transfer coefficient and both heat capacities are assumed to be constant along the flow path. Additionally, such a high number of turns is presumed that the special situation in the first and last turn does not have to be taken into account. The analytical solution of the energy balance equations yields a simple, universal formula for the log mean temperature difference correction factor F as function of NTU_I and NTU_{II}, as well as the number of channels and further geometrical parameters.

Nomenclature

$A_c = h_O b$	m	cross-sectional area of flow channel
A_O	m	total heat transfer surface area
b	m	channel spacing
$C = \sqrt{C_I C_{II}}$	W/K	mean heat capacity rate
$CN = 2\ NTU\sqrt{\pi A_c/A_O}$		Criterion Number
F	-	log mean temperature difference correction factor
$f = CN_O/CN$	-	adjustment factor for Criterion Number
h_O	m	height of exchanger
k	W/m^2K	overall heat transfer coefficient
n	-	number of channels equal to double number of turns
$NTU = kA_O/C$	-	number of transfer units (mean value)
P	-	effectiveness, dimensionless temperature change
$q = \dfrac{d\dot{Q}}{d\varphi}$	W/rad	heat flux $\dot{Q}$, related to angle φ (see Fig. 2.)
$R = C_I/C_{II}$	-	heat capacity rate ratio
$r = r'/b$	-	dimensionless radius, r' real radius
$t = \dfrac{t'\ -\ t'_{II,i}}{t'_{I,i} - t'_{II,i}}$	-	dimensionless temperature, t' real temperature of fluid I or II
$x = \psi r$	-	coordinate proportional to distance measured along main spiral

Greek letters

$\Delta(r) = t_I(r) - t_{II}(r)$	-	local temperature difference
Θ	-	mean temperature difference
$\psi = 2\pi k A_c/C$	-	cross-sectional number of transfer units (mean value)
$\kappa(t) = \psi r[t_I(r+1) - t_{II}(r-1)]$		reduced heat flux density
φ	rad	angle in polar coordinate system (see Fig. 2.)

Subscripts:

$$)_i \ , \ )_o \ - \text{inlet, outlet}$$
$$)_I \ , \ )_{II} \ - \text{fluid I,II}$$

Introduction

In the literature, the SHE has been treated analytically [1], [2], [3], [4], numerically [5], and numerically with some experiments [6].
The known approximate theoretical solutions [1],[2] have serious disadvantages concerning accuracy either for small number of channels or for large NTU.
The exact solutions [3],[4] are relatively complicated for engineering design purposes and have comparatively long calculational procedures.
The necessity of iterative adjustment for an initial vector of temperatures makes the applicability of the Runge-Kutta method to SHE with higher number of turns extremely difficult [6].
Martin at al. [7] proposed a new compact and simple formula for the calculation of the F correction factor. However, this form has a weak theoretical background and needs to be verified.

The purpose of the present paper is to provide a simple theoretically derived formula for the thermal design of SHE.

Assumptions

The outside surface of the SHE is thermally insulated. From the innermost and peripheral channel, heat is transferred only through one wall. In the channels between the centre and the periphery, heat penetrates both walls of the channel. Therefore, due to different heat transfer conditions, SHE is divided into three parts: the innermost part with two channels, the middle part with turns, which usually occupies the main volume of exchanger, and the outermost part with two channels.

The present theory fulfils the conditions for the middle part of the exchanger. Thus, the validity of the theory increases with the number of turns.

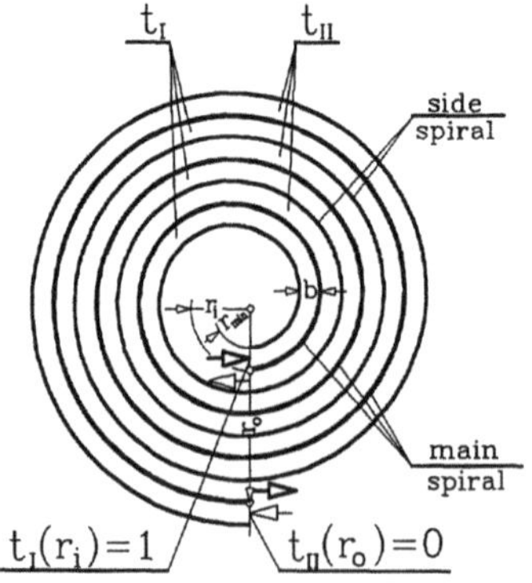

Fig. 1. Arrangement of flows in Spiral Heat Exchanger

Fluids are completely mixed in the radial and axial directions within the flow channel. At a fixed angle, the temperatures change stepwise from channel to channel (Fig. 2.). Thus, in one channel the fluid temperature is a function of the angle φ only.

Additionally the following assumptions are made:

- The shape of the spiral is optionally assumed to be the spiral of Archimedes.
- Arrangement of flows is that of Fig. 1.
- The number of channels (equal to double number of turns) is even.
- Heat capacity rates C_I, C_{II} and overall heat transfer coefficient k are constant throughout the heat exchanger.

Energy Balance Equations

A differential angle $d\varphi$ is considered as shown in Fig. 2.
For any channel, except the innermost and outermost, a set of n-2 difference differential equations may be developed and can be written, using only one set of two equations for both fluids. The components of the heat fluxes passing through the wall can be expressed by:

$$q_{j+2,j+1}=kh_O b\{(r+1)[t_I(r+\tfrac{3}{2})-t_{II}(r+\tfrac{1}{2})]\} \tag{1}$$

$$q_{j,j-1}=kh_O b\{(r-1)[t_I(r+\tfrac{1}{2})-t_{II}(r-\tfrac{3}{2})]\} \tag{2}$$

$$q_{j,j+1}=kh_O b\{r[t_I(r-\tfrac{1}{2})-t_{II}(r+\tfrac{1}{2})]\} \tag{3}$$

Changes of enthalpy along the angle φ are expressed as :

$$C_I\frac{dt_I(r-\tfrac{1}{2})}{d\varphi} \quad , \qquad C_{II}\frac{dt_{II}(r+\tfrac{1}{2})}{d\varphi} . \tag{4}$$

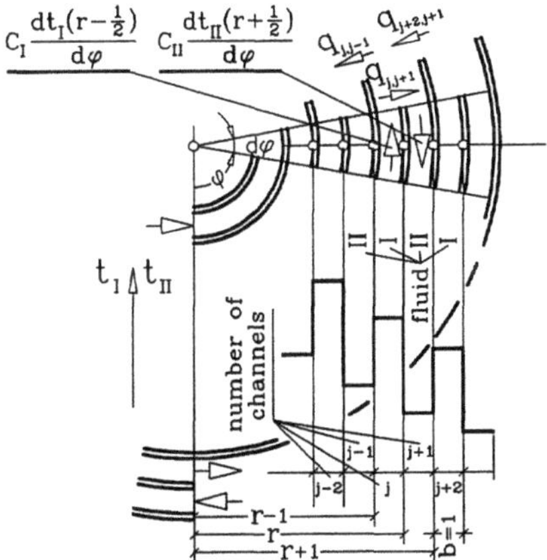

Fig. 2. Temperatures and components of energy balance in wedge of SHE

226

It is convenient to change the notation of temperatures. The local temperatures $t_I(r-\frac{1}{2})$ and $t_{II}(r+\frac{1}{2})$ will be denoted as $t_I(r)$ and $t_{II}(r)$, respectively. Considering the spiral of Archimedes yields:

$$r = r_{min} + \varphi/\pi \qquad \text{and} \qquad d\varphi = \pi dr.$$

For an incremental angle d , which includes the channels: j-1, j, j+1 and j+2 (as shown in Fig. 2.), the set of energy balance equations with new temperature notation and parameters listed in nomenclature is:

$$2\sqrt{R}\,\frac{dt_I(r)}{dr} + \psi r\Delta(r) + \kappa(r-1) = 0$$

$$\frac{2}{\sqrt{R}}\,\frac{dt_{II}(r)}{dr} + \psi r\Delta(r) + \kappa(r+1) = 0$$

(5)

with the boundary conditions: $t_I(r_i)=1$ and $t_{II}(r_o)=0$.

Both equations (5) are also valid if r is replaced by r+1 or r-1, respectively:

$$2\sqrt{R}\,\frac{dt_I(r+1)}{dr} + (r+1)\psi\Delta(r+1) + \kappa(r) = 0$$

$$\frac{2}{\sqrt{R}}\,\frac{dt_{II}(r-1)}{dr} + (r-1)\psi\Delta(r-1) + \kappa(r) = 0$$

(6)

The boundary conditions do not change.

Despite their linearity, the energy balance equations: (5) or (6) are too complicated to be solved in a simple mathematical way which could be easily accepted by engineers. Hence, the basic problem in the analysis is to find a procedure which allows one to transform these equations into either difference equations or differential equations. The second way is more promising here.

The special case of equal heat capacity rates (R=1) is considered.
Eqs. (5) and (6) are simplified by introducing R=1.

Subtracting in both systems (5) and (6) the second equation from the first, yields relations:

$$2\,\frac{d\Delta(r)}{dr} + \kappa(r-1) - \kappa(r+1) = 0$$

(7)

$$2\,\frac{d}{dr}[\frac{\kappa(r)}{\psi r}] + (r+1)\psi\Delta(r+1) - (r-1)\psi\Delta(r-1) = 0$$

with the boundary conditions:

$$\Delta_i(r_i) = \Delta_i \qquad \text{and} \qquad \kappa(r_i) = \kappa_i$$

(8)

(or for high number of channels n : $\kappa_i \approx x_i \Delta_i$).

Instead of considering the difference of temperatures or reduced heat flux densities as noted in the last terms of Eqs. (7), one can write:

$$\frac{\kappa(r+1)-\kappa(r-1)}{2} \approx \frac{d\kappa(r)}{dr}\bigg|_r \quad ; \quad \frac{(r+1)\Delta(r+1)-(r-1)\Delta(r-1)}{2} \approx \frac{d[r\Delta(r)]}{dr}\bigg|_r \tag{9}$$

Introducing formulas (9) into Eqs. (7) and integrating yields:

$$\Delta(r) - \kappa(r) = C_1$$
$$\kappa(r)/\psi r + \psi r \Delta(r) = C_2 \tag{10}$$

where C_1 and C_2 are constants of integration.

Further, the new independent variable $x = \psi r$ will be introduced in Eqs. (10). Substitution of $xD(x) = x\Delta(x) + \kappa(x)$ enables one to achieve the simple solution from set (10) and boundary conditions:

$$D(x) = 2\Delta_i \, (1+x_i x)/(1+x^2).$$

If reduced radius r_i at the inlet is small in comparison to $r_0 = r_i + n$ at the outlet (which is valid for high number of turns), the condition for difference of temperatures could be simplified $D(r_i) \approx 2\Delta_i$ and the function $D(x)$ as well:

$$D(x) = 2\Delta_i/(1+x^2). \tag{11}$$

Effectiveness of SHE

The effectiveness P for SHE can be calculated according to the definition

$$P = -\int_{r_i}^{r_0 = r_i + n} dt_I = 1 - t_I(r_0) \tag{12}$$

where r_i and $r_0 = r_i + n$ are the radii of bent walls in SHE: the smallest and the largest, respectively, through which heat is transferred between both fluids (see Fig. 1.).

For further analysis, the following geometric parameters are introduced:
- Quotient of areas A_c/A_0:
  The definition listed in the nomenclature leads to the following expression

$$2\pi A_c/A_O = \psi/NTU$$

Applying the differential $d\varphi = \pi dr$, valid for the spiral of Archimedes, and integrating the expression $rd\varphi = r\pi dr$ yields finally:

$$\frac{A_O}{\pi A_c} = \left(\int_{r_i}^{r_0} r\,dr + \int_{r_i+1}^{r_0-1} r\,dr\right) = r_0^2 - r_i^2 - r_0 - r_i = (r_0+r_i)(r_0-r_i-1) = (n-1)(n+2r_i)$$

- Another auxiliary relation, which refers to cross-sectional areas in SHE, is useful in next chapter:

$$r_0^2 - r_i^2 = n(n+2r_i) = \left(1 + \frac{1}{n-1}\right) A_O/(\pi A_c)$$

Further consideration refers to Eqs.(5) which, together with Eq.(11), allows one to express the derivative of the temperature of the first fluid:

$$\frac{dt_I(r)}{dr} = -\frac{1}{2}[\psi r \Delta(r) + \kappa(r-1)] \approx -\frac{1}{2}[\psi r \Delta(r) + \kappa(r)] = -xD(x)/2 \qquad (13)$$

Integration of the last form gives:

$$P \approx \frac{\Delta_i}{2\psi} \ln \frac{1+x_o^2}{1+x_i^2} \qquad (14)$$

Introducing expression (14) into the equation $\Delta + P = 1$, which is well known from the literature: [10],[11], one finds

$$\Delta_i = 1/(1 + \frac{1}{2\psi} \ln \frac{1+x_o^2}{1+x_i^2})$$

and

$$\frac{1}{P} - 1 = 2\psi \ / \ \ln \frac{1+x_o^2}{1+x_i^2}$$

For the case $R = 1$, the effectiveness P, NTU and the log mean temperature difference correction factor F are connected as follows [7]:

$$F = 1/[(1/P-1)NTU] \qquad (15)$$

Substituing P according to the last equation yields

$$F = \ln[(1+x_o^2)/(1+x_i^2)]/(2\psi \, NTU) \qquad (16)$$

The last relation contains two terms:

The denominator:

$$2\psi \, NTU = 4\pi \, NTU^2 A_c/A_o$$

and the square root of this term is denoted as the Criterion Number, CN:

$$CN = 2 \, NTU \sqrt{\pi A_c/A_o} \qquad (17)$$

The numerator in Eq. (16) contains an independent variable which can be expressed as follows:

$$\frac{1+x_o^2}{1+x_i^2} = 1 + \frac{x_o^2-x_i^2}{1+x_i^2}$$

and

$$\frac{x_o^2-x_i^2}{1+x_i^2} = (1 + \frac{1}{n-1}) \, CN^2/(1+x_i^2) = CN^2 \frac{1+\frac{1}{n-1}}{1+x_i^2}$$

If the number of turns in SHE is sufficiently high, the reduced radius r_i at the inlet is small in comparison to $r_o = r_i + n$. Then the fraction in the last formula can be replaced by 1, and F simplifies to:

$$F = \ln(1 + CN^2)/CN^2 \qquad (18)$$

The function F against CN is plotted in Fig. 3.

Extension of Analysis to Arbitrary Values of Heat Capacity Rate Ratio

The correction factor according to Eq. (18) turns to $F=1$ if $CN \rightarrow 0$ or $NTU \rightarrow 0$. The same result is valid in the limiting cases in which one or both heat capacity rates become infinite or if NTU_I or NTU_{II} become zero. This leads to the conclusion that in the limiting cases $NTU_I = 0$ ($C_I \rightarrow \infty$) and/or $NTU_{II} = 0$ ($C_{II} = \infty$) as well as in the special case $NTU_I = NTU_{II}$ ($R=1$) the formulas (17,18) remain valid if NTU^2 is replaced by $NTU_I NTU_{II}$ (or C^2 by $C_I C_{II}$).
The hypothesis is stated that this approach is also valid for all values $0 \leq R \leq \infty$. Thus, the formula (18), which has been derived for the special case $R=1$, is extended to general cases $0 \leq R \leq \infty$ simply by introducing the mean heat capacity rate

$$C = \sqrt{C_I C_{II}} \tag{19}$$

for the calculation of the mean number of transfer units NTU in Eq. (17).

The hypothesis is checked against calculations according to an exact theory [8] for different values of R, number of turns, and minimal radius.
For a realistic number of channels, $n > 10$, the function $F(CN)$, calculated from Eqs (17-19) and according to the exact theory [8], compare very well for values of R in the range $0.2 \leq R \leq 5$.

The mean value C according to Eq. (19) presumes a symmetric behaviour of the flow arrangement, which is approximately correct for a high number of turns.
Alternatively, the assymmetric behavior could also be taken into account by a modified mean value

$$C = C_I^\alpha C_{II}^{(1-\alpha)}$$

with an empirical exponent $\alpha = 0.5$ and $0 < \alpha < 1$.

Thermal Limits of SHE

In reference [4] it was noticed for the first time that for very high NTUs the effectiveness of SHE reaches a maximum. This phenomenon was studied in ref. [8]. Existence of a maximum was later noticed in ref. [9] as well.
It is worth emphasizing that the present theoretical result renders well these properties of SHE. Investigation of the formula $1/P = 1 + NTU/F$ regarding extremes allows one to derive the relation for the maximum of effectiveness for $R=1$

$$1/P_{max} = 1 + (c + 1/c)/\sqrt{(n-1)(n+2r_i)} \tag{20}$$

where $c = 1.980259$.
The quantity c is the coordinate where the function $CN \cdot F = \ln(1 + CN^2)/CN$ reaches its maximum. The relation $CN \cdot F$ against CN is plotted in Fig. 4. The mentioned maximum occurs for:

$$NTU_m = (c/2)\sqrt{(n-1)(n+2r_i)} \tag{21}$$

and is valid for $n \geq 6$. For $n = 4$ no maximum occurs.

In Fig. 4. the curve $CN \cdot F_M = \tanh CN$, proposed by Martin [7], does not lead to a maximum of P.

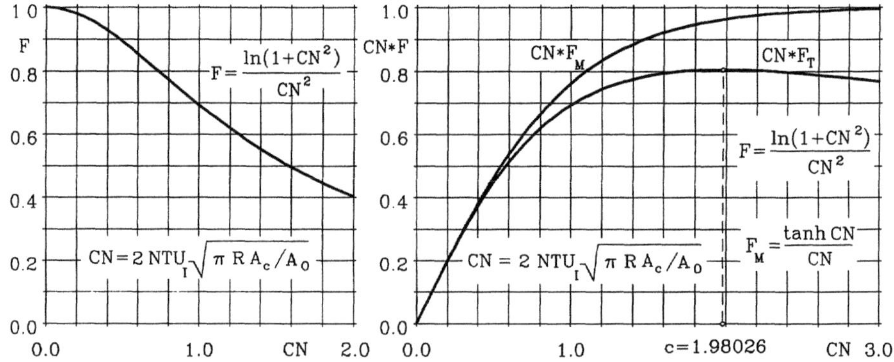

Fig. 3. Correction factor
F as a function of CN

Fig. 4. Auxiliary function CN·F
(present theory) and CN·F_M (Martin [7])

Comparison of Approximate Results to Exact Theory of SHE

The following proof of actually presented theory for different parameters NTU in SHE is mainly a comparison of effectiveness P_I achieved on the basis of the hypothesis discussed before with the data for P_I taken from ref. [8]. Actually, the evaluation of the present theoretical results is done in different ways according to the kind of parameters which are considered.

Evaluation with Regard to Different Number of Channels

The theory of this paper has been derived for a high number of turns. For a low number of turns, the Criterion Number, CN according to Eqs. (17,19) could be corrected by a factor f to yield a better value

$$CN_O = f\, CN \tag{22}$$

for Eq. (18). This adjustment factor f has been calculated using values of F from the exact theory [8] which allowes to find the value CN_O from Eq.(18). from Eq.(18). The calculation was done for the limiting case $\psi \to 0$ and R=1. The results for various radii and number of channels n are plotted in Fig. 5.

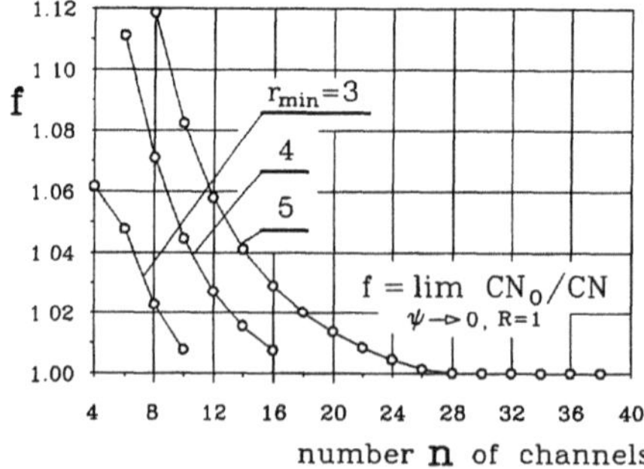

Fig. 5.
Factor $f = \lim CN_O/CN$ (on the basis of exact theory [8] in comparison to
$\psi \to 0,\ R=1$
present theory) as function of number of channels n

Evaluation for very high values of NTU and R=1

For very high values of NTU: >10-30 and $n \geq 6$, as it was investigated in refs. [4], [8], the effectiveness P achieves a maximum and with continued increase of NTU, starts to decrease, approaching finally the constant value (0.70-0.85) for unlimited NTUs or zero mean temperature difference $\Theta = 0$. The hypothesis (19) together with form (18) renders very well these properties of SHE, i.e. the maximum of P_I, which is pointed out in Fig. 6.

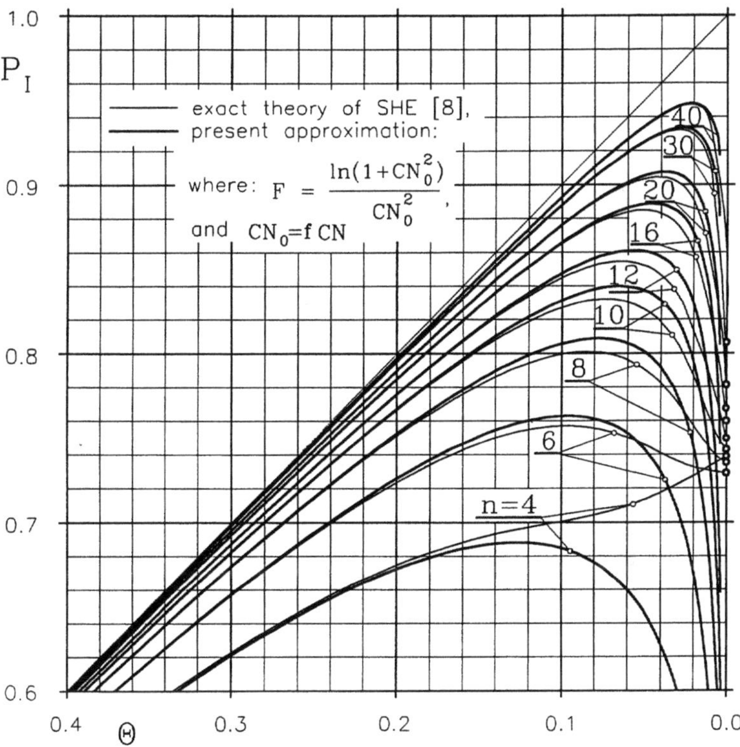

Fig. 6. Effectiveness of Spiral Heat Exchanger versus dimensionless Mean Temperature Difference for Heat Capacity Rate Ratio R=1 and different number n of channels (for n = 4,6,8; r_{min} = 2,3,4 respectively, and for $n \geq 10$, r_{min} = 5)

Conclusion

The new simple formula for the log mean temperature difference correction factor of the counterflow Spiral Heat Exchanger, derived in this paper, provides advantages over previous approches:
It has got a theoretical background and is able to describe the phenomenon of maximum of effectiveness which occurs at very high value of NTUs.

REFERENCES

1. Madejski, J.: Theory of Heat Exchange (in Polish)
   PWN Warszawa - Poznan, pp. 381-386, 1963

2. Zaleski, T.; Krajewski, W.: Method of Calculating Spiral Exchangers
   (in Polish), Inżynieria Chemiczna (Chemical Engineering)
   II, 1. 35 pp. 35-51, Warszawa, 1972

3. Cieslinski, J. P.; Bes, T.: Analytical Heat Transfer Studies in Spiral
   Plate Heat Exchangers, Reprints of XVI. Congr. of Refrigeration,
   Institut International du Froid (I.I.F.), Paris, pp. 67-72, 1983

4. Bes, Th.: Eine Methode der thermischen Berechnung von Gegen- und
   Gleichstrom-Spiralwärmeaustauschern, Wärme- und Stoffübertragung 21,
   pp. 301-309, 1987

5. Nowak, W: Calculating of Heat Exchangers Considering Heat Losses to
   Environment (in Polish), Zeszyty Naukowe Pol. Szczecinskiej
   (Scientific Journals Tech. Uni. of Stettin) No. 114, 1969

6. Buonopane, R. A.; Troupe, R. A.: Analytical and Experimental Heat
   Transfer Studies in a Spiral Plate Heat Exchanger, IV. International
   Heat Transfer Conference, Volume 1, HE 2.5, Paris, 1970

7. Martin, H.; Chowdhury, K.: Linkmeyer, H.; Bassiouny, K.M;
   Straightforward Design Formulae for Efficiency and Mean Temperature
   Difference in Spiral Plate Heat Exchanger, VIII. International Heat
   Transfer Conference, Volume 6, pp. 2793-2797, San Francisco, 1986

8. Bes, Th.; Roetzel, W.: Distribution of Heat Flux Density in Spiral Heat
   Exchangers, accepted for publication in Int. J.
   Heat Mass Transfer

9. Strenger, M.; Churchill, S.; Retallik, W.:
   Operational Characteristic of a Double-Spiral Heat Exchanger
   for the Catalytic Incineration of Contaminated Air,
   Ind. Eng. Chem. Res., 1990, 29, pp. 1977-1984

10. Roetzel, W.; Berechnung von Wärmeübertragern
    (Chapt. C of VDI- Wärmeatlas, 4th edition)
    VDI-Verlag, Düsseldorf 1984

11. Gaddis, E.S.; Spalding, D.B.; Taborek, J.:
    Heat Exchanger Theory, (vol. I of Heat Exchanger Design Handbook),
    VDI-Verlag GmbH, Hemisphere Publishing Corporation
    Washington, N. York, London 1986

Acknowledgment

The authors would like to thank the Deutsche Forschungsgemeinschaft for the financial support of this research project.

Thermal Hydraulic Performances of Plate and Frame Heat Exchangers – The CEPAJ Software

R. VIDIL, G. RATEL, J.M. GRILLOT
(Groupement pour la Recherche sur les Echangeurs Thermiques)
CENG/GRETh - 85 X - 38041 GRENOBLE CEDEX - FRANCE

1. INTRODUCTION

In the production and management of energy, 90 % of the energy used is channeled through heat exchangers. These heat exchangers have a number of applications in areas like air conditioning, the chemical and petro-chemical industries, and food-industries.

The rational use of energy largely depends on the quality of the exchangers used to optimize heat transfer between two fluids.

Plate heat exchangers hold a considerable share in the market sectors mentioned above. They offer a number of advantages, as they are compact, inexpensive, modular, and easy to disassemble and clean [14].

In this type of exchanger, the heat transfer surface is made of a series of metallic plates (equipped with joints or welded) placed vertically and at parallel distances (fig. 1). The profile of these corrugated plates are used to intensify the heat transfer and to ensure that the apparatus is mechanically sound.

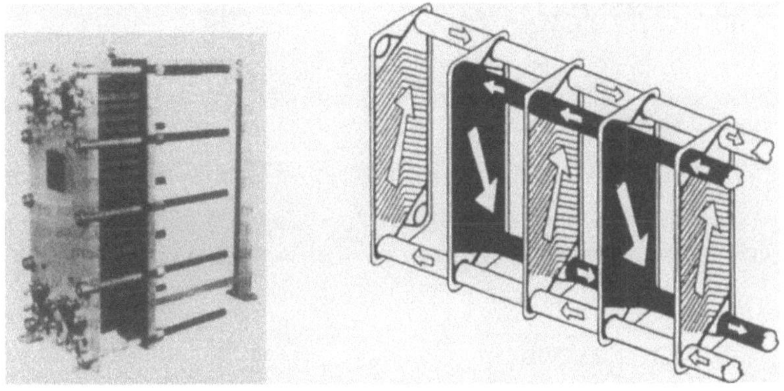

Fig. 1 : side view and cross section of a plate heat exchanger

The GRETh (Research group on heat exchanger), whose aim is to assist industry in the area of thermal heat exchangers, has established a vast research program on plate exchangers. Emphasis is placed on the study of single phase flows [1] [2] [3], two phase [7], as well as fouling [4] or the use of non-newtonian fluids.

This paper presents the main results of the thermohydraulic performances of diverse geometries. These results are integrated into the CEPAJ software presented below.

It should be reiterated here that there are 2 types of plate configurations (fig. 2) :

-   perpendicular corrugation plates,

-   inclined (or herringbone) corrugation plates.

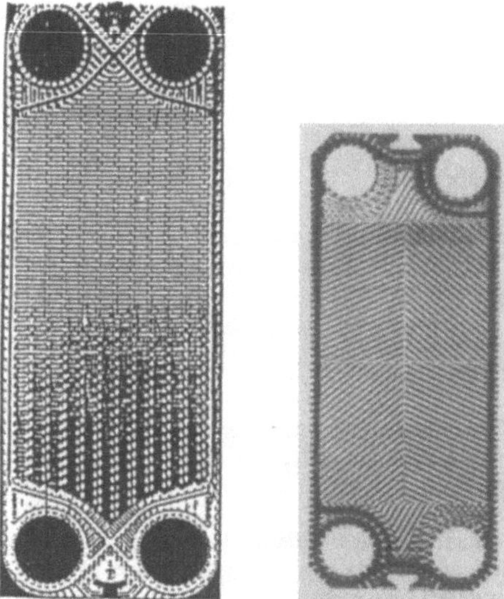

Fig. 2 : perpendicular and inclined corrugation plates

2. EXPERIMENTAL STUDY OF THE PERFORMANCE OF CORRUGATED CHANNELS

This section of the paper presents the procedure used to determine thermohydraulic performances of different corrugated channels, representative of the inner central part of plate heat exchanger.

The results are directly applied to the design of plate exchangers.

These experimental studies have allowed us to determine the influence of the corrugation angle on the friction and heat transfer correlations. Six geometrical configurations have been tested : 15°, 30°, 45°, 60°, 75° and 90°.

2.1 Experimental apparatus

The test section, made up of a channel formed by two plates, and a tightening system (fig. 3) is linked to a closed loop. Water or air are the experimental fluids used for large Reynolds numbers (Re = 1000 to 15000), and oil is used for lower Reynolds numbers (Re = 50 to 1000).

The juxtaposition of the tests has resulted in the determining of friction and heat transfer correlations using a wide range of Reynolds numbers, which cover all the industrial applications of the plate heat exchangers.

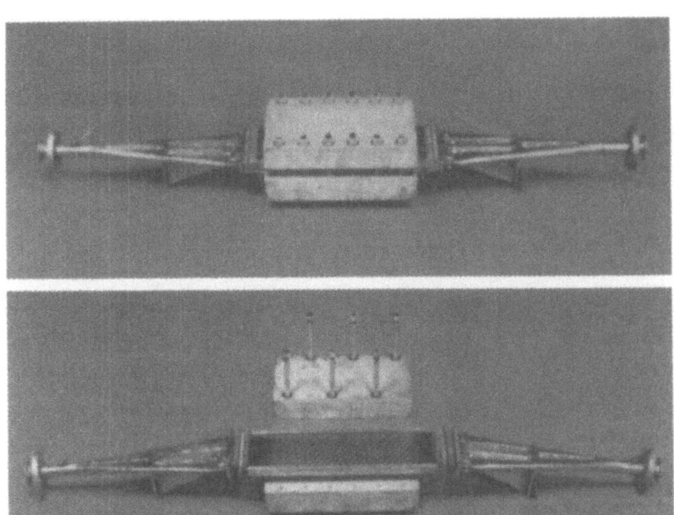

Fig. 3 : View of the test section

In order to determine the heat transfer laws [1] [3], the hot fluid is simulated by directly heating the plates electrically. The cold fluid is the experimental fluid used (water or oil). In the experimental study using air [5], the test section has three identical channels formed by the same exchange plates. The friction and heat transfer laws are obtained using global measurements.

2.2 Friction correlations

Friction correlations concern the evolution of the friction coefficient f with the Reynolds number :

$$f = g(Re) = a.Re^b$$

For each geometry, two complementary correlations are given to represent the "laminar" and turbulent flows. Using the local experimental study device, and the 60° and 90° geometries, it can be seen that the law change $(Re = Re_T)$ corresponds to the appearance of unstationary local phenomena and therefore turbulence. A physical measurement of these transitional Reynolds numbers is made in order to ensure the continuity between two calculated regressions.

236

The obtained friction laws are represented graphically for each geometry in fig. 4. The adimensional numbers used in this approach are defined as the following :

$$Re = \frac{\rho.V.D_h}{\mu}$$

$$D_h = 2 . e$$

$$V = \frac{M}{\rho . S}$$

$$S = 2 . e . la$$

$$\Delta P = 4 . f . \frac{l}{D_h} . \frac{\rho . V^2}{2}$$

D_h hydraulic diameter
V velocity
ρ mass density
μ dynamic viscosity
e corrugation depth
M mass flow rate
la channel width
f friction factor

l channel length

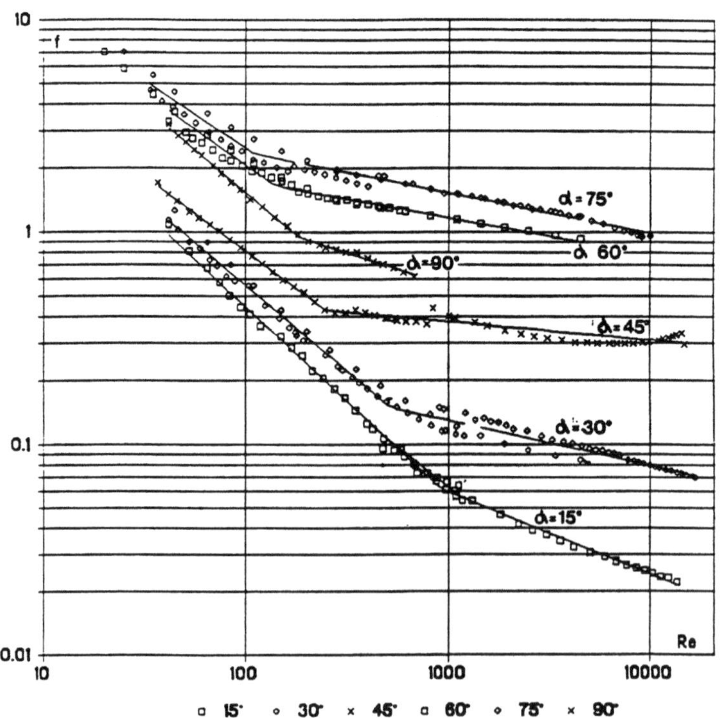

Fig. 4 : Friction factor versus Reynolds number

A comparison of the laws obtained with those of the other studies [8] [9] [10] [11] [12] has allowed us to estimate the influence of the geometric parameter e/p (with e = corrugation depth and p = corrugation pitch = distance between two peaks). It has been verified that the influence of this parameter is not significant for   < 60°.

2.3   <u>Heat transfer correlations</u>

The heat transfer coefficient uses the following non dimensional relations :

$$Nu = \frac{\alpha.D_h}{k} = a.Re^b.Pr^c.(\frac{B.\phi.D_h}{2.k})^d \qquad k \quad \text{heat conductivity}$$

$$\phi \quad \text{heat flux}$$

$$AG = \frac{B.\phi.D_h}{2.k} \qquad B \quad \text{constant}$$

The dependences between the NUSSELT and the PRANDTL numbers and the adimensional group AG (giving the influence of the fluid's thermodependence on the heat transfer, see [15]) are determined experimentally, which leads to the following values :

$$c = 0.33$$
$$d = 0.07$$

Once the two types of influence have been quantified, the regressions are calculated to predict the evolutions of the heat transfer coefficient with the Reynolds number (fig. 5).

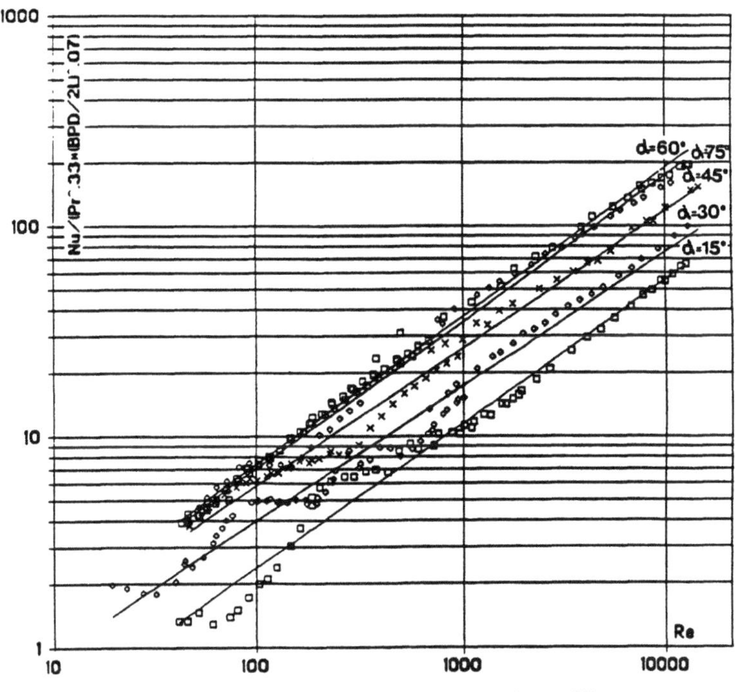

Fig. 5 : Heat transfer coefficient versus Reynolds number

A study was done on the influence of natural convection at a low Reynolds number for the 60° and 90° geometries while varying the flow direction (upward or downward, the test section is in a vertical position).

No significant difference was observed between the two flows directions in the 60° geometry.

However, this influence was significant in the 90° geometry since the downward flow transfer was found to be greater than that of the upward flow.

A physical interpretation of this influence was given using the results of the local experimental study [2].

3.   NUMERICAL CALCULATION : THE CEPAJ SOFTWARE

The numerical calculations have often been used in the GRETh plate heat exchanger program. The multidimensional flow program [2] allows the users :

- to understand and visualise the different flow regimes,

- to optimise corrugation design in order to reduce the pressure drop,

- to observe the flow distribution along a plate.

The experimental and numerical studies on the friction factor and the heat transfer coefficient laws for each type of plate tested have led to the development of a software, named CEPAJ, for the design of plate and frame heat exchangers.

The limitations of the software are :

- single phase flow with newtonian fluids,

- same pass number (maximum 5) for the cold and the hot streams.

Furthermore a perfect flow distribution is assumed for checking or sizing calculation mode.

Three design modes are available : checking, sizing and simulation.

3.1   Checking mode

3.1.a   Thermal checking of the heat exchanger

The purpose is to know whether a given heat exchanger will achieve a given heat duty.

The data are :

- the plate's geometrical characteristics and the heat exchanger geometry ;

- the process : mass flow rate, inlet and outlet temperature for both flows ;

- physical properties for both streams.

The knowledge of the physical properties leads to the building of functions relating enthalpy and temperature for each fluid :

$$H = f(T) \quad \text{or} \quad T = f^{-1}(H) \quad (1)$$

As we know, input and output thermal conditions, the energy balance between the inlet conditions of one stream and any point x inside the heat exchanger is :

$$(H^c(x) - H^c_{in}) = (H^h_{out} - H^h(x)) \cdot \frac{M^h}{M^c} \quad (3)$$

With the subscript c for cold flow and h for hot flow.

Since this relation is valid for all points inside the heat exchanger, the evolution of the enthalpy of one flow in relation to the enthalpy of the other one may be determined since enthalpy and temperature are related by the f functions, the connecting temperature curve of both flows (fig. 1) may be determined.

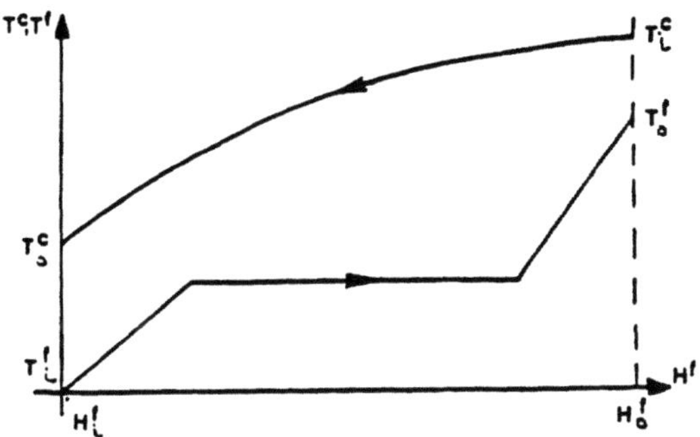

Fig. 6 : Local energy balance

The evolution curve is segmented into an appropriate number of cells to observe the linear variation of temperature in a cell. Then, in each cell limited by point x and x', the local energy balance may be obtained as follows :

$$Q = M^c (H^c(x) - H^c(x')) = U \cdot A(x,x') \cdot DTlog$$

A(x,x') is the area between point x and x'

Dtlog is the logarithmic mean temperature difference :

$$DTlog = \frac{(Tc(x) - Th(x)) - (Tc(x') - th(x'))}{\log\left(\frac{(Tc(x) - Th(x))}{(Tc(x') - Th(x'))}\right)}$$

U is the overall heat transfer coefficient.

$$U = \cfrac{1}{\cfrac{1}{\alpha_c}+\cfrac{1}{\alpha_h}+R_w+R_f}$$

α_c is the local heat transfer coefficient for the cold system

α_h is the local heat transfer coefficient for the hot system

α_c and α_h can be evaluated with physical properties, plate characteristics and velocity between plates. Since these quantities are known we can calculate $A(x,x')$. If all the mesh area between inlet and outlet is added the required area Ar to achieve the given heat duty is obtained. The comparison with the geometrical area shows the heat exchanger thermal oversizing or undersizing.

3.1.b Pressure drop evaluation

The second step of the checking mode is the pressure drop calculation for each stream. There are two ways to evaluate the pressure drop, either by calculating the pressure drop of the required heat exchanger (with the required area) or by calculating the pressure drop of the real heat exchanger assuming thermal checking agreement. In a single phase flow, and if the heat exchanger is well sized there are several differences between the two approaches.

The overall pressure drop is divided into three terms :

$$\Delta p = \Delta pfrot + \Delta Psing + \Delta pacc$$

$\Delta Pacc$ is the acceleration pressure drop. The evaluation is made between inlet and outlet manifold.

$\Delta Pfrot$ is the frictional pressure drop due to friction in the manifolds and in the channels between the plates. The first term is calculated based on the average way in the heat exchanger using the GRETh frictional pressure drop correlations.

$\Delta Psing$ is the pressure drop due to inlet or outlet effects or flow changing directions.

3.2 Sizing mode

3.2.a Logic for sizing mode

The problem of sizing heat exchanger can be seen in fig. 7. This shows the pressure drops and heat duty variations with the heat exchanger channel number. To simplify the demonstration, a constant variation of the friction factor and heat transfer coefficient with mass flow rate is assumed.

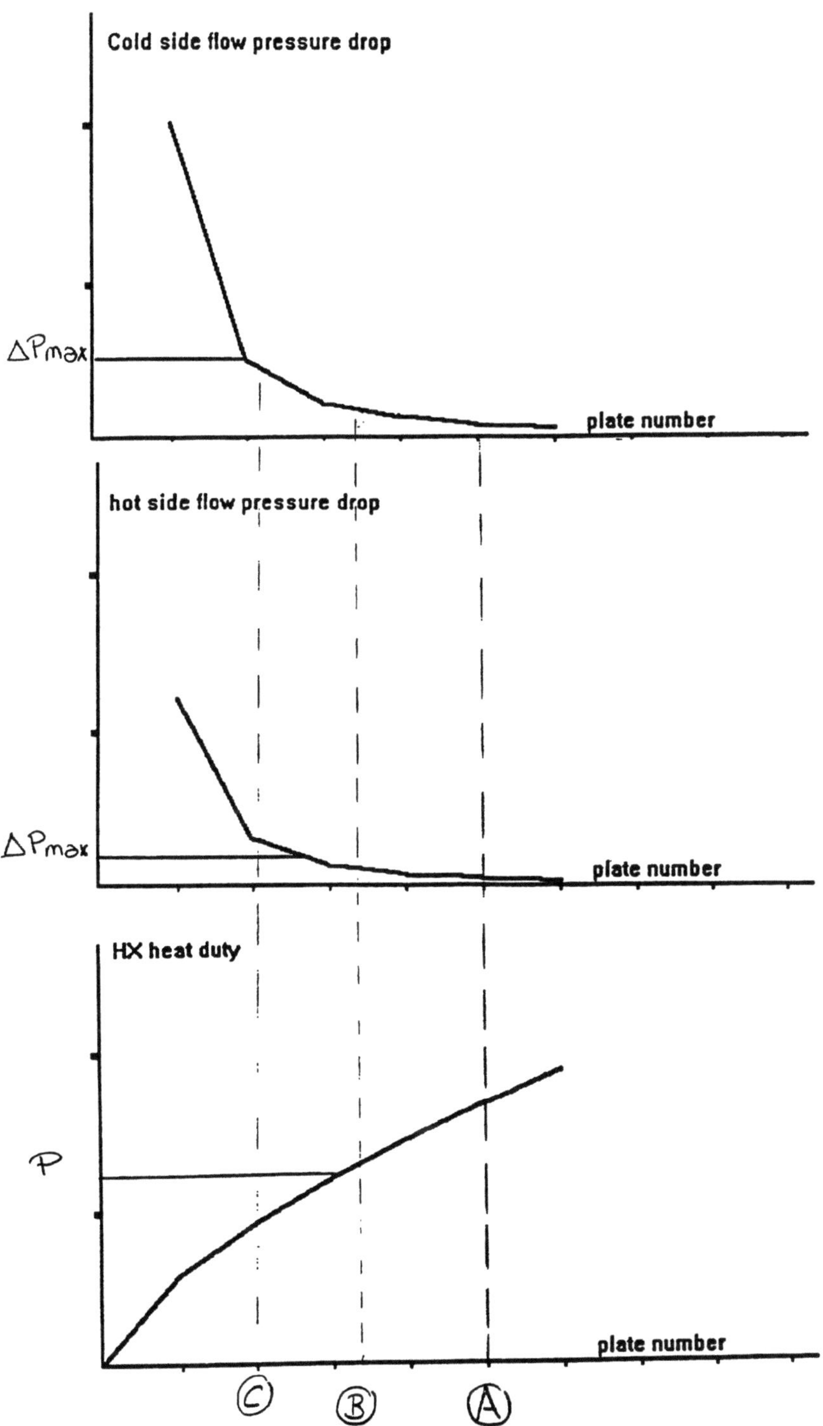

Fig. 7 : Logic for sizing calculation

If the plate number is A, the heat exchanger is oversized and there is no pressure drop limitation. The heat exchanger complies with the process specifications, but it is possible to reduce the plate number to obtain a smaller area.

If the plate number is B, the heat exchanger is well sized and there is no pressure drop limitation. The heat exchanger complies with the process specifications.

If the plate number is C, the heat exchanger is undersized and there are pressure drop limitations.

3.2.b Sizing mode with one plate type

The purpose is to find the plate number to perform a given heat duty within an imposed pressure drop limitation. Starting with a maximum plate number, the program reduces this number, evaluating thermal and hydraulical performances, until a limitation on heat duty or pressure drop appears. The calculations with a different pass number are made and then shown to the user.

3.2.c Sizing mode with different plates

The purpose is to find the plate number and the plate type needed to perform a given heat duty within an imposed pressure drop limitation. The user first selects several plate types in a library and then starting with a maximum plate number, the program searchs for the plate number for each plate type and each pass number.

All the solutions found are then shown to the user and can be classified according to area, thermal oversizing or pressure drop criteria.

3.3 Simulation mode

Using the "TRICOT" software presented in another paper [15], each pass of the plate heat exchanger can be simulated in order to determine the hydraulic and thermal behaviour of the heat exchanger taking into account the effect of maldistribution between the channels.

3.4 A practical tool

The CEPAJ software is written in FORTRAN 77 and so can be used on a personal computer or on any other computer. The data are shown by group on the screen and can be changed all together.

The plate geometries are stored in a library which can be completed by the users. Of course, all the sized heat exchanger data can also be stored in an other library.

A physical properties data base is relayed for easy access to the required thermal and hydraulic physical properties of fluids.

New or private correlations can be introduced by the user to take into account his own plate characteristics.

In checking calculation mode, plates with different corrugation angles can be used together.

In checking calculation mode, the user can observe the local characteristics of the flow (Reynolds number, Prandtl number, velocity...) in the manifolds and in the channels, and the heat transfer coefficients along the plates.

3.5 <u>Validation</u>

The CEPAJ software has been tested and validated on an plate heat exchanger performed on ESTHER test facility under industrial conditions [13]. ESTHER, with the help of seven different loops, allows for the testing of heat exchangers for various types of working fluids such as water, coolants, oils, and organic fluids.

4. CONCLUSION

The numerical and experimental studies have led to a better understanding of the local phenomena which appear in the flow inside a plate type heat exchanger. The friction factor and the heat transfer coefficient laws for each type of plate have been included in the CEPAJ software for the design of plate and frame heat exchangers [14].

The works are continuing towards two phase flow studies. These studies will increase the performance of the CEPAJ software for the design of two-phase flow plate heat exchanger.

5. REFERENCES

[1]   HUGONNOT P. : Etude locale de l'écoulement et performances thermohydrauliques à faible nombre de Reynolds d'un canal plan corrugué. Applications aux échangeurs de chaleur à plaques. Thèse de l'Université de Nancy I, Juillet 1989.

[2]   HUGONNOT P., VIDIL R. et M. LEBOUCHE : Flow regimes in a corrugated channel : Experimental and numerical approaches. Applications to plate heat exchangers, 9th Eurotherm Symposium. RFA, july 1989.

[3]   AMBLARD A. : Comportement hydraulique et thermique d'un canal plan corrugué. Application aux échangeurs de chaleur à plaques. Thèse de Docteur de l'INPG, mars 1986.

[4]   GRILLOT J.M. : Etude du dépôt de particules en phase gazeuse dans les canaux d'échangeurs thermiques à plaques. Thèse de Docteur de l'INPG, avril 1989.

[5]   SOURLIER P. : Contribution à l'étude des propriétés convectives de fluides thermodépendants lors de l'écoulement en canal de section rectangulaire. Thèse de Docteur de l'Université de Nancy I, juin 1988.

[6]   LEULIER J.C. : Comportement hydraulique et thermique des échangeurs à plaques traitant des produits non-newtoniens. Thèse de Docteur de l'Université de Nancy I, janvier 1988.

[7]   SACLIER F. : Etude hydraulique et thermique de l'ébullition convective du fluide R22 dans des canaux de section rectangulaire. Thèse de l'Institut National Polytechnique de Grenoble, 1989.

[8]   BASSIOUNY M. K. : Experimentelle und theoretische unter suchungen über mengenstrauverteilung, druckverlust und wärmeübergang in plattenwareaustauschern. Fortschrittberichte, VDI Verlag, 1985.

[9]   FOCKE W. W., ZACHARIADES J., OLIVIER I. : The effect of the corrugation inclination angle on the thermohydraulic performance of plate heat exchangers. Int. J. of heat Mass Transfer, Vol. 28, n°8, pp.1469-1479, 1985.

[10]   OKADA et al. : Design and heat transfer characteristics of new plate heat exchanger. Heat Transfer Japanese Research, Vol.1, n°1, pp.90-95, 1972.

[11]   KULLENDORFF A. : Local mass transfer studies over a sine-shapped wave in a plate heat exchanger. Extrait de "Transfer studies on heat exchangers with a mercury evaporation method", 1974.

[12]   GAISER G., KOTTKE V. : Enhancement of heat transfer in plate heat exchangers and regenerators with corrugated passages. 9th Eurotherm Symposium, RFA, july 1989.

[13]   VIDIL R., ICART G. : La plate-forme d'essai ESTHER du GRETh. Un outil de qualification des échangeurs à la disposition des constructeurs. Journées MEI 90, Echangeurs et récupérateurs de chaleur, 1990.

[14]   VIDIL R., GRILLOT J.M., MARVILLET C., MERCIER P., RATEL G. : Les échangeurs à plaques - Description et éléments de dimensionnement. 2e édition, septembre 1990. Editions LAVOISIER.

[15]   THONON B., P. MERCIER, M. FEIDT (LEMTA Nancy) : Flow distribution in plate heat exchangers and consequences on thermal and hydraulic performances. Eurotherm 18, Février 1991.

Flow Distribution in Plate Heat Exchangers and Consequences on Thermal and Hydraulic Performances

B. THONON * ; P. MERCIER * , M. FEIDT **

* GRETh   CEN Grenoble BP 85X   38041 GRENOBLE CEDEX   FRANCE
** LEMTA   Université de NANCY I   URA-CNRS 875
   2 Av de la Forêt de Haye   54504 Vandoeuvre Les Nancy CEDEX   FRANCE

The misunderstanding of the consequences of a non-uniform flow distribution on thermal and hydraulic performances generally involves a poor design of the heat exchanger. Thus, a numerical model is developped to predict the flow distribution, and is validated by experiences. It appears that the overall pressure drop is closely related to the flow distribution, and that the thermal performances are less affected. But heterogeneity in temperature is observed at the outlet of the channels.

INTRODUCTION

The conventional approach, to the thermal and hydraulic design of heat exchangers, is commonly used under the assumption of there beeing uniform flow distribution between all the channels.

The maldistribution of flow deteriorates the thermal and hydraulic performances, and induces a local heterogeneity in the thermal and pressure fields, MUELLER [1] has reviewed the causes and effects of non uniform flow distribution in heat exchangers.

If numerous investigations have been made to determine pressure drop and heat transfer characteristics of corrugated channels [2] [3], very few studies closely examine the problem of flow distribution [4] [5], especially in the hydraulic area. Only the edge effect and the pass arrangements have been studied [6] in regard of the thermal efficiency.

In plate heat exchangers, the non uniform flow distribution originates from the difference in the pressure profile at the outlet and inlet port.

The overall pressure drop of the heat exchanger is also a function of the mean flow rate in each channel.

In each channel we use for one phase flow the relation:

$$\Delta P_1 = 4 \ f \ \rho \ \frac{U_1^2}{2} \frac{L}{D_H}$$

Where f, the friction factor, is a function of the Reynolds number : $f = a \ Re^b$
The coefficients a and b, depend on the angle of the corrugation [2] [3].
If we consider the physical properties of the fluid to be constant inside the heat exchanger, we obtain the overall pressure loss coefficient:

$$K_H = \frac{\Delta P_{maximum}}{\Delta P_{nominal}} = \left[\frac{U_{maximum}}{U_{nominal}} \right]^{2+b}$$

In some cases, K_H can reach values of about 10.

For heat transfer, the decrease of the transfered heat power depends on the flow distribution of the two fluids (hot and cold), and the arrangement of the heat exchanger.
The sensitivity of the heat exchanger to maldistribution is proportional to the NTU of the exchanger. But in most cases the thermal effect of maldistribution is low, and will be hidden in the design by the safety fouling factor.
The present study concerns the problem of single phase flow distribution in plate heat exchangers, whith both thermal and hydraulic investigations.
A numerical model has been developped and validated using the experimental results.

NUMERICAL MODEL : The CEPAJ Software

The model studies the effect of flow distribution on thermal and hydraulic performances. Determining the pressure profile between the ports allows us to predict the flow rate in each channel and calculate the heat power exchanged between the two fluids.
The heat exchanger is described as a network of channels. The pressure at each node is calculated, as well as the flow rate for each branch (fig 1).

General hypothesis :   Steady state regime
                           The two fluids are incompressible
                           No outside thermal losses

Flow in the ports :
The flow is considered one dimensional and the mass and momentum balances are established. The control volume used is similar to BAJURA's one [7] (fig 2).

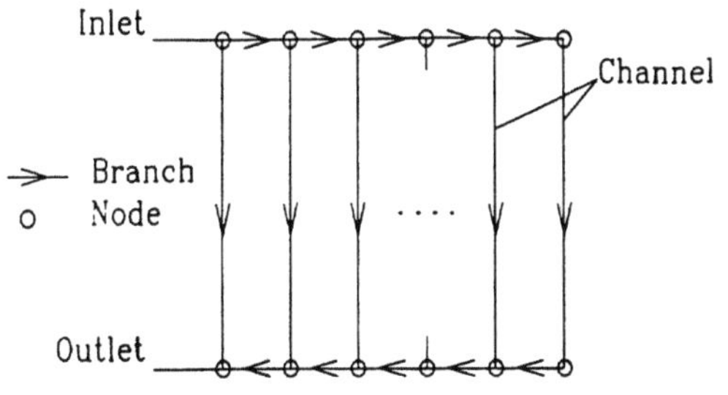

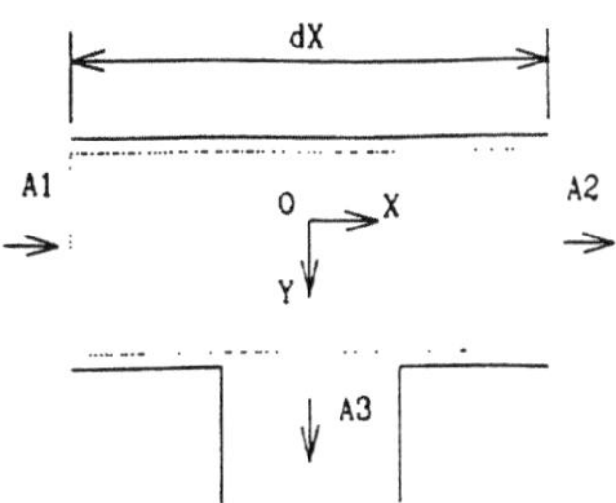

Figure 1 : Network representation
of the exchanger

Figure 2 : Control volume

The mass balance is classicaly:

$$\sum M_i = 0$$

We express a general momentum balance in projection on the Ox axis, where the pressure and the velocity are averaged on the control volume :

$$\frac{1}{\rho}\frac{dP}{dX} + \theta\, U\, \frac{dU}{dX} + 4\, f\, \frac{1}{D_H}\, \frac{U^2}{2} = 0$$

The momentum correction factor θ takes into account the velocity profile on the passage area and the losses due to the branching flow.

Therefore, to solve the momentum balance, θ and f are to be determined from specific experiments.

Flow in the channels :

We establish the momentum balance for the channel, including the regular and the singular pressure losses at the inlet and the outlet.

$$P_{in} - P_{out} = \frac{1}{2}\, \rho\, K_{in}\, U_{pin}^2 + 4\, f\, \rho\, \frac{U_b^2}{2} + \frac{1}{2}\, \rho\, K_{out}\, U_{pout}^2$$

The loss coefficients K_{in} and K_{out} have been choosed in the litterature [8].

Heat transfer :

The knowledge of the velocity field of the fluids allows the thermal field to be calculated.

A step by step method is used to solve the energy balance of each channel.

$$V_b\, \rho\, C_p \left(\frac{\partial T}{\partial t} + U_b\, \frac{\partial T}{\partial y} \right) = \phi\, A_b$$

The heat flux density ϕ received by the control volume Vb, comes from the two adjacents channels (fig 3).

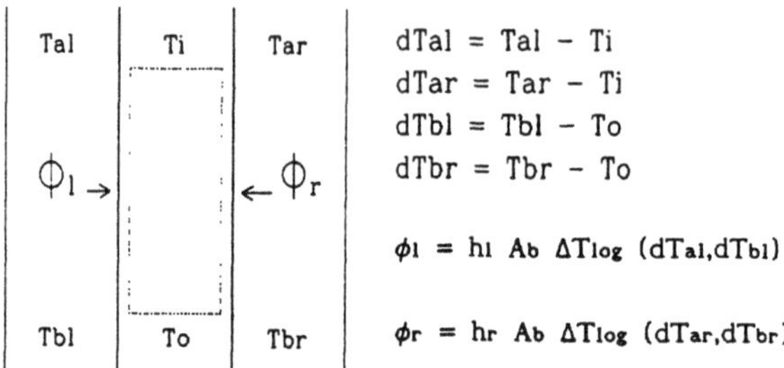

$$dTal = Tal - Ti$$
$$dTar = Tar - Ti$$
$$dTbl = Tbl - To$$
$$dTbr = Tbr - To$$

$$\phi_l = h_l \, A_b \, \Delta T_{log} \, (dTal, dTbl)$$

$$\phi_r = h_r \, A_b \, \Delta T_{log} \, (dTar, dTbr)$$

Figure 3 : Thermal control volume

The equivalent heat transfer coefficients h_l and h_r are calculated from experimental data [2] [3].

Physical properties :

The physical properties (ρ, Cp, μ, λ) of the two fluids are calculated over the entire network, and are reinjected in the hydraulic module of the model.

EXPERIMENTATION

Experiments has been done on an industrial plate heat exchanger. Pressures and temperatures are measured by inserting hypodermic needles into the gaskets (between two plates), of the outlet and inlet ports and of several channels of the cold fluid (fig 4).

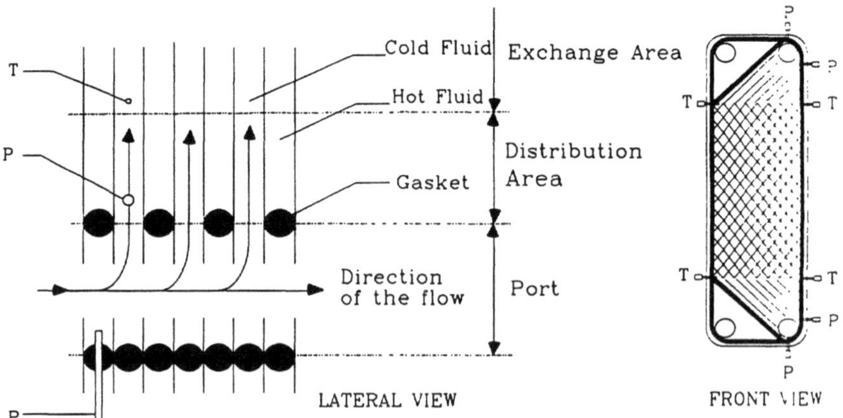

Fig 4 : Exchanger instrumentation

Different parameters of the numerical model are established independantly, this for a large range of physical properties and for different flow rates.

Friction factor in the ports :
A uniform flow is established in the port (no branching flow). And the friction factor f_p is calculated from the pressure loss.

$$\Delta P_p = 4\ f_p\ \rho\ \frac{U_p^2}{2}\ \frac{L}{D_H}$$

It appears that the different values of the friction factor are not significantly related to the Reynolds number. It is thus possible to assimilate the port as a rough duct in turbulent flow.

Momentum correction factor :
An isothermal flow is considered, and for a given total mean flow rate, the pressure loss in the channels and the pressure profile in the two ports are measured.
The momentum balance in the domain can be expressed in a general form:

$$\Delta P_{pl} = \alpha\ M_l\ dM_l + \beta\ M_l^2$$

With $\alpha\ M_l\ dM_l$ is an acceleration term (α is a function of θ and $\frac{A_b}{A_p}$), and $\beta\ M_l^2$ represents the regular pressure loss due to friction in the port (the value of β is known).
The coefficient α is to be calculated with a step by step method (fig 5).

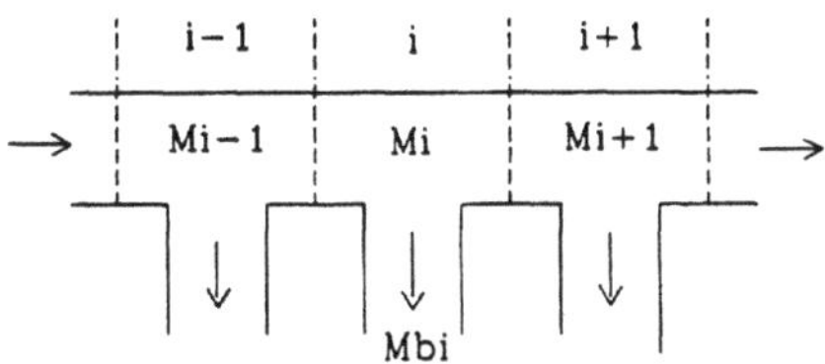

Figure 5 : Description of the inlet port

The flow rate M_{bl} is known (pressure loss in the channels)
With $\quad M_l = M_{l-1} - M_{bl}\quad$ (mass balance)
And $\quad M_l$ = Total inlet mass flow rate of the exchanger
Thus, the coefficient α can be identified for the inlet and the outlet ports (fig 6). It appears that the total flow rate and the physical properties do not significantly influence the value of the momentum correction factor.

The consequences of flow maldistribution on thermal and hydraulic performances are studied using a large range of non dimensionnal numbers (70<Re<10000 , 2.3<Pr<55.0 , 17<Nu<200) and for "U" type and "Z" type exchangers.

RESULTS

By introducing the different parameters in the numerical model, we can simulate the flow distribution in a plate heat exchanger.
A good agreement is found between the numerical results and the experimental data (fig 6-7)

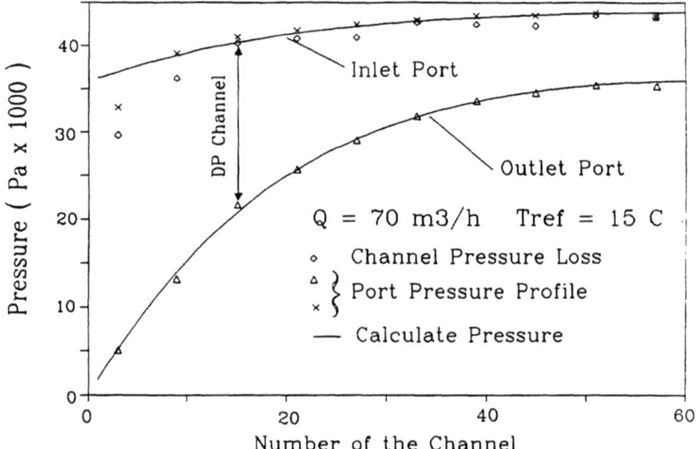

Figure 6 : Pressure field of the exchanger

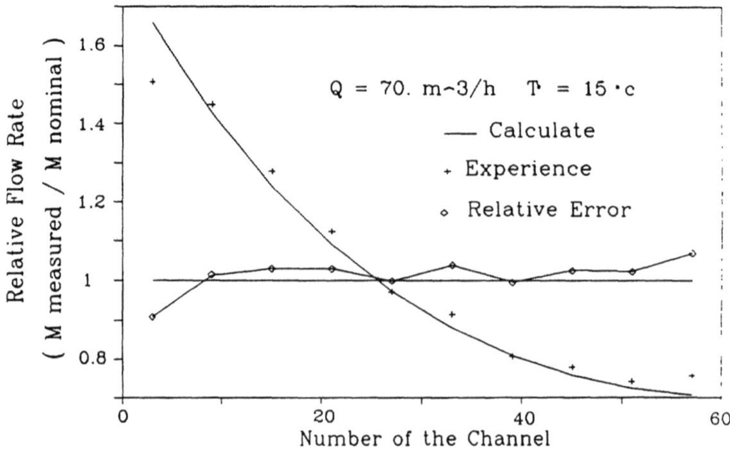

Figure 7 : Comparaison of numerical and experimental data

To measure the quality of the distribution, several non dimensionnal criteria are introduced:

Overall pressure loss $\quad K_H = \dfrac{\Delta P \ \text{overall}}{\Delta P \ \text{nominal}}$

Overall heat power loss $\quad K_P = \dfrac{\text{exchanged heat power}}{\text{nominal heat power}}$

The nominal pressure loss and heat power are calculated for a uniform flow distribution.

We can also define mean quadratic criteria:

$$SIGM = \frac{1}{N} \sum_{i=1}^{n} \left(\frac{M_i}{M_{avg}} - 1 \right)^2$$

$$SIGT = \frac{1}{N} \sum_{i=1}^{n} \left(\frac{\Delta T_i}{\Delta T_{avg}} - 1 \right)^2$$

Influence of the flow rate conditions :

The flow conditions are represented by a Reynolds number which is estimated in the channel for an average flow rate. The heat exchanger is adiabatic and the number of channels is constant. The variation of the Reynolds number is obtained by changing the total mean flow rate and the physical properties of the fluids (water, aqueous glycol solution).

This section will be limited to the study of the global hydraulic criteria K_H and SIGM (fig 8), for two different exchangers, one with 40 channels the other with 60 channels.

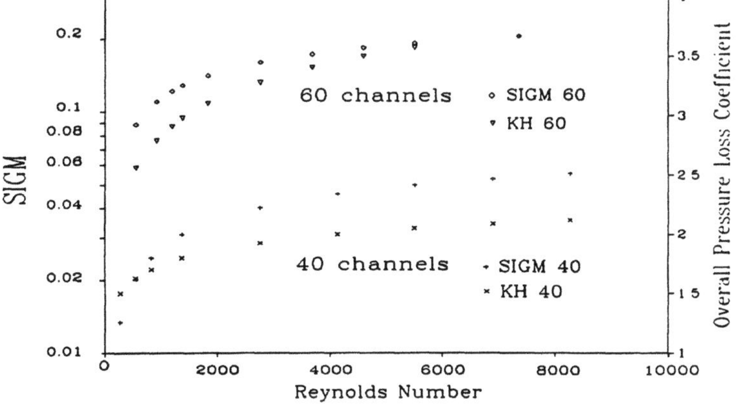

Figure 8 : Influence of the flow conditions

It appears that the two criteria increase with the Reynolds number in a similar way (exponential function), and that they are extremely dependant on the number of channels.

Influence of the configuration ("U" type or "Z" type) :
We will compare two exchangers with the same nominal conditions, one in an "U" configuration, the other in an "Z" configuration.
The local flow rate and outlet temperature are studied (fig 9-10).

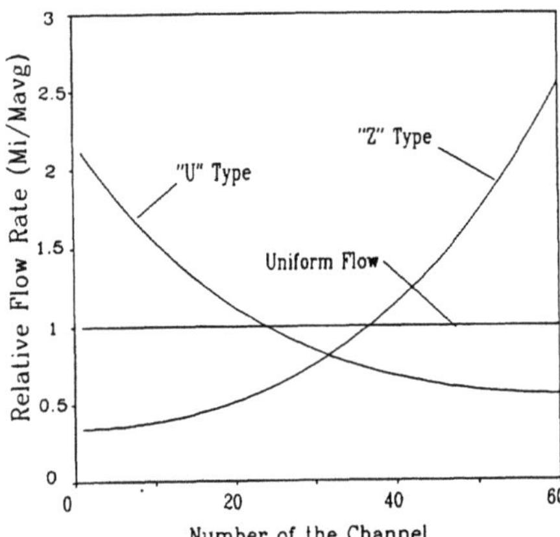

Figure 9 : Influence of the configuration on the flow distribution

Figure 10 : Influence of the configuration on the temperature distribution

We compare the two configurations for both fluids to the reference exchanger with uniform flow distribution, for the four criteria.

		KP	KH	SIGT	SIGM
HOT FLUID	Uniform	1.	1.	2.87 e-3	0.
	"U" type	0.97	3.91	14.83 e-3	2.00 e-1
	"Z" type	0.94	4.70	22.78 e-3	4.14 e-1
COLD FLUID	Uniform	1.	1.	3.06 e-3	0.
	"U" type	0.97	3.73	10.26 e-3	1.91 e-1
	"Z" type	0.94	4.47	23.55 e-3	3.97 e-1

As is shown, the loss of transfered heat power is poor (3% and 6%). But heterogeneity in the outlet temperature of the channel is observed (approximatively ± 20%).

As for the hydraulic aspect, the overall pressure loss coefficient is extremely high leading to a poor design of the heat exchanger.

It also appears that, the performances of the "Z" configuration are poorer than those of the "U" configuration, both on the thermal and hydraulic aspects.

CONCLUSIONS

This study put forward consequences of maldistribution on the performances of a plate heat exchanger.

If the thermal efficiency is less affected by a non uniform flow distribution than the overall pressure drop, a significant difference in temperature is observed at the outlet of the channels.

Thus it seems important, to take in account the flow distribution for the design of plate heat exchangers.
This is particularly true for chemical and food industries, where the residence time and the temperature are to be controlled all over the transfer surface.
All these results are included in the CEPAJ software [9].

The optimal design of an heat exchanger depends on physical and economic criteria, but some problems of maldistribution may be avoided by using flows at low Reynolds numbers in the exchanger, and by reducing the number of channels if possible.

NOMENCLATURE

A	Area	m^2		θ	Momentum Coefficient	
Cp	Heat Capacity	J/kg K		λ	Thermal Conductivity	W/m K
DH	Hydraulic Diameter	m		ρ	Fluid Density	kg/m^3
f	Friction Factor			ϕ	Heat Flux Density	W/m^2
h	Heat Transfer Coefficient	W/m^2 K		μ	Dynamic Viscosity	kg/m s
K	Loss Coefficient					
L	Length	m			Subscript	
M	Mass Flow Rate	kg/s		b	branch	
N	Number of channels			H	Hydraulic	
P	Pressure	N/m^2		P	Power	
T	Temperature	K		in	inlet	
U	Velocity	m/s		out	outlet	
V	Control Volume	m^3		p	port	

REFERENCE

[1]   A.C. MUELLER & J.P. CHIOU
"Review of Various Type of Flow Maldistribution in Heat Exchanger"
Heat Transfer Engineering Vol 9 n°2 pp 36-50 1988

[2]   W.W. FOCKE, J. ZACHARIADES & I. OLIVER
"The Effect of the Corrugation Inclination Angle on the Thermohydraulic
Performances of a Plate Heat Exchanger"
Int Journal of Heat & Mass Transfer Vol 28 n°8 pp 1469-1478 1985

[3]   P. HUGONNOT
"Etude Locale de l'Ecoulement et Performances Thermo-Hydraulique à Faibles
Nombres de Reynolds, Application aux Echangeurs à Plaques"
Thèse Université de NANCY I Chapitre VI pp 164-201 1989

[4]   M.F. EDWARDS, D.I. ELLIS & M. AMOOIE-FOUMENY
"The Flow Distribution in Plate Heat Exchanger"
First UK National Conference on Heat Exchanger pp 1288-1302 1984

[5]   M.K. BASSIOUNY & H. MARTIN
"Flow Distribution and Pressure Drop in Plate Heat Exchanger"
Chemical Engineering Science Vol 39 n°34 pp 693-704 1984

[6]   S.G. KANDLIKAR & R.K. SHAH
"Multipass Plate Heat Exchanger - Effectiveness NTU
Results and Guidelines for Selecting Pass Arrangements"
Transaction of the ASME
Journal of Heat and Mass Transfer Vol 111 pp 300-313 1989

[7]   R.A. BAJURA & E.H. JONES JR.
"Flow Distribution in Manifolds"
Transaction of the ASME
Journal of Fluid Engineering pp 654-666 DEC 1976

[8]   Heat Equipment Design Handbook   Section 2.2.7
Hemisphere Publishing Corporation 1983

[9]   R. VIDIL, G. RATEL, J.M. GRILLOT
"Thermal-Hydraulic Design Performances of Plate and Frame Heat Exchangers
the CEPAJ Software"
EUROTHERM SEMINAR N°18   HAMBOURG   Fev 1991

Welded Plate Heat Exchangers as Refrigerants Dry-Ex Evaporators

Ch.MARVILLET

Groupement pour la Recherche sur les Echangeurs Thermiques
Centre d'Etudes Nucléaires de Grenoble
85X   38041 GRENOBLE CEDEX

1/BACKGROUND

The subject of enhanced heat transfer has developped to the point that it is of serious interest for heat exchanger application in *heat pumps, air conditionning systems and refrigeration units.*

Enhanced heat transfer surfaces are used for different reasons:

- *to obtain less expensive* heat exchangers in reducing the number of heat exchangers or plates, the shell diameter and, on the whole, the heat exchanger size.
- *to obtain more compact heat exchangers,*anyone designing heat pumps or refrigeration units may try and keep the amount of refrigerant as small as possible, the more so as an "ozone friendly" substitute is not yet commercially available.
- *to reduce the mean temperature difference* for the heat exchangers and to provide increased thermodynamic process efficiency which yields savings in operating costs.

The plate heat exchangers are made of high performance heat transfer surfaces: the corrugated or the studded plates may dramatically increase the heat transfer coefficient with single phase fluids. This kind of heat exchanger has a very large number of applications with viscous or turbulent flows.
Nowadays the plate heat evaporators are increasingly used in refrigeration plants. The welded or brazed plate heat exchangers may partially replace the conventional bundle heat exchangers designed for pool boiling or forced flow boiling because of their greater compactness and smaller refrigerant capacity.

The design, the conception and the optimization of plate evaporators require precise, reliable and, if possible, complete thermohydraulic data and correlations. Available information on convective boiling and two phase flow characteristics is very limited in the case of corrugated and studded geometries: in tables 1 and 2, data of different sources is given. A distinction is made between references that provide:
    - "global" thermohydraulic performances of plate evaporators caracterised by a global heat transfer coefficient and total pressure drop (table 1)
    -"local" thermohydraulic performances of corrugated plates caracterised by a local heat transfer coefficient and frictionnal pressure drop (table 2).

TABLE 1 :LIST OF REFERENCES		
Authors	evaporator type	working fluids
JACOBSEN (1983) /1/	Plate and frame H.E	R22 R500 R114
DUTTO (1990) /2/	Brazed plate H.E	R142b
HAUKAS (1983) /3/	Welded plate H.E	R12 R22
PANCHAL (1983) /4/	plate H.E	R717

TABLE 2 :LIST OF REFERENCES		
Authors	evaporator type	working fluids
UEHARA (1988) /5/	Plate and frame H.E	R22 R717
OEHARA (1988) /6/	Brazed plate H.E	R12
ENGEL- HORN /7/	Brazed plate H.E	R22
CAREY (1987) /8/	Corrugated test sections	R113

This paper presents experimental work on four different plate geometries during forced convection boiling of refrigerant R22. The heat transfer coefficient and pressure drop correlations have been established in wide operation conditions (mass velocity, vapor quality, heat flux and saturated pressure) and the comparisons have been made between theplates' performances.

2/METHODOLOGY

To evaluate the performances of the different geometries, stainless steel test sections have been built. Table 3 shows the main features of the plates. The hydraulic diameter is defined as a volumetric one and the heat transfer surface as a projected one.

These test sections (in vertical position) are composed of three parallel channels and, of course, of four welded plates:

- inside the central channel, the refrigerant R22 flows upwards and is partially evaporated.

- inside the two lateral channels, water flows downwards and is cooled.

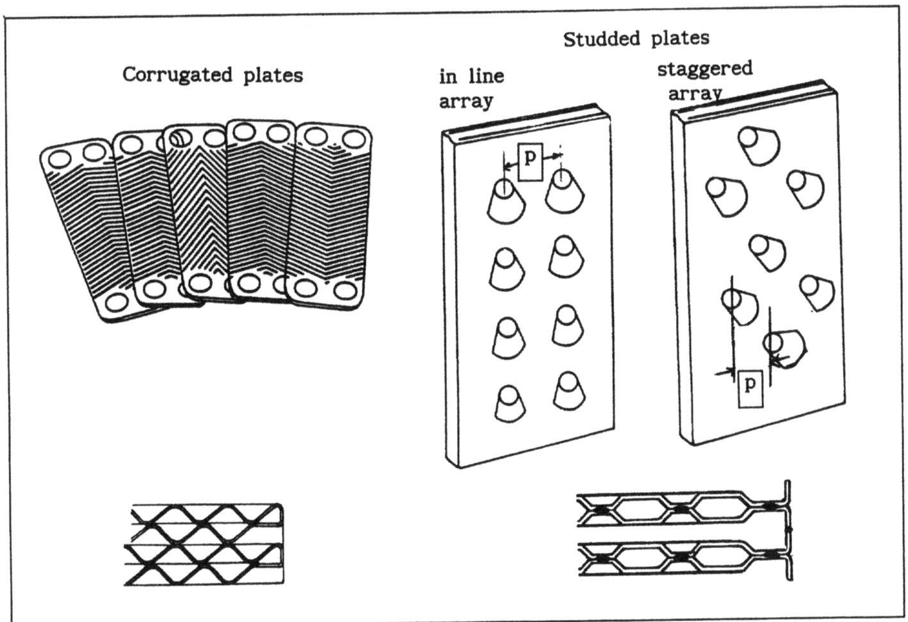

FIGURE 1: CORRUGATED AND STUDDED PLATES
(FRONT AND CROSS-SECTIONNAL VIEWS)

	Plate n°1	Plate n°2	Plate n°3	Plate n°4
Geometry	smooth	studded	studded	corrugated
length l	1284 mm	1284 mm	1284 mm	950 mm
width w	146 mm	146 mm	146 mm	170 mm
hydraulic diameter	7.6 mm	8.35 mm	9.03 mm	6.7 mm
pitch p between studs or corrugations	-	30 mm in-line array	21 mm staggered array	15 mm angle with flow direction: 30°
heat transfer surface	0.194 m^2	0.250 m^2	0.250 m^2	0.323 m^2

TABLE 3: GEOMETRICAL CARACTERISTICS OF PLATE GEOMETRIES

The objective of the present work is to evaluate the heat transfer coefficient between the wall of the internal plates and the evaporating refrigerant fluid flowing upwards. The main parameters having a significant influence on this heat transfer coefficient are:

- the refrigerant mass velocity G (kg/m2.s), defined as the ratio of refrigerant mass flowrate to the **mean flow** area (in the cases of studded plates channels, flow area does obviously not have a constant value)

- the mean thermodynamic vapor quality x (%).At the inlet of the central channel of the test section, the vapor quality of the refrigerant flow may be imposed from 10% up to 80% when the vapor quality does not increase by more than 15% from the inlet to the outlet of the central channel.

- the heat flux ϕ (W/m2) defined with projected heat transfer surface (and not the developped heat transfer surface).

- the saturated pressure Psat (or temperature Tsat) of the refrigerant.

The test loop whose flow sheet is represented in figure 2 provides for testing in a wide range of conditions: table 4 summarizes the explored test conditions of the different geometries.

The test loop is made of four main circuits:

-the refrigerant circuit's components are: a variable speed circulation pump used to control the refrigerant mass flowrate, two volumetric flowrate sensors BOPP&REUTER, a system of three coaxial preevaporators to control the vapor quality at the inlet of the test section, the four welded plates test section, a liquid/vapor separator, and a refrigerant horizontal condenser.

-the warm water circuit on the test section whose components are a regulated electrical heater (maximal output:24kW), a recirculation pump and an electromagnetic massflowrate sensor (WAFERMAG BROOKS).

-the warm water circuit on the system of preevaporators whose components are a regulated electrical heater (maximal output:100 kW), a recirculation pump and an electromagnetic massflowrate sensor (WAFERMAG BROOKS).

- the iced glycol/water circuit on the condenser whose components are a refigeration unit CIAT TK600, a recirculation pump, a heat exchanger and an electrical heater both hold to control the glycol/water temperature and refrigerant saturated pressure.

The inlet and outlet temperatures of fluids flowing through the test section are measured with 100 ohm platinium resistances, the pressure drop on refrigerant flow is measured with a differential manometer ROSEMOUNT 1151 DP.

	x (%)	G(kg/m2.s)	Tsat(°C)	ϕ(kW/m2)
plate n°1	10-90	200-650	5-17	10-60
plate n°2	10-90	200-700	4-16	10-60
plate n°3	10-85	200-600	2-14	10-50
plate n°4	10-80	200-900	3-20	5-60

TABLE 4: TEST CONDITIONS WITH R22 EVAPORATION

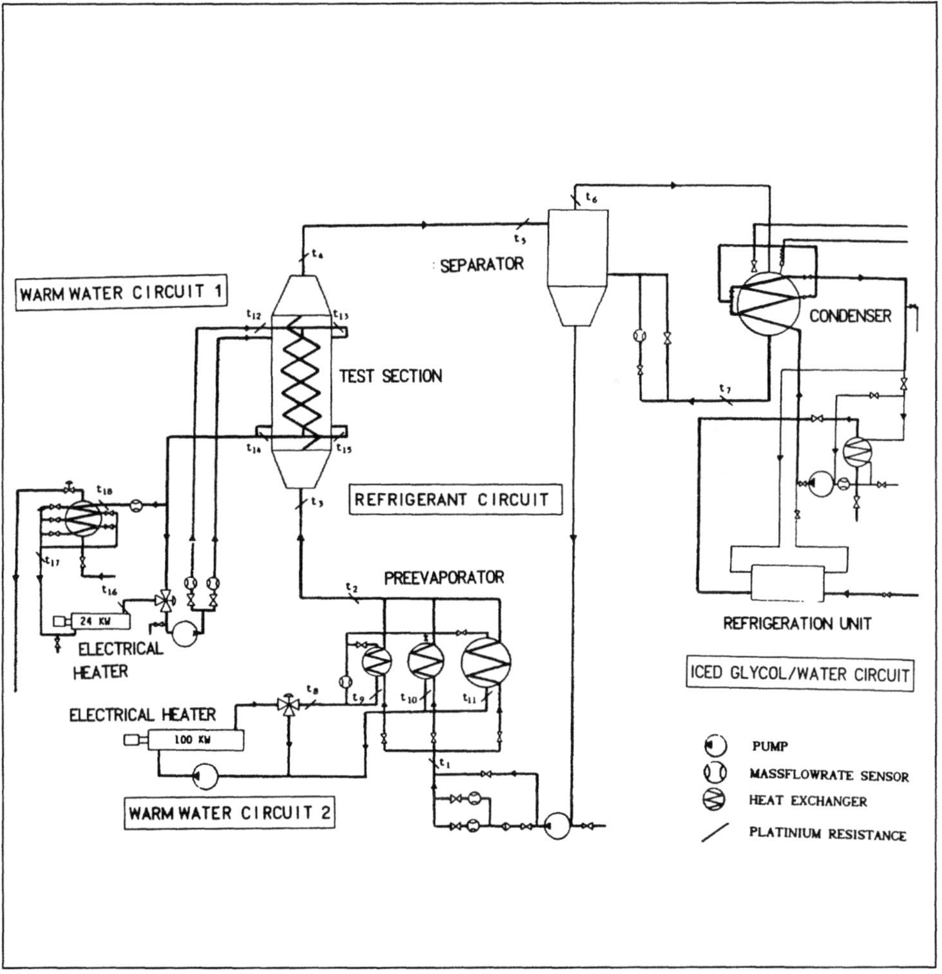

FIGURE 2: FLOW SHEET OF EVAPORATION TEST LOOP

260

The inlet and outlet vapor quality and the global heat transfer coefficient U are calculated from temperatures and massflowrate measurements.

From this experimental data are deduced the two parameters which caracterize the thermohydraulic performances of the plates:

- the mean heat transfer coefficient (αdp) between the wall and the evaporating fluid R22. From a previous test, the mean heat transfer (αw) between water flow and internal plate is evaluated by the Wilson plot method. αdp is deduced from the following relation (e and λ are the thickness and thermal conductivity of plates) :

$$\alpha dp = \cfrac{1}{\cfrac{1}{U} - \cfrac{e}{\lambda} - \cfrac{1}{\alpha w}} \qquad (1)$$

- the frictionnal pressure drop of the refrigerant two-phase flow (Δpf) which is deduced from the measured pressure drop Δp:

$$\Delta pf = \Delta p - \Delta pg - \Delta pa \qquad (2)$$

The gravitationnal term Δpg and acceleration term Δpa are evaluated with a local void fraction calculated from the LOCKART-MARTINELLI correlation.

3/CORRELATION OF EXPERIMENTAL DATA

3.1/HEAT TRANSFER COEFFICIENT DATA

For each geometrical parameters, more than one hundred different test conditions are explored as can be seen in figure 3.

To correlate this experimental data, the thermal performance of the smooth, studded and corrugated channels is represented by the factor R defined as the ratio of the measured evaporation heat transfer coefficient αdp and the calculated pool boiling heat transfer coefficient αpool :

$$R = \cfrac{\alpha dp}{\alpha pool} \qquad (3)$$

The coefficient αpool is calculated with the correlation proposed by reference /9/. The expression for refrigerant R22 is as follows:

$$\alpha pool_{(W/m^2)} = 2200*(\phi/20000)^{(0.9-0.3*Pr^{0.3})}$$
$$*(2.1*Pr^{0.27}+Pr*(4.4+\frac{1.8}{1-Pr})) \quad (4)$$

where Pr is the refrigerant reduced pressure
Φ is the heat flux (W/m2)

This factor R is represented in figures 4 and 4bis, for each four geometrical parameters , against the product of the two adimensional numbers Bo (Boiling Number) et Xtt (Lockart-Martinelli Parameter) whose definitions are:

$$Bo = \phi \, / \, G.\Delta h_{lv} \quad (5)$$

$$Xtt = \left[\frac{1-x}{x}\right]^{0.9} * \left[\frac{\mu_l}{\mu_v}\right]^{0.1} * \left[\frac{\rho_v}{\rho_l}\right]^{0.5} \quad (6)$$

The fluid parameters Δh_{lv}, μ_v, μ_l, ρ_v, ρ_l are the latent heat, the vapor and liquid viscosity, the vapor and liquid density.

As can be seen in figures 4 and 4bis, it may be concluded for the four tested geometrical parameters that:

-when the product Xtt.Bo is larger than 0.00015, the factor R is equal to a constant value: R=1. Nucleate boiling is the predominant mecanism between the wall and refrigerant during R22 evaporation inside a vertical channel.**Under these conditions, the four different geometrical parameters have very similar heat transfer coefficients and no heat transfer intensification has been caused by corrugations and studs.**

-when the product Xtt.Bo is less than 0.00015, the factor R is higher than 1, and increases with the decrease of product Xtt.Bo: Two phase forced convection becomes the predominant mecanism during R22 evaporation. **Under these conditions, no significant heat transfer intensification can be noted between smooth and studded plates. However the corrugated plates have higher heat transfer coefficient than the three other plates.**

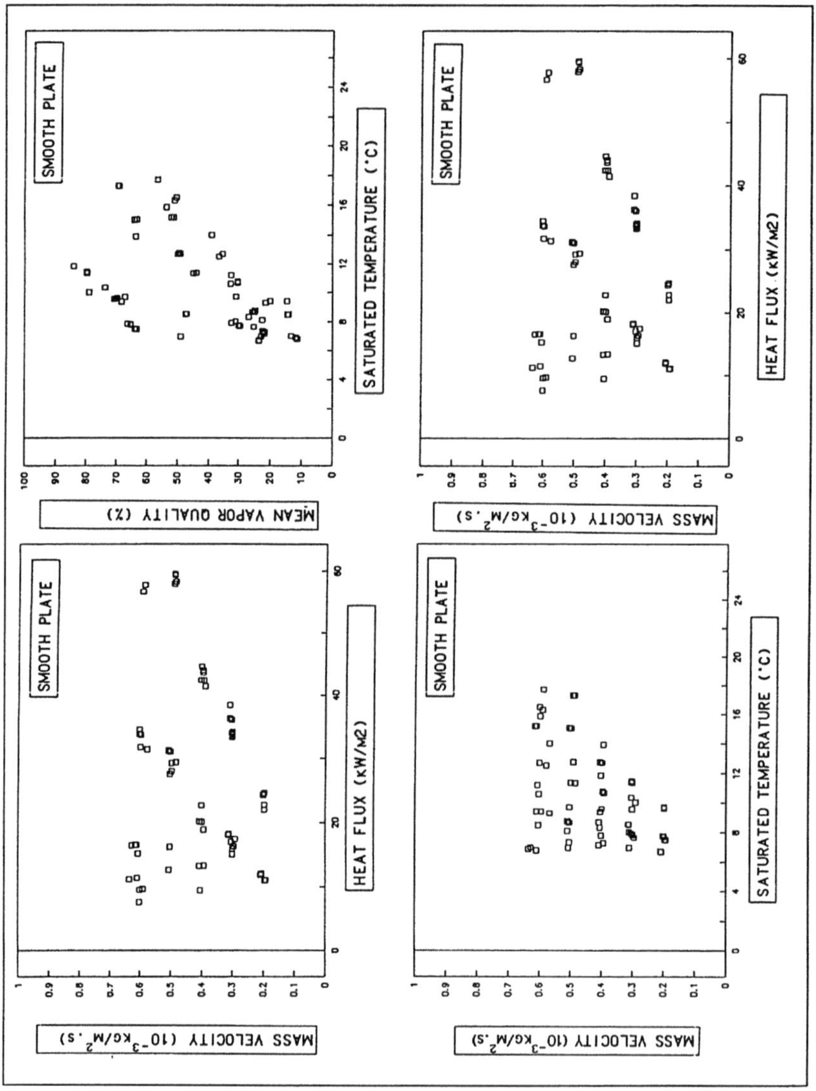

FIGURE 3: SMOOTH PLATE TEST CONDITIONS

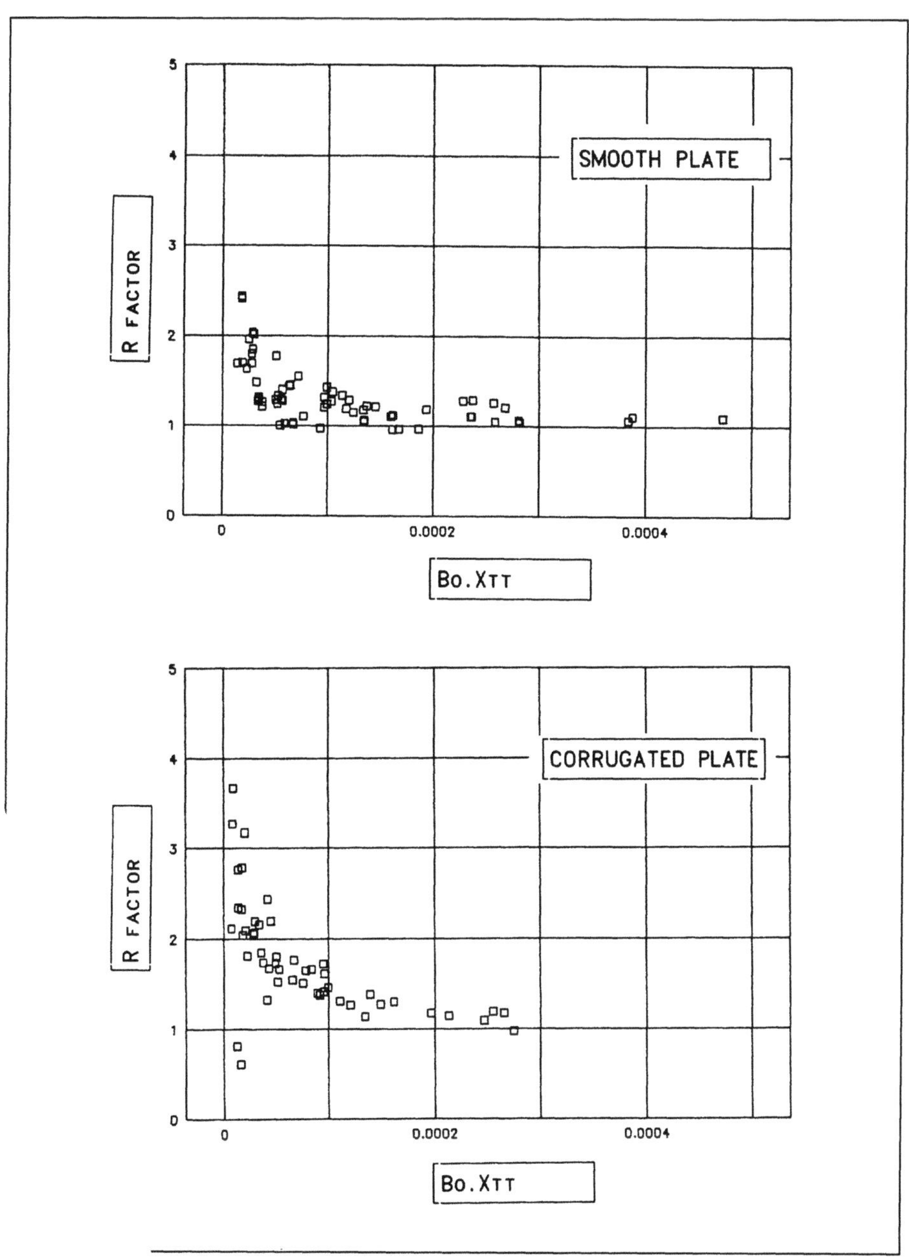

-+: Relation between R factor and the product Bo.Xtt

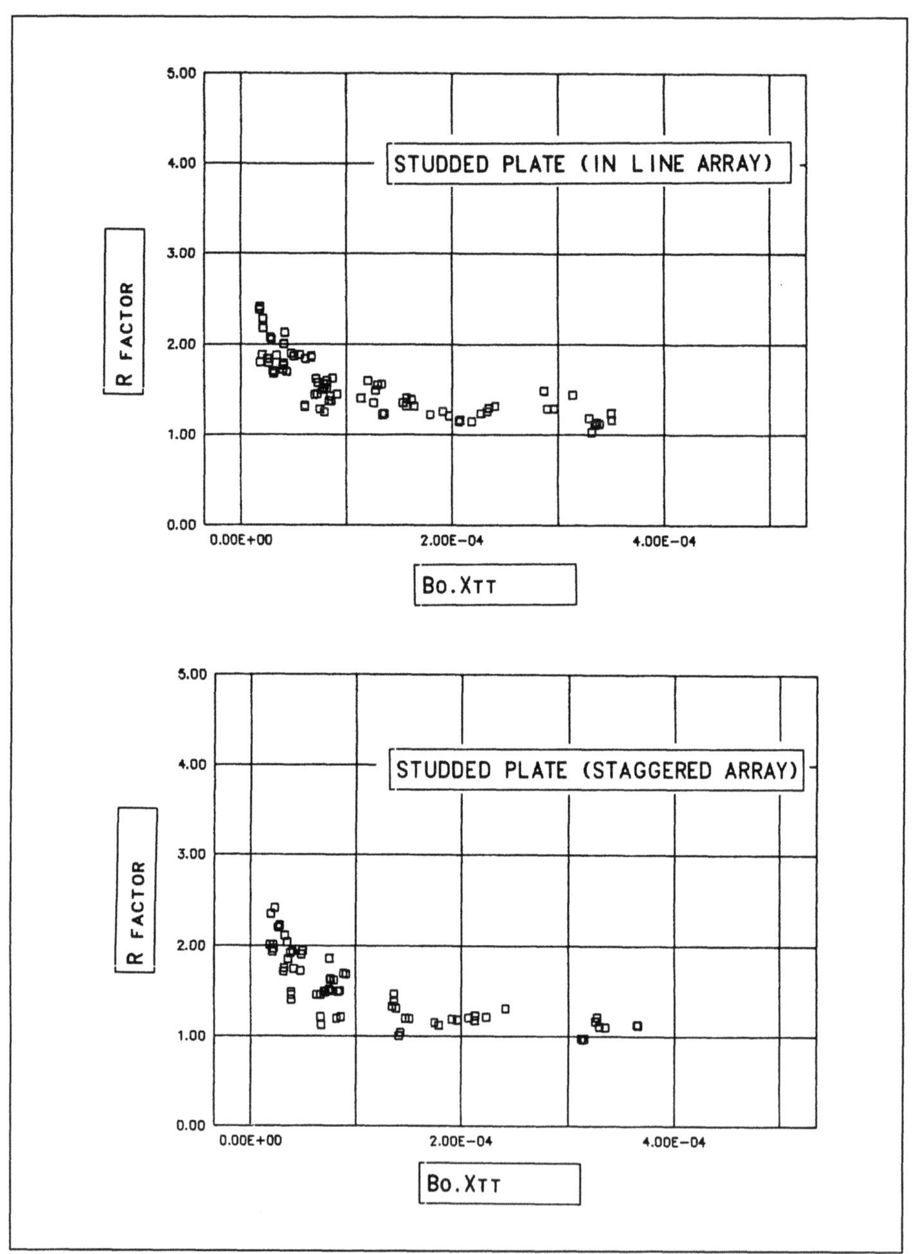

FIGURE 4BIS: RELATION BETWEEN R FACTOR AND THE PRODUCT BO.XTT

3.2/FRICTIONAL PRESSURE DROP DATA

The frictional pressure drop in smooth, corrugated and studded channels is strongly dependant on the Lockart Martinelli parameter Xtt as shown in figures 5 and 5bis. For each of the four different plate geometrical parameters we can propose a frictional pressure drop correlation of similar expression:

$$\frac{\Delta pf}{\Delta p_{smooth}} = A * Xtt^n \qquad (7)$$

where $\Delta psmooth$ is the frictionnal pressure drop of the liquid phase flowing in a smooth plate channel of similar hydraulic diameter . The mass velocity of liquid phase in this smooth plate channel is $G*(1-x)$ where is G is the total mass velocity as defined previously.

Contrary to heat transfer coefficient data, a very strong difference can be observed between the frictional pressure drop data in the four different channels: frictional pressure drop in studded plate channels are much higher than in smooth and corrugated plate channels with similar vapor quality, mass velocity and refrigerant saturated pressure. In figure 6, a comparison of frictional pressure drop in the four different channels has been evaluated with calculated values from previous correlations.

4/CONCLUSION

Four different plate geometrical parameters have been tested in a wide range of working conditions (from 200 to 900 kg/m2.s and from 10 to 60 kW/m2.K) during R22 forced convective boiling. Wall-refrigerant heat transfer coefficient and frictional pressure drop have been deduced from experimental data.

We conclude that:

- the frictional pressure drop is strongly dependant on the plate geometry and for each geometry a specific correlation has been proposed.The studded plates present a frictional pressure drop much higher than the smooth ,and corrugated plate channels.

- the heat transfer coefficient (in a nucleate boiling regime) is not dependant on the plate geometry. Pool boiling heat transfer correlations can be used for the prediction of heat transfer. Pure nucleate boiling is predominant when the product Xtt.Bo is higher than 0.00015.

- When the product Xtt.Bo is less than 0.00015, two phase forced convection make a significative contribution. Corrugated plates have higher heat transfer coefficients than the others. To elaborate heat transfer correlation, more complete experimental data are needed.

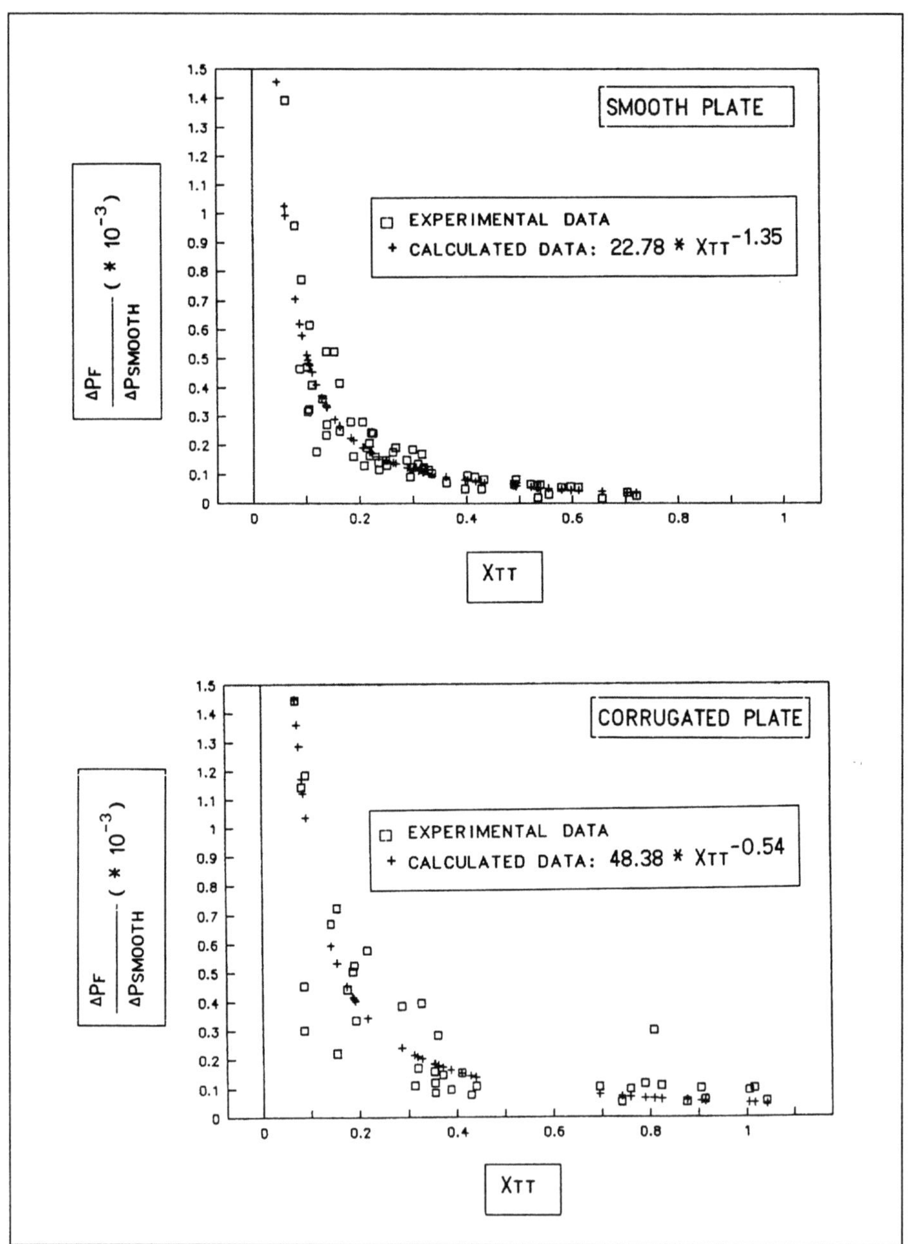

FIGURE 5: RELATION BETWEEN THE PRESSURE DROP RATIO $\Delta P_F/\Delta P_{SMOOTH}$ AND THE LOCKART MARTINELLI PARAMETER X_{TT}.

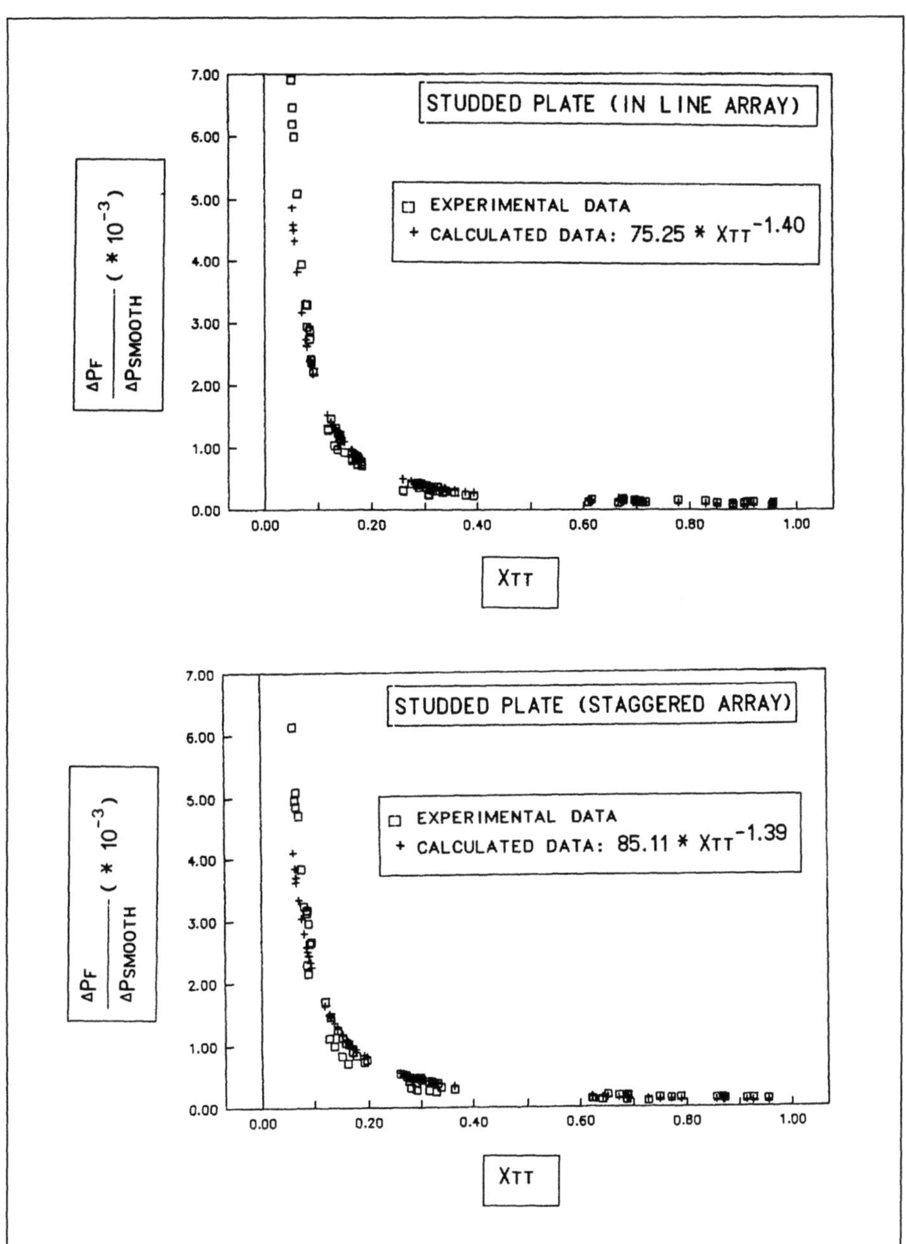

FIGURE 5BIS: RELATION BETWEEN THE PRESSURE DROP RATIO $\Delta P_F/\Delta P_{SMOOTH}$ AND THE LOCKART MARTINELLI PARAMETER X_{TT}.

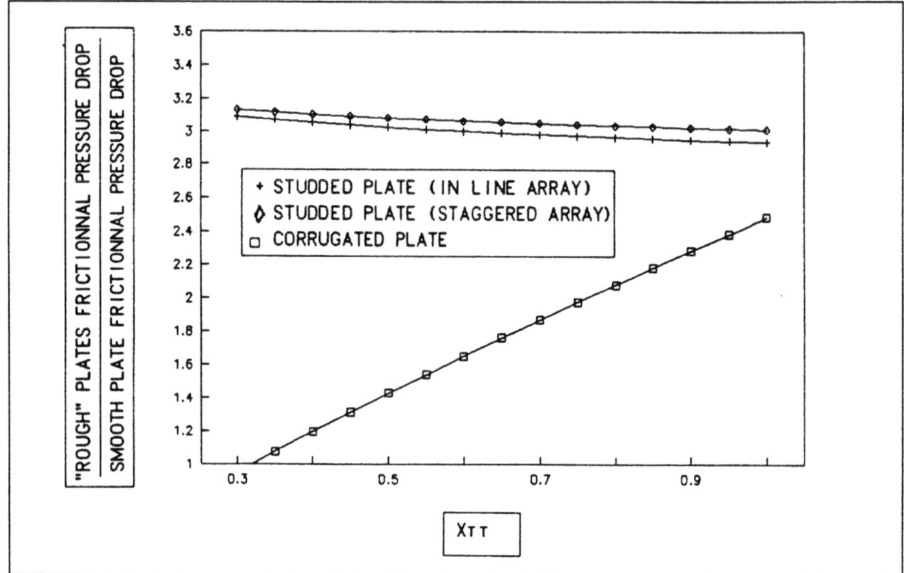

FIGURE 6: COMPARISON OF THE DIFFERENT PLATES FRICTIONNAL PRESSURE DROP

5/REFERENCES

1. JACOBSEN, C; Advanced heat pump techniques, XVIth international congress of refrigeration (1983)

2. DUTTO, T; Essais d'un évaporateur et d'un condenseur à plaques brasées. Report EDF HE 14/W 2903 (1989).

3. HAUKAS, HT; Design of a plate type evaporator for heat pumps. Int.J.Refr, vol.7, 1984.

4. PANCHAL, C.B; HILLIS, D.L; THOMAS, A; Convective boiling of ammonia and freon 22 in plate heat exchangers. Proc. ASME/JSME. Therm.Eng.Jt.Conf.,1983.

5. UHEARA, H; KUSUDA, H; MONDE. M; NAKAOKA, T; Shell and plate type heat exchanger for an OTEC plant. Proc.ASME/JSME Therm.Eng.Conf.1983.

6. OHEARA, T; TAKAHASHI, T; High performance Evaporator Development. SAE International Congress (1988).

7. ENGELHORN, H.R; REINHART ,A.M; Investigations in heat transfer of a plate evaporator. Eurotherm Seminar N°8, Paderborn (1989).

8. CAREY, V.P; COHEN, M; A comparison of the flow boiling performance characteritics of partially-heated cross-ribbed channels with different rib geometries. Int.J.Heat.Mass Transfer. Vol atesN°12, pp.2459-2474, 1989.

9. VDI Wärmeatlas ; Behältersieden. Ha1-Ha22 (1984).

Development of a Compact Heat Exchanger
for Gas Turbine Heat Recovery

A. BONTEMPS
Groupement pour la Recherche sur les Echangeurs Thermiques (G.R.E.Th.)
85 X – 38041 GRENOBLE Cedex (France)
and
Université Joseph Fourier
B.P. 53 X – 38041 GRENOBLE Cedex (France)

M. BRUN
Société TURBOMECA
Bordes
64320 BIZANOS (France)

SUMMARY

The recovery of exhaust gas energy from gas turbines to heat air before it
enters the combustion chamber can reduce specific consumption significantly.
This can be achieved by using a heat exchanger operating at higher pressures
and temperatures. The design concept of a lightweight compact recuperator is
presented together with prototype testing leading to the full scale heat
exchanger. Thermal and aeraulic performances were measured and cyclic duty and
fouling tests were carried out.

INTRODUCTION

Gas turbines are widely used in generating power plants as well as in
transportation. Compared to diesel engines their specific consumption can
reach up to 30 % higher. This can be reduced significantly by adding a heat
exchanger which recovers heat in the exhaust gas and releases it into the
combustor inlet compressed air-stream.

This paper deals with the selection, realization and testing of a recuperator
type compact heat exchanger developed by TURBOMECA and G.R.E.Th. (Groupement
pour la Recherche sur les Echangeurs Thermiques) with the support of A.F.M.E.
(Agence Française pour la Maîtrise de l'Energie).

The design features of a metalic plate recuperator are described. Dimensioning
procedure was established from basic heat transfer analysis and aeraulic
characteristics (flow arrangement, pressure drop) were determined from
computer simulation.

An elementary heat exchanger (3 channels) was built and tested and
performances allowed us to demonstrate feasibility. Consequently a full scale
heat exchanger was realized. An extensive laboratory test programme was
undertaken to determine aeraulic and thermal performances which were compared
to those of the elementary prototype.

Transient thermal regime tests were carried out by simulating an entire range
of turbine operations (start up, steady state, shut down).

PRELIMINARY CHOICE

The two basic types of heat exchangers are regenerators or recuperators according to whether heat storage takes place or not in a heat exchanger component. Both types can be classified as dynamic or static heat exchangers. Among the dynamic heat exchangers, rotary regenerators (or thermal wheels) are frequently used with gas turbine exhaust. Using a metal disk allows the heat exchanger to operate successfully up to 800°C [1]. Ceramic disks are under development to withstand higher temperatures (1000°C) [1,2]. For practical purposes a static recuperator was selected. To develop such a recuperator operating conditions of concerned engines must be taken into account :

- material resistance to temperatures of 700°C,
- withstanding pressures of 10 bar,
- lowest volume and weight,
- cyclic study,
- best compromise between effectiveness and pressure drop.

From a technological point of view, static heat exchangers can be classified as tube or plate heat exchangers. Both are used for heat recovery, however plate heat exchangers offer high surface area densities and subsequent low volume requirement. Moreover such compact plate heat exchangers differ from secondary surface heat exchanger in which the surface steering the fluid is not the same as the one where heat transfer takes place, or primary surface heat exchangers, in which these two surfaces are identical. The latter (primary surface) has been selected because it seems better adapted to a smaller volume.

BASIC STRUCTURE

The heat exchanger consists of a stack of corrugated metal plates, pressed together in a frame and arranged in a U-layout (figure 1). Each plate has two central ports whose edges are sealed by welding in order to provide two independent channels for air and for exhaust gases called hereafter "air" channels and "gas" channels. This choice produces counter-current flows leading to maximum effectiveness. Details of construction are given in reference [3] and are briefly recalled.

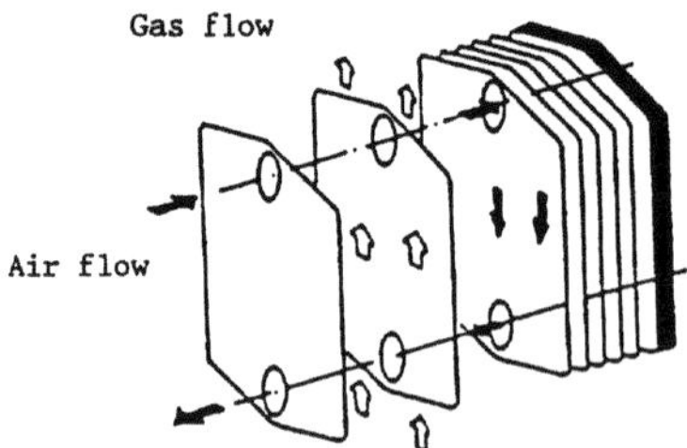

Fig.1   Basic design

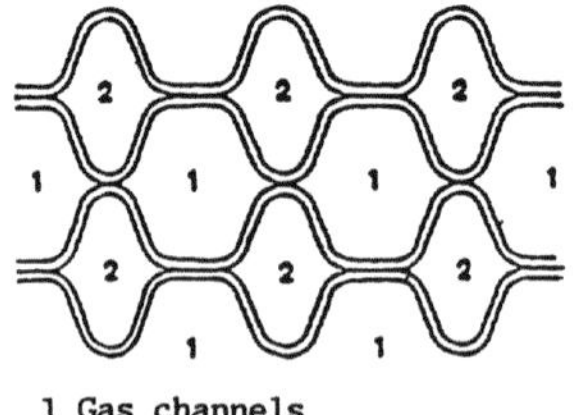

Fig. 2   Air an gas channel cross sections

PLATE DESIGN

The corrugated patterns consist of chevron design allowing the plates to support each other. However due to the strong difference in pressures, the flow passages must have different cross-sections to obtain comparable gas velocities. This is achieved by choosing dissymetric geometry as seen in figure 2.

To ensure an even flow of fluids into air channels a plate is divided into three parts, two distribution surfaces (near the front) and a current part (figures 3 and 4).

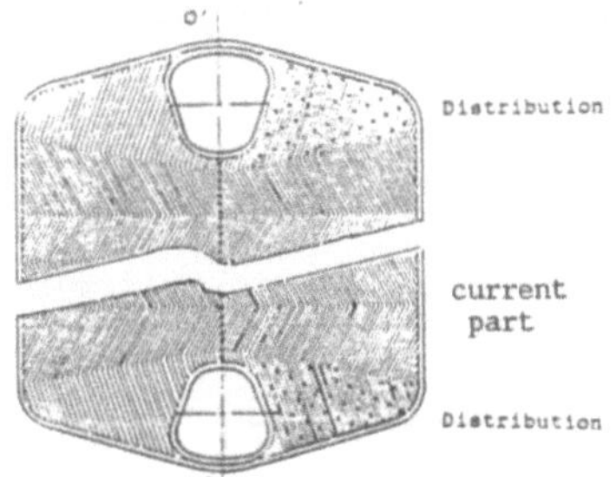

Fig. 3 Plate view with chevron
patterns

Fig. 4 Plate view after stamping
and before laser cutting

PREDIMENSIONING : AERAULIC AND THERMAL SIZING

Requirements

Present conditions related to concerned turbines are :

- mass flow rate : up to 5 kg/s
- maximum gas inlet temperature $T_{gi} = 700°C$
- air inlet temperature $T_{ai} = 340°C$
- heat exchanger effectiveness more than 75 %

In our case effectiveness is defined by :

$$E = \frac{T_{ao} - T_{ai}}{T_{gi} - T_{ai}} \tag{1}$$

T_{ao} being the outlet air temperature

- Total relative pressure drop less than 10 %
  this pressure drop is defined by

$$\frac{\Delta P}{P} = \frac{\Delta P_g}{P_g} + \frac{\Delta P_a}{P_a} \tag{2}$$

where ΔP_a and ΔP_g are the air and gas pressure drop respectively.

- Flow between channels as uniform as possible.

Flow distribution in air-side passages

The study of air and gas flows allows us to determine the form and dimensions of plates. For air flow, this was achieved in two steps :

- by calculating the pressure and the flow rate repartition along the air
  circuit considered as a one-dimensional loop,
- by determining the air streamlines between plates to avoid dead or
  recirculation zones.

The air circuit inside the heat exchanger is modeled by an aeraulic network
which can be represented by a manifold whose headers, made of the successive
plate ports, are connected by air channels between plates (figures 5 and 6).

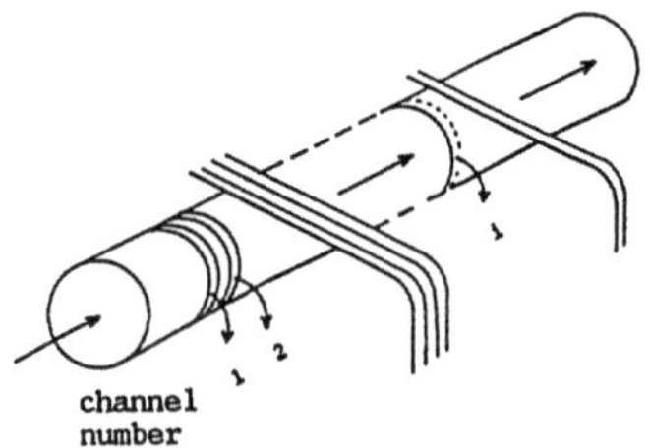

Fig. 5   Analogy between an air
         flow header and a manifold

Fig. 6   One-dimensional air flow
         circuit for flow rate
         distribution and pressure
         drop analysis

One-dimensional mass and momentum conservation equations are solved by the
TRICOT code [4] with the following hypothesis :

- steady state conditions
- temperatures in the headers are inlet and outlet temperatures and between
  the plates are the average between these two temperatures.

To fulfill the requirements, the following parameters have been adapted :
- cross sections of inlet and outlet headers, S_0 and S_i
- friction factors of air channels

The influence of S_0 and S_i can be observed in figure 7 where the ratio q/q_m is
reported as a function of the channel position, q being the calculated air
flow rate and q_m being the mean air flow rate obtained by dividing the total
air flow rate by the number of channels. It is shown that $q = q_m$ when the
cross sections are related as :

$$\frac{S_0}{S_i} = \left(\frac{\rho_{ai}}{\rho_{ao}}\right)^{1/4} \tag{3}$$

where ρ_{ai} and ρ_{a0} are inlet and outlet fluid densities respectively.

Taking temperature values into account, it has been choosen

$$S_0 = 1.23\, S_i \tag{4}$$

In figure 8, results form the computer code TRICOT are shown for heat
exchangers with 360 plates. It is noted that channel flow rates are identical
within 5 % and that pressure drop is $\Delta P_a = 18340$ Pa ($\Delta P_a/P_a = 2.3$ %). Inside
each channel (within the space between two plates), the fluid repartition over

a whole plate is determined from the ARMOR code which solves two-dimensional Navier-Stokes equations with a friction term. Two friction factors Λ_x and Λ_y are defined depending on chevron angles, the total pressure drop over the plate being the same found with the TRICOT code.

Results can be observed in figure 9 where air streamlines and velocity vectors are reported showing a good repartition of flow.

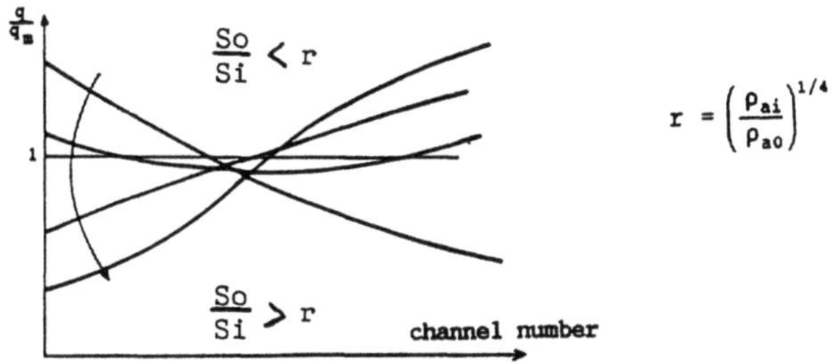

Fig. 7   Effect of inlet and outlet port area
ratio on air flow rate distribution

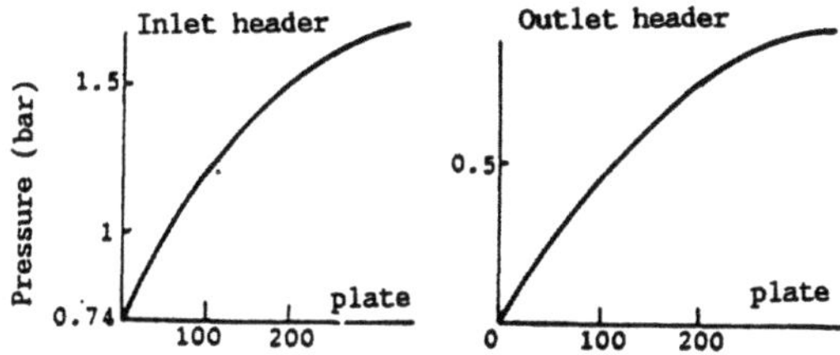

Fig.8   Pressure variations in inlet and outlet headers

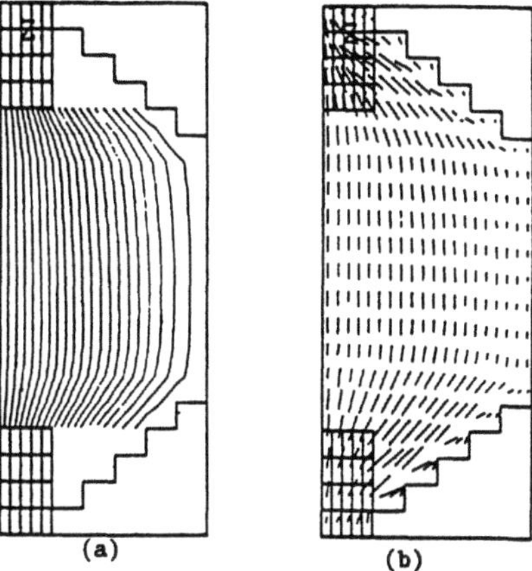

Fig. 9   Air flow distribution over a plate (a) stream lines (b) velocity
vectors

Flow distribution in gas-side passages

Gases enter at a parallel angle to the plates and the distribution between passages is assumed uniform as a first approximation. Pressure drop is calculated from classic correlations used for plate heat exchangers [5]. For a 1200 kW turbine, with maximum gas flow rate, it is obtained $\Delta P_g/P_g = 4$ %.

Thermal sizing

Thermal design has been carried ut by using an LMTD (logarithmic mean temperature difference) method with an iterative process. The hypotheses are those commonly used :

- steady state conditions,
- constant viscosity, thermal conductivity and specific heat of each fluid within the corresponding temperature range,
- constant heat transfer coefficient,
- uniform fluid flow rate repartition through each channel,
- negligible heat loss to the surroundings.

Heat power is calculated from :

$$W = K \cdot A_g \cdot \Delta T_{ml} \tag{5}$$

where ΔT_{ml} is the logarithmic mean temperature difference defined as

$$\Delta T_{ml} = \frac{(T_{gi} - T_{ao}) - (T_{go} - T_{ai})}{\ln \dfrac{T_{gi} - T_{ao}}{T_{go} - T_{ai}}} \tag{6}$$

and where A_g is the gas side surface area.

The overall heat transfer coefficient is calculated with respect to the surface A_g by

$$\frac{1}{K} = \frac{1}{h_g} + \frac{1}{h_a \dfrac{A_a}{A_g}} + r \tag{7}$$

where r is the thermal conduction resistance. The convection coefficients h_g (gas side) and h_a (air side) are given by

$$h_g = \frac{Nu_g \, \lambda_g}{d_g} \quad \text{and} \quad h_a = \frac{Nu_a \, \lambda_a}{d_a} \tag{8}$$

λ_g and λ_a are the thermal conductivities of gas and air respectively, d_g and d_a are the volumetric hydraulic diameters of the gas and air channel defined by

$$d_k = \frac{4 \, V_k}{A} \quad k = a, g \tag{9}$$

V_k being the void volume between plates and A the wetted surface area.

The gas Nusselt number is calculated from a turbulent fluid flow correlation :

$$Nu_g = B \, Re^n \, Pr^{1/3} \tag{10}$$

The B and n coefficients are derived from studies on plate heat exchangers [5] and with the plate patterns considered are choosen as B = 0.074 and n = 0.75. These values will be ajusted with the experimental results.

The flow regime of the air side can be either laminar or turbulent. As a first approximation, the air passages can be considered as cylindrical tubes and classical correlations can be used [3].

When K is determined, effectiveness is deduced by (1).

The obtained results allowed us to determine the heat exchange surface area for turbines with power ranging from 400 kW to 1200 kW. To adapt the heat exchanger to these different powers, a reference plate of area $A_r = 0.18$ m^2 will be used. This plate associated with the chosen geometry leads to a compactness of 1000 m^2/m^3.

<u>Heat exchanger construction</u>

Recuperator design features are shown in figure 10 and a complete module is presented in figure 11. It can be observed that pressure is withstood by two thick cover plates.

Plates of 0.2 mm thickness are made by stamping a corrugated surface pattern on sheet metal (figure 4). Each plate is welded to the next by means of a CO_2 laser beam allowing the deformations to be minimized and leading to high welding velocities (several meters per minute).

The plates are stacked one by one. The external edge is first welded to the preceeding plate then the port edges are welded to the following one. After each welding, the laser beam is shifted 1 mm in order to cut the two welded metal sheets.

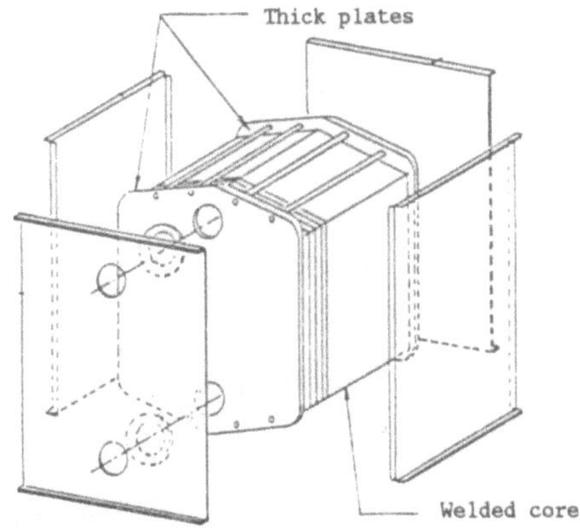

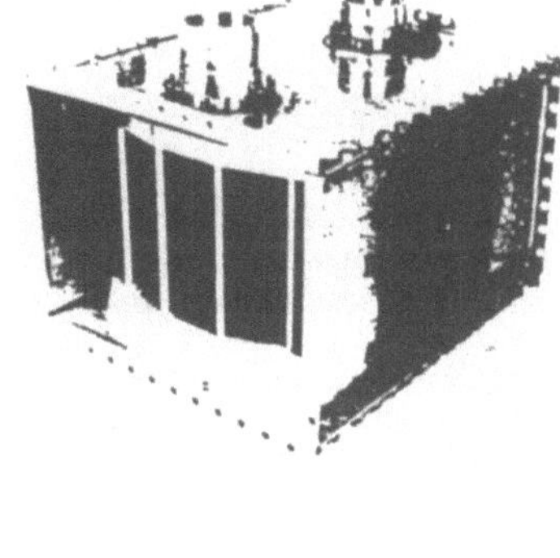

Fig. 10 Heat exchanger parts   **Fig. 11** View of the heat exchanger

Experimental study of an elementary heat exchanger

An elementary heat exchanger (3 channels) with an air channel under pressure (1 to 8 bar) between two heated air channels at atmospheric pressure has been made and mounted on a test loop (figure 12). Channel flow rates are of the same order or magnitude as those of the full scale heat exchanger. However, inlet temperatures are less than those of the real case ($T_{ig} \simeq 70°C$ for gas ant T_{ia} is room temperature for air).

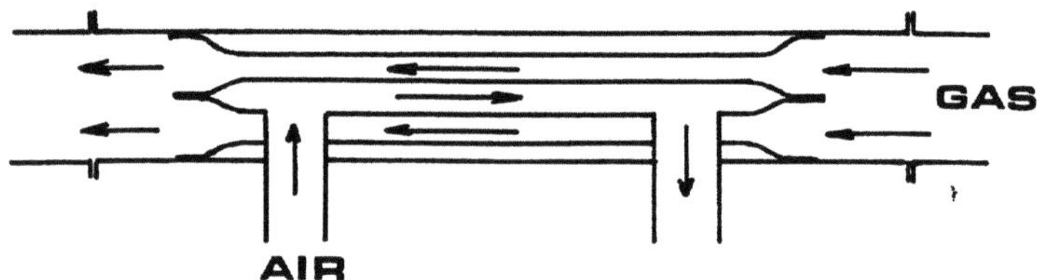

Fig. 12 Schematic drawing of the elementary heat exchanger

Pressure drop has been measured on each side and friction factors Λ have been deduced from :

$$\Delta P = G^2 \left(\frac{1}{\rho_0} - \frac{1}{\rho_i} \right) + \Lambda \frac{L}{d_k} \frac{G^2}{2\rho_m} \tag{11}$$

where G is the mass velocity, d_k the hydraulic diameter, L a characteristic dimension of the channel, ρ_i and ρ_0 are fluid densities at inlet and outlet of the heat exchanger and ρ_m is the average density defined by [6]

$$\frac{1}{\rho_m} = \frac{1}{2} \left(\frac{1}{\rho_i} + \frac{1}{\rho_0} \right) \tag{12}$$

Results are presented in figures 13 and 14 together with those concerning the full scale heat exchanger. Friction factors are given under a reduced form for reasons of confidentiality.

The overall heat transfer coefficient K was determined by measuring inlet and outlet temperatures and fluid flow rates. Results can be seen in figure 15 together with the results of the complete heat exchanger.

Experimental study of a full scale heat exchanger

A heat exchanger of significant size (200 plates, 100 "gas" channels and 100 "air" channels, figure 11) was built and mounted on the GAZTON test loop which is a general facility for gas–gas heat exchangers. The system consists of two circuits, a primary cooling loop with pressurized air flow (1 bar to 8 bar) at room temperature and a secondary heating loop with atmospheric air flow heated by means of a gas burner. In our case the temperature of the atmospheric air flow was limited to 150°C. Mass flow rates were varied from 0.1 to 0.7 kg/s.

Pressure drop analysis

Friction factors are given in figure 13 (air side) and in figure 14 (gas side). It can be noticed that experimental points of the elementary heat exchanger line up with those of the complete heat exchanger for the gas passages and do not for the air passages. This is explained by the fact that the total pressure drop results from the pressure drops in the air channels plus the pressure drop in the inlet and outlet headers. Such headers do not exist in the elementary heat exchanger.

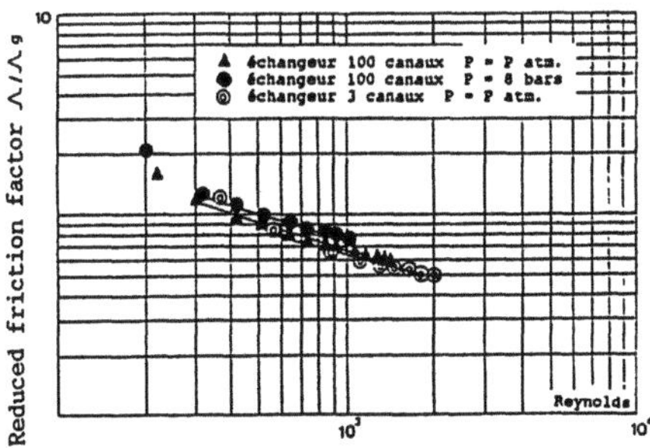

Fig. 13 Gas side friction factor as a function of Reynolds number

Fig. 14 Air side friction factor as a function of Reynolds number

Heat transfer analysis

Heat transfer coefficients have been determined by measuring heat flow rate, inlet and outlet temperatures and by using formula (5). In figure 15, are reported experimental values of the overall heat transfer coefficient K for the elementary and the full scale heat exchangers. Moreover, in the same figure are shown theoretical curves calculated from the formula (7) and the following correlations :

- air flow
$$Nu_a/Nu_{a0} = 0.01 \ Re^{0.8} \ Pr^{1/3} \tag{13}$$

- gas flow
$$Nu_g/Nu_{g0} = 0.01 \ Re^{0.74} \ Pr^{1/3} \tag{14}$$

Nu_{a0} and N_{g0} being coefficients specific to the given geometry. Compared to the preliminary thermal design, the obtained correlation is almost the same in the air case but it is rather different in the gas case. This is attributed to maldistribution effects which were not taken into account in preliminary dimensioning. This is clearly shown in figure 15 where the theoretical curve underestimates the experimental values found for the elementary heat exchanger where maldistribution effects of the gas flow are negligible.

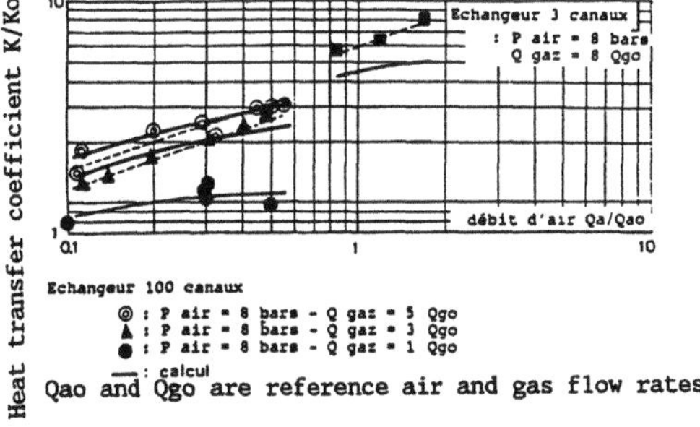

Fig. 15   Overall heat transfer coefficient as a function of air flowrate

Transient thermal regime tests

Resistance to thermal cycling is a major concern for high temperature heat exchangers. Tests were carried out by simulating as entire range of turbine operation (start up, steady state, shut down). Several series of 40 cycles of 30 mn duration were realized with the following conditions :

- Cycles with constant gas and air flow rates during which burners heating the gas side fluid were started up and stopped. Temperature variations are given in figure 16.
- Cycles with constant gas flow rates, burners working, during which air compressor was started up and stopped. Temperature variation are given in figure 17.

Ten cycles per day were carried out and pressure tests were realized every morning at room temperature. No significant leak was detected.

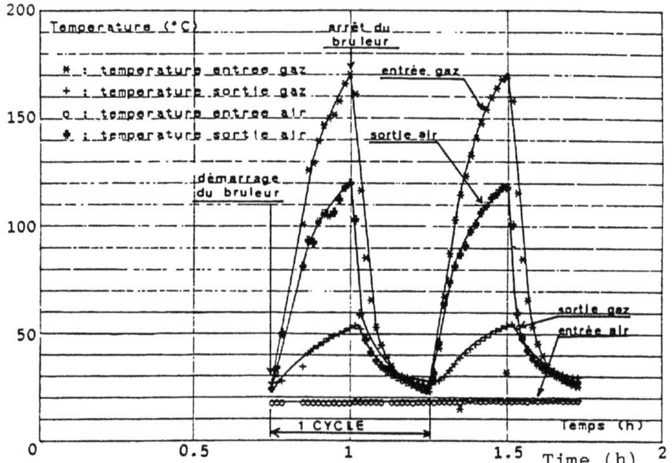

Fig. 16 Thermal cycling by starting up and stopping burner

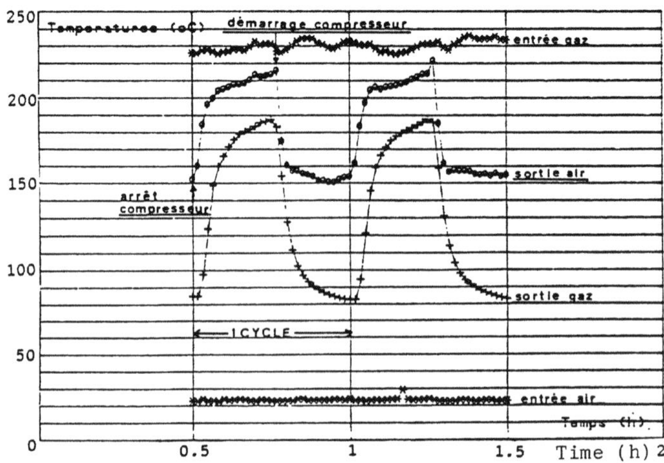

Fig. 17 Thermal cycling by starting up and stopping air compressor

Heat exchanger fouling tests

Fouling control and removal of deposits are of particular concern for transportation applications. Deposits can significantly deteriorate heat exchanger efficiency both through reduction in heat transfer and blockage of gas passages.

Exhaust gases from a diesel engine were used as fouling fluid in the elementary heat exchanger (3 channels). Tests showed that the channels were never destroyed and could easily be cleaned.

CONCLUSION

Designing, construction and tests of a heat exchanger for gas turbine heat recovery have been carried out. Technological characteristics of the proposed heat exchanger are :
- welded plate recuperator
- primary surface
- compactness of 1000 m^2/m^3
Following tests were carried out :
- feasibility tests on an elementary heat exchanger
- thermal and aeraulic performance measurements on a full scale heat exchanger
- reliability tests (duty cycling and fouling)

Overall heat transfer coefficients were deduced and the experimental effectivenss was found to be about 75 %. Modifications to improve this value are in progress. Use of this heat exchanger on TURBOMECA turbines would lead to an increase in efficiency of up to 20 %.

REFERENCES

[1]   SHAH, R.K. :
      Advances in compact heat exchanger technology and design theory.
      Heat Transfer 1982, Vol. 1, 123/142, Hemisphere Publishing Corporation, Washington (1982).

[2]   RAHNKE, C.J.R. :
      Structural design of ceramic rotary heat exchangers in compact heat exchangers, H.I.D. Vol. 10. Shah, R.K., Mac Donald, C.F. and Howard, C.P. Editors, ASME, New York (1980).

[3]   BONTEMPS, A., LAURO, F., VIDIL, R., PERBOS, A. and BRUN, M. :
      Echangeur de chaleur compact pour la récupération d'énergie sur les turbines à gaz. Definition, realisation et essais. Revue Génerale de Thermique, 348 (1990).

[4]   BONTEMPS, A., MERCIER, P., SOLECKI, J.C., TREILLE, P. and VIDIL, R. :
      Etude de la distribution de débit dans un échangeur à plaques et joints. Présentation et validation du logigiel TRICOT.
      GRETh Technical Report 87-119 (1987) (Unpublished).

[5]   KUMAR, H. :
      The plate heat exchanger : construction and design 1st U.K. heat exchanger symposium, Leeds (1984).

[6]   KAYS, W.M. and LONDON, A.L. :
      Compact heat exchangers 2nd edition, Mac Graw Hill, New York (1984).

High Performance Titanium Plate Fin Heat Exchanger Using a Novel Manufacturing Process

C I ADDERLEY
Rolls-Royce and Associates Limited

J O FOWLER
Rolls-Royce plc

INTRODUCTION

There is increasing interest in compact forms of heat exchange
equipment. In the offshore industry, for example, there is an
obvious incentive to reduce both the weight and the volume of
topsides plant. Onshore, weight is generally less important but
volume is often the main factor in installation costs. The plate
fin heat exchanger (PFHE) is a compact form which offers
significant advantages over shell and tube exchangers in terms of
weight, volume and thermal effectiveness. It also has a
multistream capability; a single unit can therefore replace
several shell and tube exchangers adding further to its
attractiveness.

Although PFHEs have been in use for about 50 years they have been
confined mainly to specialist applications, such as cryogenics
and transport, rather than general process use. There are
several reasons for this reticence on the part of plant designers
- some of them unrelated to the equipment capabilities. What is
clear, however, is that PFHEs can only replace shell and tube
exchangers in applications where they can be operated reliably at
the same pressures and temperatures.

Current PFHE technology is centred on brazed aluminium units.
This choice of material limits the operating temperature to a
maximum of about 50°C and the method of manufacture constrains
pressures to 80 - 90 bar. Higher temperatures can be
accommodated by the use of other materials (such as stainless
steels, titanium and nickel alloys) but the pressure constraint

still applies. Moreover, the tooling costs for forming the finned sections in the conventional manufacturing process are generally higher for these alternative materials.

The superplastic forming/diffusion bonding (SPF/DB) manufacturing process overcomes the limitations imposed by the conventional PFHE manufacturing route and allows greater flexibility in the design of PFHEs. The titanium SPF/DB process developed by Rolls-Royce for the cost-effective manufacture of critical aero-engine components is now being applied to the manufacture of PFHEs. Using this process the primary and secondary heat transfer surfaces are joined by solid state diffusion bonds. This results in a high integrity exchanger which can operate at pressures in excess of 200 bar and at temperatures up to 300°C. The SPF/DB manufacturing process, together with the excellent corrosion resistance of titanium, therefore greatly extends the range of application of the PFHE. This paper describes the SPF/DB process and its application to two designs of PFHE currently under development.

THE SPF/DB PROCESS

Superplastic forming and diffusion bonding are well known metallurgical phenomena, their development into cost effective manufacturing processes however is a relatively recent advance.

A titanium alloy SPF/DB process has been developed by Rolls-Royce plc primarily for the manufacture of high integrity aero-engine parts. The ability of the process to produce complex hollow geometries, associated with verifiable high integrity solid state joints in a reliable manner at high volume throughputs suggested that the basic technology might find further application in the field of high performance heat exchangers. Rolls-Royce has initiated a development programme based on the principles of simultaneous engineering in order to develop the technologies for such applications.

Superplasticity is a deformation phenomenon which allows some materials to strain to high elongation without the initiation of tensile instability or necking. This enables the generation of high volume fractions of hollowness in a heat exchanger matrix allowing designs of good mechanical and thermal performance, together with low weight and high material yield.

Above about 800°C Ti alloys have the ability to absorb their own oxides, so, provided clean metal surfaces are protected from surface contamination by the provision of a suitable joint face environment, and sufficient pressure is applied to the mating surfaces then solid state diffusion bonding takes place, to the extent that no interface can be detected. No macroscopic deformation takes place during bonding and therefore shape and size stability is maintained during the operation. Solid state diffusion bonding produces a joint with parent metal properties without the presence of a heat affected zone or other material such as flux or bond promoter. Its use within a heat exchanger therefore eliminates the possibility of chemical interaction with the process fluids.

In the manufacturing process titanium sheet of near net shape and controlled surface finish is cleaned to a high standard and a bond inhibitor is deposited onto the joint faces. The deposit specifies the ultimate internal configuration of the heat exchanger, including the inlet zone, distributors and the secondary surfaces.

Although the internal geometry is fixed at this stage, the deposition process allows considerable flexibility of design, to satisfy both mechanical and thermal requirements. Heat exchanger panels are then bonded by solid state diffusion. This results a high integrity joint of known and reproducible properties, which allows the construction of a high performance yet weight efficient element, taking full advantage of the mechanical properties of titanium.

The diffusion bonded sandwich is then superplastically formed in a closed die as shown schematically in Figure 1. The element is finally ironed against the die form at high pressure. This results in the production of smooth flat surfaces which ensures good conformance of each element to its neighbours in the heat exchanger matrix.

After Bonding

After SPF

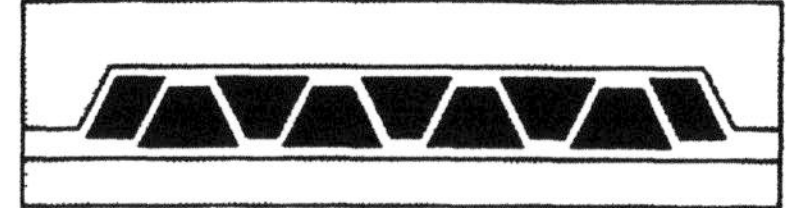

After Ironing

FIGURE 1: The SPF Process

For its aero-engine business, Rolls-Royce has developed this manufacturing process specifically for titanium alloys. Superplasticity is exhibited by a number of commercially significant alloys and the method is currently being developed to allow its use with both stainless steel and nickel alloys.

<u>THE PFHE ELEMENT</u>
The basic element of the Rolls-Royce titanium PFHE consists of two parting sheets separated by the secondary heat transfer surface (ie the fins) formed from three sheets of titanium by the SPF/DB process described above. The secondary surface is joined to the parting sheets by solid state diffusion bonds. The flexibility of the SPF/DB process allows a wide range of secondary surface geometries to be formed.

Three examples of relatively simple geometries are shown in
Figure 2. These are straight corrugations and a uniform pitch
dot core. The internal form of the dot core is shown with the
parting sheets removed in Figure 3. Conventional finning
arrangements such as herringbone, serrated and perforated can be
easily produced.

FIGURE 2: Examples of Element Internal Geometries

FIGURE 3:   Dot Core Geometry

A further degree of flexibility exists in the ability to vary the
secondary surface geometry within the element. This allows flow
distributors at each end of the element to be formed integrally
with the secondary heat transfer surface in a single SPF/DB
operation. Furthermore the distributor geometry may be tailored
to achieve a uniform flow distribution across the element. In
the two designs of exchanger currently under development, simple
straight corrugations are used for the fins with a dot core
geometry for the distributors. The transition between these
geometries is shown in Figure 4.

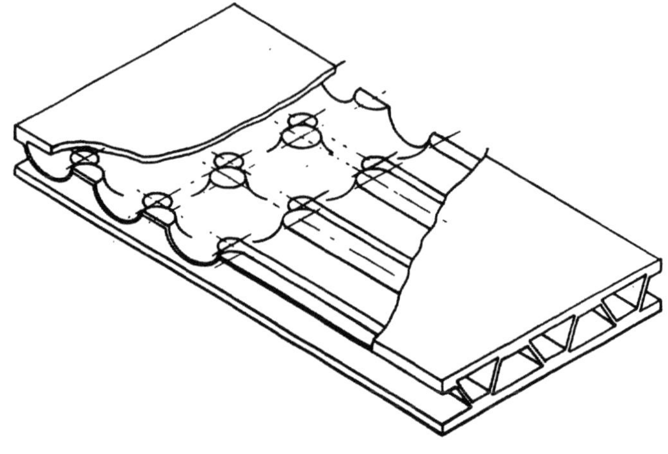

FIGURE 4:   Distributor - Fin Transition

The dimensions of the internal geometry (ie, fin pitch, etc) are
constrained only by requirements to meet the pressure duty of the
element.   Here the reduced thickness of the secondary surface and
the unsupported area of the parting sheets are of primary concern
since the solid state diffusion bonds exhibit parent metal
properties.   Passage height may be chosen to meet the thermal-
hydraulic requirements of the exchanger.   The exchangers
currently under development have a passage height of 2 mm and a
fin pitch of 8 mm.   This gives a considerable margin on the
operating pressure of 200 bar.

In conventional PFHEs, bars are brazed to the parting sheets to
seal the gap at the edges of an element.   In the Rolls-Royce
titanium PFHE, separate edge bars are not required since the
three sheets of titanium are diffusion bonded together at the
edges to form a seal.

THE PFHE MATRIX

The PFHE matrix is constructed as an assembly of individual elements. The form of this assembly depends on the process requirements. The PFHEs currently under development are intended for offshore application as seawater/methane compressor intercoolers. Typical operating parameters for this duty are given in Table 1. For this two stream liquid/gas duty we have considered two matrix designs: a jacketed matrix and a bonded matrix.

Jacketed Matrix

In this design the SPF/DB elements are used only for the high pressure methane stream. The elements are separated and held in position as a bundle by toothed tie-bars. The seawater flows through passages between elements. The bundle is then inserted into a steel jacket. Figure 5 shows the element design and arrangement and Figure 6 shows the jacketed bundle.

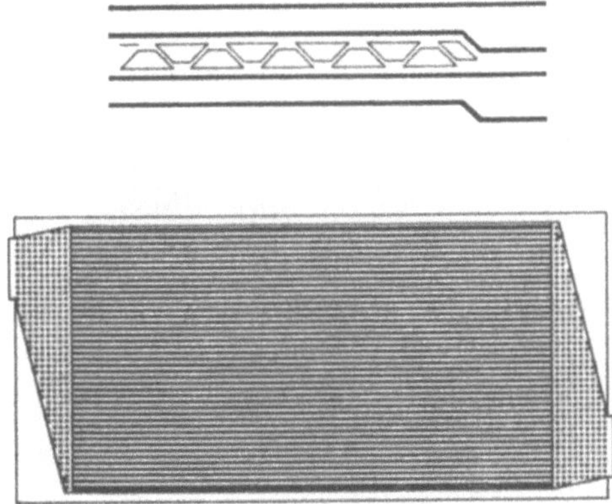

FIGURE 5:   The Jacketed Matrix Element

The gas header is formed by inserting the open ends of elements into slots in a flat plate to which a tank header is welded as shown in Figure 7. The floating head arrangement for the gas stream allows removal of the bundle from the jacket.

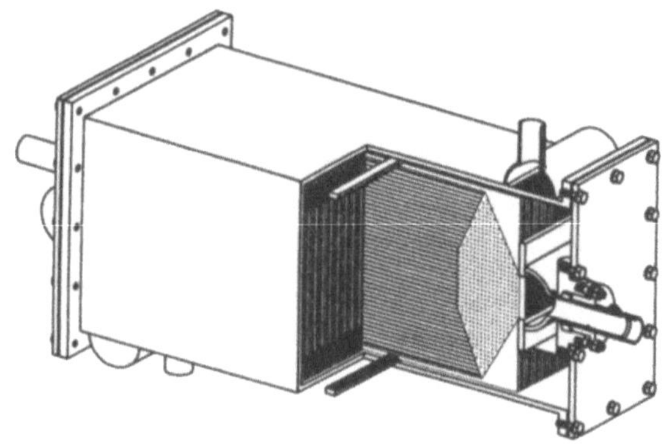

FIGURE 6:   The Jacketed Matrix Exchanger

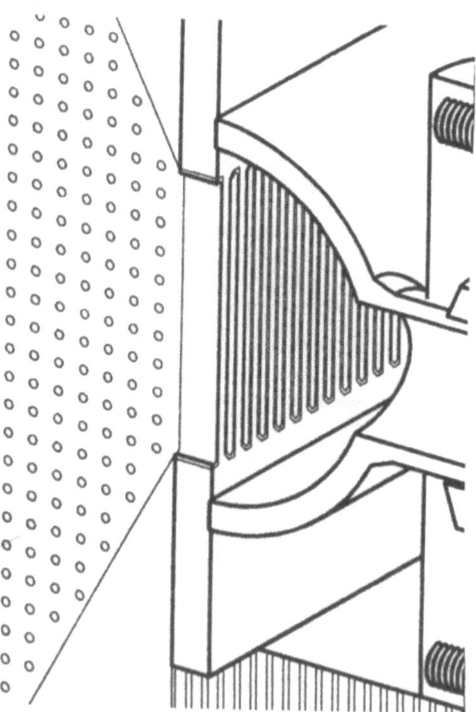

FIGURE 7:   Gas Header Detail

In order to achieve the required seawater flow distribution on
the 'shell' side of the matrix surface features can be provided
in the seawater passages on the outside of the element parting
sheets. These may be formed during the SPF phase of the element
manufacture by corresponding shapes on the SPF dies.
Alternatively, chemical etching may be used or baffles may be
welded to the parting sheets.

The advantages of this design are:

- the ability to remove the matrix for cleaning of the seawater
  passages where biological fouling may occur.
- reduced cost compared with an exchanger manufactured wholly
  from titanium.

For the seawater/methane duty, the main parameters of this matrix
design are given in Table 2.

Bonded Matrix

In this design both the process streams flow through the SPF/DB
elements. The parting sheets of adjacent elements are joined
together using an activated diffusion bond to form the matrix.
Manifold-type headers are provided by holes which pass through
each element in the matrix. Figure 8 shows an element for the
two stream methane/seawater application.

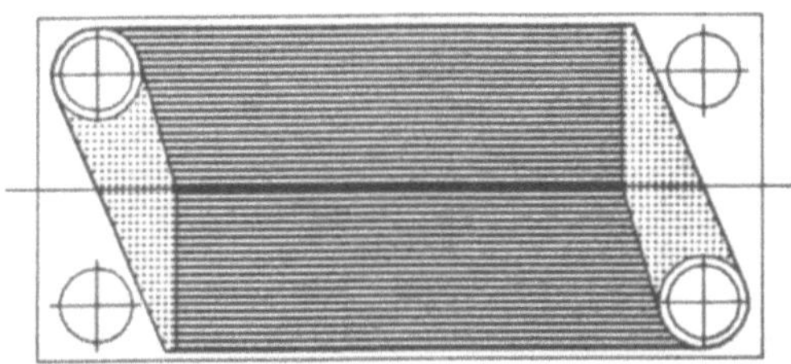

FIGURE 8: Bonded Matrix Element

The internal cavity formed during the SPF process is shaped so that header holes for the stream which does not enter the element are drilled through the solid metal formed by diffusion bonding of the three titanium sheets. The end elements of the matrix are manufactured with a thicker parting sheet on one side to form cap sheets to which nozzles, supports, etc may be welded. Figure 9 shows a cut-away view of this exchanger.

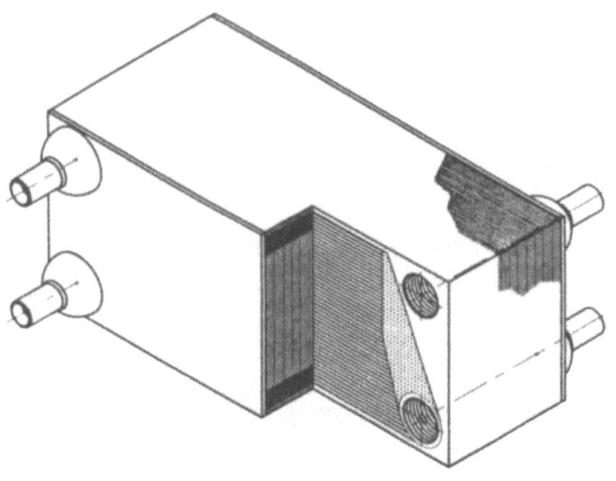

FIGURE 9:   The Bonded Matrix Exchanger

The activated diffusion bond between the elements holds them together in good thermal contact. Exposure of the activator to the process fluids at the joints between elements within the headers is prevented by autogeneous seal welds.

The main parameters of this matrix design for the seawater/ methane duty are given in Table 3.

Unlike the jacketed matrix, this design can easily be adapted for multistream operation by allowing for more header holes in the elements during the SPF/DB process. Furthermore, the elements may be stacked for cross flow heat exchange and the secondary surface geometry may be tailored to each process stream.

CONCLUDING REMARKS

Rolls-Royce has developed a unique titanium forming technology
capable of producing high performance critical aero parts with
the level of product assurance necessary for flight. This
process is now being applied directly to the manufacture of high
integrity PFHEs which can be operated at pressures and
temperatures greatly in excess of the capabilities of
conventional PFHE technology. This allows PFHEs to be used for
process duties which are currently the preserve of shell and tube
exchangers with consequent savings of both weight and space.
Further benefits may be realised due to the multistream
capability of the PFHE which enables a single unit to perform the
duties of two or more shell and tube exchangers. An on-going
development programme will extend the manufacturing process to
stainless steel and nickel alloys thereby increasing the range of
application of the PFHE further still.

TABLE 1.   Seawater/Methane Application Operating Conditions

Medium		Primary Methane	Secondary Seawater
Flowrate	Kg/sec	6	60
Pressure	bar	200	20
Inlet Temperature	°C	150	10 - 30
Pressure Drop Limit	bar	0.6	0.7

TABLE 2:   Jacketed Matrix Parameters

Number of elements		50
Gas passage height	mm	2
Fin pitch	mm	8
Water passage height	mm	2
Overall dimensions	mm	1,000 x 500 x 500
Thermal duty	MW	1.5 - 2.0
Exchanger dry weight	Kg	360
Exchanger wet weight	Kg	420

TABLE 3:   Bonded Matrix Parameters

Number of elements		100
Passage of height	mm	2
Fin pitch	mm	8
Overall dimensions	mm	800 x 400 x 320
Thermal duty	MW	2.0
Exchanger dry weight	Kg	370
Exchanger wet weight	Kg	430

Regenerators

Optimal Thermal Control of Regenerative Heat Exchangers

E. Van den Bulck

Department of Mechanical Engineering
Katholieke Universiteit Leuven
Celestijnenlaan 300A, B-3001 Heverlee BELGIUM

Summary

The performance of regenerative heat exchangers with active control of the local heat transfer coefficient h or the local transfer surface area a is reported. It is found that within each regenerator pass (ha) should be enhanced in a narrow zone which shifts from the inlet face of the regenerator at the start of the regeneration period to the exit face at the end of the regeneration period. Outside this zone, (ha) may be small so as to limit the pressure drop of the fluid flow. When applying this control measure, the effectiveness of the regenerator approaches the counterflow exchanger effectiveness for rotational speeds which are only half of those conventionally used. Carryover losses and seal wear may thus be significantly reduced.

1 Introduction

Regenerative heat exchangers — configured either as an arrangement of parallel packed beds or as rotary regenerators — are widely used in the air-conditioning and process industries where they can be found in a variety of gas-to-gas heat recovery applications. The design of regenerative heat exchanger cores follows the conventional design procedures for compact heat exchangers, and involves computation of the core dimensions to yield a prescribed heat exchanger efficiency and independently prescribed pressure drops for both fluid streams. The design procedures for sizing compact heat exchangers are well established [1,2]. The dimensions of the heat exchanger are determined by the geometric characteristics of the selected flow passages. High performance regenerators often employ laminar-flow narrow passages in either a parallel-plate or a honeycomb arrangement. These passages achieve high heat transfer rates for moderate pressure drops.

Although the thermal design procedures of regenerator cores may be similar to those of recuperative heat exchangers, it is well known that the heat exchange in regenerators follows a dynamic wave type behavior which is very different from the stationary heat exchange in recuperators. At any moment, the heat exchange between the fluid stream and the regenerator core is primarily confined to a narrow zone which moves through the

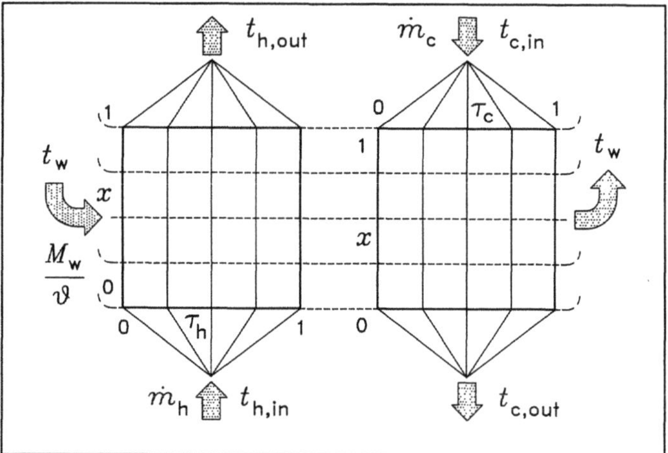

Fig. 1   Schematic representation of regenerative heat exchangers.

exchanger core in the direction of the fluid flow at a low velocity. Outside of this narrow zone the temperatures of the exchanger core and the fluid stream are almost equal, and the heat exchange is negligible. Nonetheless, the thermal design of regenerator cores in general does not take into account the dynamics of heat transfer in regenerators.

In order to optimize the regenerator performance, the flow passages are constructed so as to maximize the heat transfer coefficient with the constraint of limited pressure drops. This optimization leads to flow passages with uniform dimensional characteristics in the axial flow direction. For such designs, increases in heat transfer rates are usually obtained at the expense of substantially increased pressure drops. If the regenerator core were to be constructed so as to maximize the heat transfer coefficient only in the regions where the heat transfer is taking place, the thermal performance of the regenerator may be improved without a corresponding increase in pressure drop. With heat transfer regions that are moving with time, maximizing the heat transfer rate in these regions leads to optimal dynamic control of the transfer rate characteristics of the regenerator core. Such a study is presented in this paper.

2   Performance objective

Figure 1 illustrates the flow configuration of rotary or valved types of regenerative heat exchangers. Fluid stream h flows in the x-direction with a thermal capacity rate C_h. Fluid stream c flows in the opposite direction and has a thermal capacity rate C_c. The regenerator matrix *'flows'* from left to right with a thermal capacity rate C_r. The matrix material passes sequentially through each of the regenerator periods and is then recirculated. All three flows are unmixed. The length of the heat exchanger in both flow directions and the time period of each regenerator pass are scaled to 1. Figure 1 also shows the remarkable similarity between regenerators and crossflow heat exchangers arranged as liquid-coupled indirect-transfer-type exchanger systems.

The performance of the regenerator is modeled following the conventional assumptions which are detailed in [3]. The resulting system of conservation and transfer-rate equations reads as follows:

$$\left.\begin{array}{ll}\dfrac{\partial t_h}{\partial x} + C^*_{r,h}\dfrac{\partial t_w}{\partial \tau_h} = 0 & \dfrac{\partial t_c}{\partial x} + C^*_{r,c}\dfrac{\partial t_w}{\partial \tau_c} = 0 \\[2mm] \dfrac{\partial t_h}{\partial x} = \dfrac{(ha)_h}{C_h}(t_w - t_h) & \dfrac{\partial t_c}{\partial x} = \dfrac{(ha)_c}{C_c}(t_c - t_w)\end{array}\right\} \quad \text{for } 0 \leq x, \tau_h, \tau_c \leq 1, \quad (1)$$

where the thermal capacity rate ratios are defined as

$$C^*_{r,h} = \dfrac{C_r}{C_h}, \quad C^*_{r,c} = \dfrac{C_r}{C_c}, \quad \text{with} \quad C_h = (\dot{m}c_p)_h, \quad C_c = (\dot{m}c_p)_c, \quad C_r = \dfrac{(Mc_p)_w}{\theta}, \quad (2)$$

and θ the period of rotation. The periodic boundary conditions can be formulated as

$$\left.\begin{array}{l}t_w(x, \tau_h = 1) = t_w(x, \tau_c = 0) \\[1mm] t_w(x, \tau_h = 0) = t_w(x, \tau_c = 1)\end{array}\right\} \quad \text{for } 0 \leq x \leq 1. \quad (3)$$

and with the inlet conditions specified as

$$\begin{array}{ll}\text{at } x = 0: & t_h(\tau_h) = 0, \quad \text{for } 0 \leq \tau_h \leq 1, \\[1mm] \text{at } x = 1: & t_c(\tau_c) = 1, \quad \text{for } 0 \leq \tau_c \leq 1,\end{array} \quad (4)$$

the effectiveness of the regenerator is given by

$$\epsilon_r = \dfrac{C_h}{C_{min}} \int_0^1 t_h(\tau_h)|_{x=1}\, d\tau_h = \dfrac{C_c}{C_{min}} \int_0^1 (1 - t_c(\tau_c)|_{x=0})\, d\tau_c \quad (5)$$

where $C_{min} = \min\{C_h, C_c\}$.

The Numbers of Transfer Units are defined separately for each period

$$N_{TU,h} = \dfrac{(ha)_h}{C_h} \quad \text{and} \quad N_{TU,c} = \dfrac{(ha)_c}{C_c} \quad (6)$$

where h is the local heat transfer coefficient and a is the local transfer area density such that for each period

$$A = \int_0^1 \int_0^1 a\, d\tau\, dx \quad \text{and} \quad (\bar{h}A) = \int_0^1 \int_0^1 (ha)\, d\tau\, dx \quad (7)$$

where A is the total transfer surface area. In the following discussion both a and h, and thus $N_{TU,h}$ and $N_{TU,c}$ may vary with time and axial flow position.

The *objective* of the design analysis can now be formulated as follows. Find the optimal distributions $(ha)_h$ and $(ha)_c$ so as to maximize the heat exchanger effectiveness

$$(\epsilon_r)_{max} = \max_{(ha)_h, (ha)_c} \epsilon_r(N_{TU,h}(x, \tau_h), N_{TU,c}(x, \tau_c), C^*_{r,h}, C^*_{r,c}) \quad (8)$$

subject to the constraint that $(\bar{h}A)_h$ and $(\bar{h}A)_c$ are finite. Maximizing the exchanger effectiveness by dynamically actuating the local transfer capacity (ha) is referred to as *optimal thermal control*.

298

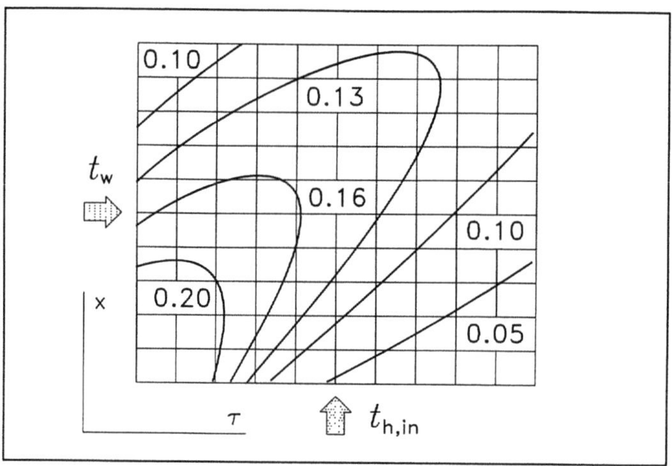

Fig. 2   Contours of constant temperature difference between the hot fluid stream and the matrix core for the inlet conditions as specified in equation (4). $C^*_{r,h} = C^*_{r,c} = 1$, $N_{TU,h} = N_{TU,c} = 5$, and $(ha)_h$ and $(ha)_c$ are constant.

The optimal distributions of $(ha)_h$ and $(ha)_c$ are suggested in figure 2. This figure shows contours of constant temperature difference between the hot fluid stream and the regenerator matrix for the conditions as specified in the figure caption. This figure indicates that the heat exchange in regenerators follows a dynamic wave type behavior which is very different from the stationary heat exchange in recuperators. At any moment, the heat exchange between the fluid stream and the regenerator core is primarily confined to a narrow zone which moves through the exchanger core in the direction of the fluid flow at a low velocity. This moving zone is called the 'thermal wave'. Temperature differences between the fluid stream and the matrix are appreciable only within this wave. Outside of the thermal wave the temperatures of the exchanger core and the fluid stream are almost equal, and the heat transfer is negligible.

It may be worthwhile to promote the heat transfer in the zones of high temperature difference while saving on (ha) in the areas of low temperature difference. Such an approach may lead to savings in material costs and, more significantly, to savings in operating costs associated with pressure drops of flows through the regenerator matrices.

3   Analysis I.

The analysis presented in appendix A shows that the effectiveness of the heat exchange in a regenerator matrix, with a step change of the fluid inlet temperature and a uniform initial temperature of the matrix core, increases for a constant $(\bar{h}A)$ if the local heat transfer capacity (ha) is actuated as indicated in figure 3. The heat transfer is promoted in a narrow zone which shifts along the trajectory $x = \tau$ such that there is negligible heat transfer outside of this zone. The effectiveness reaches a maximum for a

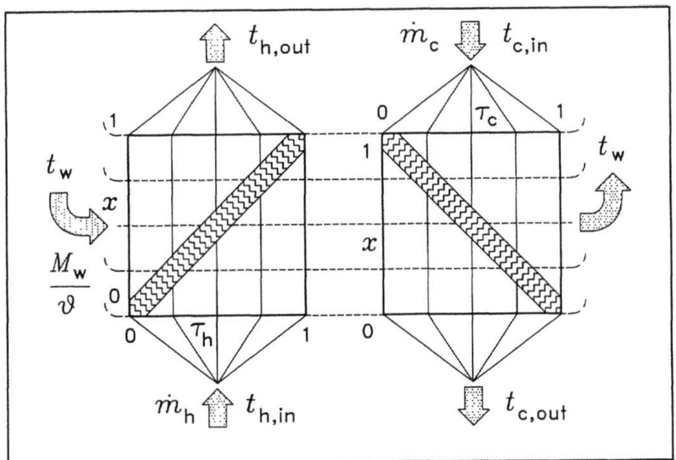

Fig. 3 Optimal actuation of the heat transfer rate (ha) in a regenerator pass. The heat transfer is promoted in a narrow zone which shifts along the trajectory $x = \tau$.

Dirac δ-distribution of (ha) aligned along the 'diagonal' of each regenerator pass. The effectiveness is then equal to the counterflow heat exchanger effectiveness with the same capacitance-rate ratio and the same average ($\bar{h}A$). The outlet temperature of the fluid stream is constant with time and the matrix temperature at the end of the regeneration period is uniformly distributed with the axial flow position. For each pass holds

$$\epsilon_{f,\delta} = \frac{t_{f,out} - t_{f,in}}{t_{w,initial} - t_{f,in}} = \epsilon_{cf}(N_{TU}, 1/C_r^*) \quad \text{and} \quad t_{w,initial} - t_{w,final} = \frac{t_{f,out} - t_{f,in}}{C_r^*}. \tag{9}$$

For the two-pass system illustrated in figure 1, the regenerator effectiveness with two δ-distributions of (ha) in each pass can then be expressed as

$$\epsilon_{r,\delta} = \frac{\dfrac{\epsilon_{h,\delta}}{C_{r,h}^*} \dfrac{\epsilon_{c,\delta}}{C_{r,c}^*} C_{r,min}^*}{1 - \left(1 - \dfrac{\epsilon_{h,\delta}}{C_{r,h}^*}\right)\left(1 - \dfrac{\epsilon_{c,\delta}}{C_{r,c}^*}\right)} \quad \text{where} \quad \begin{cases} \epsilon_{h,\delta} = \epsilon_{cf}(N_{TU,h}, 1/C_{r,h}^*) \\[2mm] \epsilon_{c,\delta} = \epsilon_{cf}(N_{TU,c}, 1/C_{r,c}^*) \\[2mm] C_{r,min}^* = C_r/C_{min} \end{cases} \tag{10}$$

This expression is similar to the effectiveness of liquid-coupled counterflow heat exchangers. It is shown in [4] that the effectiveness of crossflow heat echangers, with δ-distributions of (ha) aligned along the diagonal, equals the counterflow heat exchanger effectiveness.

The effectiveness $\epsilon_{r,\delta}$ reaches a maximum with respect to $C_{r,min}^*$,

$$\text{at} \quad \left(\frac{C_r}{C_{min}}\right)_{opt} = \frac{2}{1 + C^*} : \quad (\epsilon_{r,\delta})_{max} = \epsilon_{cf}(N_{TU,0}, C^*) \tag{11}$$

where

$$C^* = \frac{C_{min}}{C_{max}} \quad \text{and} \quad N_{TU,0} = \frac{1}{C_{min}}\left[\frac{1}{1/(hA)_h + 1/(hA)_c}\right]. \tag{12}$$

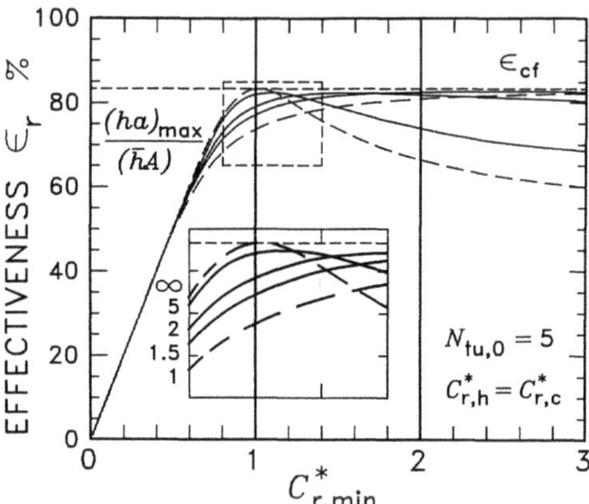

Fig. 4 Regenerator effectiveness as a function of $C_{r,min}^*$ with $(ha)_{max}/(\bar{h}A)$ as parameter.

Thus, the effectiveness of regenerators with optimal actuation of (ha) approaches the counterflow exchanger effectiveness and there exists an optimum operating condition for the regeneration period or rotational speed.

4 Analysis II and Discussion

In practice, the (ha) δ-distributions can be approximated as gaussian distributions with finite width and amplitude. As the width of these (ha)-distributions vary, their peak values vary accordingly such that the average $(\bar{h}A)$ remains constant for each of the regenerator passes. To study the effect of gaussian (ha)-distributions with finite width, the effectiveness of the regenerator is computed from equation (5) using a numerical solution of the conservation and transfer rate equations (1) with the inlet conditions and periodic boundary conditions as specified by equation (4-3). The numerical method is a central-central explicit finite difference scheme of second order, with Richardson's extrapolation to zero grid size. The trapezium rule is used for the integration of the outlet temperature in equation (5).

Figure 4 shows the regenerator effectiveness as a function of $C_{r,min}^*$ with the ratio $(ha)_{max}/(\bar{h}A)$ as parameter. High values of this parameter reflect narrow distributions as more of the heat transfer capacity is packed along the optimal control trajectory. The effectiveness curve corresponding to a ratio of 1 corresponds to the curve of a conventional regenerator with a constant (ha) in each period. The effectiveness of these regenerators increases monotonously with $C_{r,min}^*$ and reaches asymptotically the counterflow effectiveness for $C_{r,min}^* \to \infty$. Regenerators are generally operated at values of $C_{r,min}^* \approx 3 \dots 5$. This value is a compromise between high regenerator effectivenesses at high-speed operation and small carryover losses and reduced seal wear at low-speed operation. $C_{r,min}^*$ is proportional with the rotational speed of the regenerator.

Figure 4 shows that the effectiveness curves exhibit a maximum with respect to $C^*_{r,min}$ as $(ha)_{max}/(\bar{h}A)$ is increased by thermal control. The optimum condition of operation is the value of $C^*_{r,min}$ which yields a maximum regenerator effectiveness. Figure 4 indicates that the optimum condition of operation shifts from a value of infinity for an uncontrolled regenerator to a value of 1 for an optimally controlled regenerator. Applying thermal control to the regenerator allows the regenerator to be operated at lower rotational speeds than conventional, thereby reducing the carryover losses and the seal wear without the otherwise corresponding drop in effectiveness. With $(ha)_{max}/(\bar{h}A) = 2$, the regenerator could be operated at $C^*_{r,min} \approx 1.5$ with the same thermal performance as a regenerator with uniform (ha) and which is operated at $C^*_{r,min} \approx 3$.

5 Conclusion

A study of the optimal distribution of the (ha)-value within regenerative heat exchangers is presented. It is shown that the optimal distribution is a narrow distribution aligned along the optimal control trajectory $x = \tau$ of the exchanger and that high exchanger effectivenesses are obtained at lower rotational speeds than conventional. Operating the regenerator at lower rotational speeds reduces the carryover losses and the seal wear. Although not discussed in this paper, applying optimal control could also lead to reducing the pressure drops of the fluid flows.

The existing methods of enhancing the heat transfer coefficients in forced convection flow cannot be applied for the active control of regenerators. Alternative active methods for heat transfer enhancement need to be investigated.

A Optimal thermal control of a regenerator matrix

The following study applies to a porous matrix heated or cooled by an axial fluid stream as well as to an insulated duct with a step change of the inlet temperature of the fluid flow. The transient response of a porous matrix is modeled by:

$$\frac{\partial t_f}{\partial x} + C^* \frac{\partial t_w}{\partial \tau_h} = 0$$
$$\frac{\partial t_f}{\partial x} = N_{TU}(t_w - t_f) \tag{A.1}$$

supplemented with the initial condition for the matrix temperature and the inlet condition for the fluid stream:

$$t_f(x = 0, \tau) = 0 \quad \text{and} \quad t_w(x, \tau = 0) = 1. \tag{A.2}$$

To find the optimal trajectory of the N_{TU} distribution, a coordinate transformation as suggested by the discussion in the paper is used,

$$w = cx + \tau \quad \text{and} \quad z = x - c\tau, \tag{A.3}$$

302

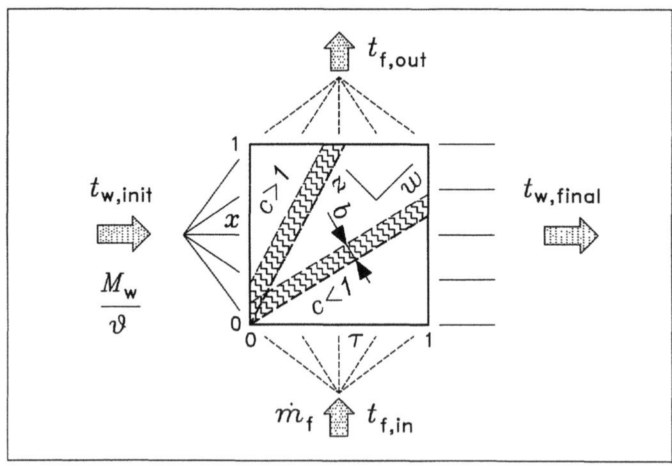

Fig. A.1   Regenerator matrix with a shifting transfer area of width b and following the trajectory $z = c\tau$.

where c is a parameter. This transformation is illustrated in figure A.1 for cases where $c < 1$ and $c > 1$. The heat transfer zone is a rectangular area with its base aligned along the w-axis and with width b. The N_{TU} is constant in this zone.

The conservation and transfer rate equation in terms of these coordinates can be written as:

$$cC^*\frac{\partial t_w}{\partial z} - \frac{\partial t_f}{\partial z} = c\frac{\partial t_f}{\partial w} + C^*\frac{\partial t_w}{\partial w} \tag{A.4}$$

$$\frac{\partial t_f}{\partial z} = N_{TU}(t_w - t_f) - c\frac{\partial t_f}{\partial w} \tag{A.5}$$

with initial conditions

$$\text{at } w = 0: \quad t_f(z) = 0 \quad \text{and} \quad t_w(z) = 1 \quad \text{for } 0 \leq z \leq b, \tag{A.6}$$

and boundary conditions

$$\left.\begin{array}{l} \text{at } z = 0: \quad t_f(w) = 0 \\ \text{at } z = b: \quad t_w(w) = 1 \end{array}\right\} \quad \text{for } 0 \leq w. \tag{A.7}$$

Substituting equation (A.5) in equation (A.4) and taking the Laplace transform $t(w) \to \bar{t}(p)$ yields

$$\left.\begin{array}{l} \dfrac{\mathrm{d}\bar{t}_w}{\mathrm{d}z} = \dfrac{N_{TU} + pC^*}{cC^*}\bar{t}_w - \dfrac{N_{TU}}{cC^*}\bar{t}_f - \dfrac{1}{c} \\[2ex] \dfrac{\mathrm{d}\bar{t}_f}{\mathrm{d}z} = N_{TU}\bar{t}_w - (N_{TU} + pc)\bar{t}_f \end{array}\right\} \tag{A.8}$$

The solution of this system of ordinary differential equations is

$$\left.\begin{array}{l} \bar{t}_w(z) = A_1(N_{TU} + pc + \lambda_1)e^{\lambda_1 z} + A_2(N_{TU} + pc + \lambda_2)e^{\lambda_2 z} + \dfrac{N_{TU} + pc}{p(N_{TU} + pc + cN_{TU}/C^*)} \\[3ex] \bar{t}_f(z) = A_1 N_{TU}e^{\lambda_1 z} + A_2 N_{TU}e^{\lambda_2 z} + \dfrac{N_{TU}}{p(N_{TU} + pc + cN_{TU}/C^*)} \end{array}\right\} \tag{A.9}$$

where λ_1 and λ_2 are the roots of the characteristic equation:

$$\lambda^2 + [N_{TU}(1 - 1/C^*c) + p(c - 1/c))]\lambda - p(p + N_{TU}/c + N_{TU}/C^*) = 0 \qquad (A.10)$$

The coefficients A_1 and A_2 need to be determined from the boundary conditions in equation (A.7):

$$\left. \begin{array}{l} \text{at } z = 0: \quad \bar{t}_f = 0 \\ \text{at } z = b: \quad \bar{t}_w = 1/p \end{array} \right\} \quad \text{for } 0 \le w, \qquad (A.11)$$

which leads to the system of equations:

$$\left. \begin{array}{l} A_1 N_{TU} + A_2 N_{TU} = -\dfrac{N_{TU}}{p(N_{TU} + pc + N_{TU}c/C^*)} \\[3mm] A_1(N_{TU} + pc + \lambda_1)e^{\lambda_1 b} + A_2(N_{TU} + pc + \lambda_2)e^{\lambda_2 b} = \dfrac{N_{TU}c/C^*}{p(N_{TU} + pc + N_{TU}c/C^*)} \end{array} \right\} \qquad (A.12)$$

The effectiveness is computed from the outlet temperature t_f at $z = b$. The case of interest here is a *Dirac δ-distribution* of N_{TU}, which is modeled by taking the limit for $b \to 0$ where $bN_{TU} \to N_{TU,0}$. Taking the limit in equations (A.10) and (A.12), solving for the roots λ_1 and λ_2 and coefficients A_1 and A_2, inserting the results in equations (A.9), and taking the inverse Laplace transform yields a constant outlet temperature for the fluid stream:

$$\text{at } x = 1 \quad \left\{ \begin{array}{ll} t_f = \dfrac{1 - e^{-N_{TU,0}(1 - 1/cC^*)}}{1 - \dfrac{1}{cC^*}e^{-N_{TU,0}(1 - 1/cC^*)}} = \epsilon_{cf}(N_{TU,0}, 1/cC^*) & \text{for } \tau < 1/c \\[6mm] t_f = 0 & \text{for } \tau > 1/c \end{array} \right. \qquad (A.13)$$

where ϵ_{cf} is the effectiveness of a counterflow heat exchanger with the number of transfer units equal to $N_{TU,0}$ and with the capacity-rate ratio equal to $1/cC^*$. The effectiveness of the transient response of a porous matrix with the N_{TU} distributed as a delta spike shifting along the trajectory $x = c\tau$ follows from figure A.1:

$$\begin{array}{l} \text{for } c \le 1: \quad \epsilon_{f,\delta} = t_f \\ \text{and for } c \ge 1: \quad \epsilon_{f,\delta} = t_f/c \end{array} \qquad (A.14)$$

This effectiveness reaches a maximum for a given $N_{TU,0}$ at $c = 1$.

The results of the study presented in this appendix can be formulated as follows:

The optimal thermal control of a regenerator matrix requires that the heat transfer capacity (ha) between the fluid stream and the matrix material can be actuated with time and position within the flow direction. The control consists of locally increasing (ha) within a narrow transfer zone, and to let this zone shift in time from the entrance face of the matrix at the start of the regeneration period to the the exit face of the matrix at the end of the period. The thermal effectiveness of the regenerator will approach the effectiveness of a counterflow heat exchanger, with the same dimensionless parameters N_{TU} and C^*, as the width of the transfer zone shrinks to zero while the average $(\bar{h}A)$ remains constant.

304

NOMENCLATURE

a	local transfer area density	**Greek symbols**	
A	coefficient or total transfer surface area	δ	Dirac δ-function
		ϵ	heat exchanger effectiveness
b	width of transfer zone	ϵ_{cf}	counterflow heat exchanger effectiveness
c	slope of transfer zone		
C	fluid stream or matrix thermal capacity rate	ϵ_r	regenerator effectiveness
		λ	root of characteristic equation
C^*	capacity-rate ratio	τ	dimensionless time variable
$\dot{m}$	mass flow rate		
M	mass of regenerator matrix		
N_{TU}	local or average number of transfer units	**Subscripts**	
		c	cold fluid stream or cold pass of regenerator
h	heat transfer coefficient		
p	Laplace variable	cf	counterflow
t	temperature	f	fluid stream
$\bar{t}$	Laplace transformed of temperature	h	hot fluid stream or hot pass of regenerator
w	dimensionless time variable	r	regenerator
x	dimensionless flow coordinate	w	matrix material
z	dimensionless flow coordinate	δ	for a Dirac δ-distribution

References

[1] Shah, R.K.: Compact Heat Exchanger Design Procedures, in *Heat Exchangers, Thermal-Hydraulic Fundamentals and Design*, edited by S. Kakaç, A.E. Bergles and F. Mayinger, pp. 495-536, 1981, Hemisphere Publishing Corp., Washington.

[2] Kays, W.M., and A.L. London, *Compact Heat Exchangers*, McGraw-Hill Inc., New York, 1984.

[3] Shah, R.K., Thermal design theory for regenerators. In *Heat Exchangers, Thermal-Hydraulic Fundamentals and Design* (Edited by S. Kakac, A.E. Bergles and F. Mayinger), pp. 721-763, Hemisphere Publishing Corp., Washington D.C. (1981).

[4] Van den Bulck, E., Optimal Design of Crossflow Heat Exchangers, *J. Heat Transfer*, in press, 1991.

A Simplified Model for Helical Heat Exchanger for Long-Term Energy Storage in Soil

Y. RABIN,*§ E. KORIN** and E. SHER*

*Mechanical Engineering Department and **The Institutes for Applied Research,
Ben-Gurion University of the Negev, Beer- Sheva 84105, Israel

Summary
A simplified numerical code for thermal analysis of a helical heat exchanger for use in long-term thermal energy storage in soil was developed. The model was verified for a particular case for which an analytical solution was available from the literature and was validated with experimental data obtained from field experiments. The differences between predicted and measured data were in the range of ±1°C, which is considered satisfactory for engineering design purposes. The model was prepared for use with an personal computer and thus provides a convenient and reliable design tool for such a system. The computer code may be easily modified for the study of the influence of incorporating phase-change material elements in the storage well.

Introduction

The development of alternate energy sources to fossil fuels has acquired high priority in view of the environmental problems associated with the extraction and consumption of these fuels and the instability of oil prices. The utilization of waste heat and renewable energy sources as viable alternatives depends on the development of cost-effective thermal energy storage systems for the short term and, particularly, for the long term. Among the various techniques proposed for long-term thermal energy storage, a method based on soils as the heat storage medium is considered one of the most promising possibilities. Most of the R&D effort on this method has been devoted to cold and moderate zones and usually involves vertical multiple well storage. The different heat exchanger models previously proposed include the U-shaped exchanger of Reuss et al. [1], in which 25-mm diameter polypropylene pipes are inserted into the soil in vertical bores about 0.15 m in diameter and 10 m in depth. For warm and arid regions, a helical heat exchanger for seasonal heat storage has been proposed by Nir et al. [2]. Earlier work on this idea included the investigation of the general concept and the development of theoretical model for the design of such a system for application in unsaturated soils, which constitute the norm in arid zones [2]. Recently, an experimental field system, based on this model, was built and operated at the Institutes for Applied Research of Ben-Gurion University of the Negev. The system was used to obtain experimental data for validation testing of the model and for investigating certain engineering and operational issues [3]. It was shown that for clay soils with a water content above 20% and operational temperatures in the range of 20-80°C the effect

---

§ Y.R. is an M.Sc. student in the Mechanical Engineering Department, Ben-Gurion University of the Negev.

of soil drying on the heat transfer process may be neglected. Based on this conclusion the objective of the present work was to develop a simplified and reliable code for use in a personal computer for the simulation of thermal energy storage in a helical heat exchanger inserted into the soil. The paper describes the simplified theoretical model, the verification testing, and the results of a comparison with experimental data.

<u>Theoretical Model</u>

A schematic description of the system and the geometric parameters of the helical heat exchanger are given in Fig. 1.

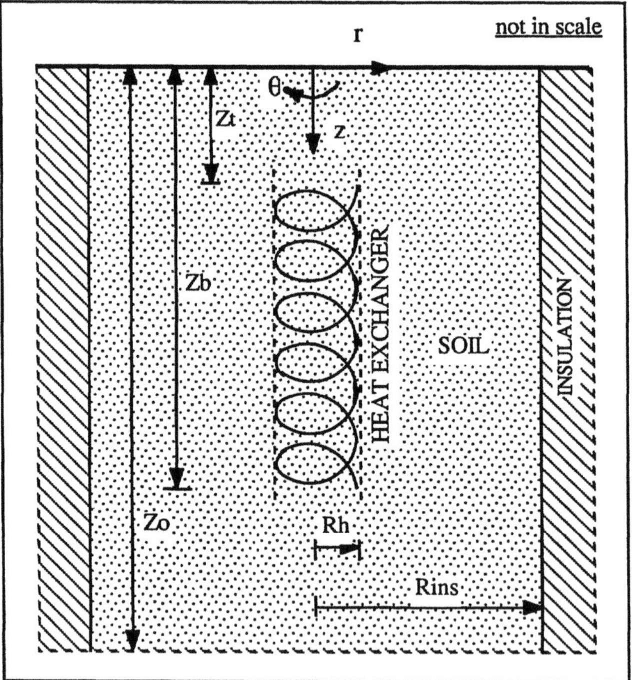

Fig. 1: Schematic description of the system

where r, θ, z are the coordinates of the cylindrical system, Zt and Zb are the distance of the upper and lower sides of the heat exchanger from the surface of the ground, respectively, Zo is the total height of the system, Rh is the radius of the heat exchanger and Rins is the radius of the system.

A Simplified Model for Helical Heat Exchanger for Long-Term Energy Storage in Soil

Y. RABIN,*§ E. KORIN** and E. SHER*

*Mechanical Engineering Department and **The Institutes for Applied Research,
Ben-Gurion University of the Negev, Beer- Sheva 84105, Israel

Summary

A simplified numerical code for thermal analysis of a helical heat exchanger for use in long-term
thermal energy storage in soil was developed. The model was verified for a particular case for
which an analytical solution was available from the literature and was validated with
experimental data obtained from field experiments. The differences between predicted and
measured data were in the range of ±1°C, which is considered satisfactory for engineering
design purposes. The model was prepared for use with an personal computer and thus provides
a convenient and reliable design tool for such a system. The computer code may be easily
modified for the study of the influence of incorporating phase-change material elements in the
storage well.

Introduction

The development of alternate energy sources to fossil fuels has acquired high priority in view
of the environmental problems associated with the extraction and consumption of these fuels
and the instability of oil prices. The utilization of waste heat and renewable energy sources as
viable alternatives depends on the development of cost-effective thermal energy storage systems
for the short term and, particularly, for the long term. Among the various techniques proposed
for long-term thermal energy storage, a method based on soils as the heat storage medium is
considered one of the most promising possibilities. Most of the R&D effort on this method has
been devoted to cold and moderate zones and usually involves vertical multiple well storage.
The different heat exchanger models previously proposed include the U-shaped exchanger of
Reuss et al. [1], in which 25-mm diameter polypropylene pipes are inserted into the soil in
vertical bores about 0.15 m in diameter and 10 m in depth. For warm and arid regions, a helical
heat exchanger for seasonal heat storage has been proposed by Nir et al. [2]. Earlier work on
this idea included the investigation of the general concept and the development of theoretical
model for the design of such a system for application in unsaturated soils, which constitute the
norm in arid zones [2]. Recently, an experimental field system, based on this model, was built
and operated at the Institutes for Applied Research of Ben-Gurion University of the Negev. The
system was used to obtain experimental data for validation testing of the model and for
investigating certain engineering and operational issues [3]. It was shown that for clay soils
with a water content above 20% and operational temperatures in the range of 20-80°C the effect

---

§ Y.R. is an M.Sc. student in the Mechanical Engineering Department, Ben-Gurion University of the Negev.

$$k\frac{\partial T}{\partial z}(0,r,t) = q_{surface}(T(o,r,t),T_{air}(t),q_{solar}(t),P_{aw}(t)) \tag{1e}$$

where $q_{surface}$ is the heat flux through the upper surface of the system, T_{air} is the air temperature, q_{solar} is the flux of the solar radiation absorbed by the surface, and P_{aw} is the partial pressure of the water vapor in the air in mm hg.

A further assumption is made that the ground surface is effectively "gray":

$$q_{surface}(r,t) = q_{solar}(t) + h_c(T_{air}(t) - T(0,r,t)) + \varepsilon\sigma(T_{ef}(t)^4 - T(0,r,t)^4) \tag{2}$$

where h_c is the convection heat transfer coefficient, ε is the emissivity of the soil and σ is the Stefan-Boltzman constant.

The convective heat transfer coefficient can be determined from the correlation [4]:

$$h_c = 6.2 + 1.4 \cdot u_{wind} \quad [w/m^2 \cdot K] \tag{3}$$

where u_{wind} is the wind velocity in [m/s].

The effective sky temperature T_{ef} is defined as the temperature of a black body which would radiate to the ground the same flux as actually reaches it from the sky [5]:

$$T_{ef}(t) = \left(0.55 + 0.065\sqrt{P_{aw}(t)}\right)^{0.25} \cdot T_{air}(t) \quad [K] \tag{4}$$

It is interesting to note that eq. 4 suggests that in an arid zone at or near sea level (as is the case at the experimental region) the effective nocturnal sky temperature can be expected to be approximately 15°C below the ambient air temperature.

At temperatures in the range of the ambient temperature it is possible to approximate the fourth power temperature difference of the thermal radiation by a linear difference:

$$q_{surface}(r,t) = q_{solar}(t) + h_c(T_{air}(t) - T(0,r,t)) + h_r(T_{ef}(t) - T(0,r,t)) \tag{5}$$

where the radiative heat transfer coefficient h_r is defined as:

$$h_r = 4\varepsilon\sigma T_{av}^3 \tag{6}$$

and the average temperature, as:

$$T_{av} = (T_{ef} + T(0,r,t))/2 \quad [K] \tag{7}$$

For the working fluid stream:

$$a \cdot \int_0^{2\pi} k \cdot \frac{\partial T}{\partial r^*}\bigg|_{r^*=a} \cdot d\theta^* = \pi \cdot a^2 \cdot \rho_f \cdot Cp_f \cdot \left(\frac{Cv_f}{Cp_f} \cdot \frac{\partial T_f}{\partial t} + u_{av} \cdot \frac{\partial T_f}{\partial z^*}\right) \tag{8}$$

where θ^*, r^*, z^* are coordinates related to the pipe (Fig. 2), T_f is the temperature of the fluid, a is the radius of the pipe, u_{av} is the average velocity of the fluid, Cp_f is the constant-pressure specific heat of the fluid and Cp_v is the constant-volume specific heat of the fluid.

Considering the fact that for liquid heat transfer the term $\partial T_f/\partial t$ is about three orders of magnitude smaller than $u_{av} \cdot \partial T_f/\partial z^*$ [6], equation (8) of the heat balance becomes:

$$a \cdot \int_0^{2\pi} k \cdot \frac{\partial T}{\partial r^*}\bigg|_{r^*=a} \cdot d\theta^* = \dot{m} \cdot Cp_f \cdot \frac{\partial T_f}{\partial z^*} \tag{9}$$

where $\dot{m}$ is mass flow rate of the fluid.

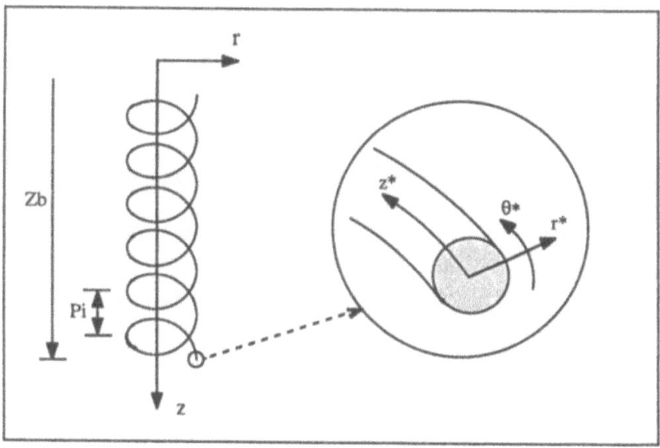

Fig. 2: The system cylindrical coordinates and the pipe cylindrical coordinates.

The boundary conditions are:
$$T_f(z,t) = T(z,r,t) \qquad r,z \in \Omega \qquad (9a)$$
where Ω is the surface of the heat exchanger pipe.
$$T_f(0,t) = T_{inlet}(t) \qquad (9b)$$
where T_{inlet} is the inlet temperature of the working fluid to the heat exchanger, and determined by the heat source of the system.

The initial condition is:
$$T_f(z^*,0) = T_0(z^*) \qquad (9c)$$
where the transformation between the pipe coordinates system and the cylindric ground coordinates system is defined as:
$$z = Zb - \frac{z^*}{2\pi} \cdot Pi \qquad (10)$$
where Pi is the pitch of the helical heat exchanger (Fig 2).

The theoretical model can be solved by the finite differences method. The system is separated into small ring elements, and in each element a small lump system behavior is assumed. The temperature field is calculated by the following implicit equation:
$$T_{ij}^{p+1} = T_{ij}^p + \frac{\Delta t}{C_{ij}} \left[\sum_n \frac{T_n^p - T_{ij}^p}{R_{ijn}} + \dot{q}_{ij}^p \right] \qquad (11)$$
where the index i,j refers to the discretization with respect to space, and the index p, to time, Δt

is a time interval, C_{ij} is the heat capacity of an element ij defined by (12) and R_{ijn} represents the thermal resistance between element ij and its neighbor n, calculated from eq. 13 for pure conduction and from eq. 14 for convection.

$$C_{ij} = \rho_{ij} \cdot Cp_{ij} \cdot V_{ij} \tag{12}$$

$$R_{ijn} = \begin{cases} \dfrac{\Delta r}{(r_{ij} + 1/2 \cdot \Delta r) \cdot \Delta z \cdot k} & n = i, j+1 \\[3mm] \dfrac{\Delta r}{(r_{ij} - 1/2 \cdot \Delta r) \cdot \Delta z \cdot k} & n = i, j-1 \\[3mm] \dfrac{r_{ij}}{\Delta r \cdot \Delta z \cdot k} & n = i-1, j \\[3mm] \dfrac{r_{ij}}{\Delta r \cdot \Delta z \cdot k} & n = i+1, j \end{cases} \tag{13}$$

$$R_{ijn} = \frac{1}{A_{ij} \cdot h_c} \tag{14}$$

The criterion for the solution stability is:

$$\Delta t \le \left[\frac{C_{ij}}{\sum\limits_n 1/R_{ijn}} \right]_{min} \tag{15}$$

The numerical scheme (eq. 11) was selected so as to enable the calculation of the temperature field associated with phase change (liquid-solid) processes in the system, which are an important aspect that we intend to study in the next step.

In order to solve eq. 9 numerically the cross section of the pipe is considered as to have a square cross section having the same perimeter as the pipe to simulate an identical heat transfer surface area. The temperature profile in z* direction, in the working fluid, is determined by:

$$T_{in,e+1}^{p+1} = T_{in,e}^p + \frac{1}{\dot{m} \cdot Cp_f} \cdot \sum_n \frac{T_n^p - T_{b,e}^p}{R_{ijn}} \tag{16}$$

where $T_{in,e}$ and $T_{out,e}$ are the inlet and outlet temperatures of the fluid in element e, respectively, and T_b is the bulk temperature of the fluid in the element:

$$T_{b,e}^p = \frac{1}{2} \left(T_{out,e}^p + T_{in,e}^p \right) \tag{17}$$

Verification testing of the theoretical model

The first step in testing the numerical computer code was to check the overall energy balance of the system. The overall energy transfer from the working fluid (water) was calculated from:

$$E = \int_0^t \dot{m} \cdot Cp_f \left[T_{inlet}(t) - T_{outlet}(t) \right] \cdot dt \tag{18}$$

where E is the stored energy.

The total energy absorbed by the soil is:

$$E(t) = \iiint_V \rho \cdot Cp \cdot [T(r,z,t) - T(r,z,0)] \cdot dv \qquad (19)$$

The energy balance was calculated from the results of a simulation for a six-month energy storage period. The testing was performed with the following numerical values: the initial temperature of the soil was uniform and equal to 20°C, the inlet fluid temperature was 70°C and the mass flow rate was 40 kg/h. In order to assess the effect of neglecting the time dependent term $\partial T_f/\partial t$ in eq. (8), energy conservation was defined as the ratio of the total energy (eq. 19) to the total energy storage (eq. 18). It was observed that 88.0% of the energy conservation was obtained after the first hour, 97.8% after 6 h, 98.9% after 24 h and 99.94% after six months. From these results it can be seen that neglecting the time-dependent $\partial T_f/\partial t$ term in eq. 9 may effect the results only during the very short time after the step change in the fluid input temperature.

The computer code results have also been examined against an exact solution for a simplified problem. In this case a one-dimensional heat conduction problem in an infinite cylinder with constant thermophysical properties was solved. The domain of solution was subjected to a step wall temperature at time t>0. The governing equation for this case is therefore:

$$\frac{\partial^2 T}{\partial r^2} + \frac{1}{r} \cdot \frac{\partial T}{\partial r} = \frac{1}{\alpha} \cdot \frac{\partial T}{\partial t} \qquad (20)$$

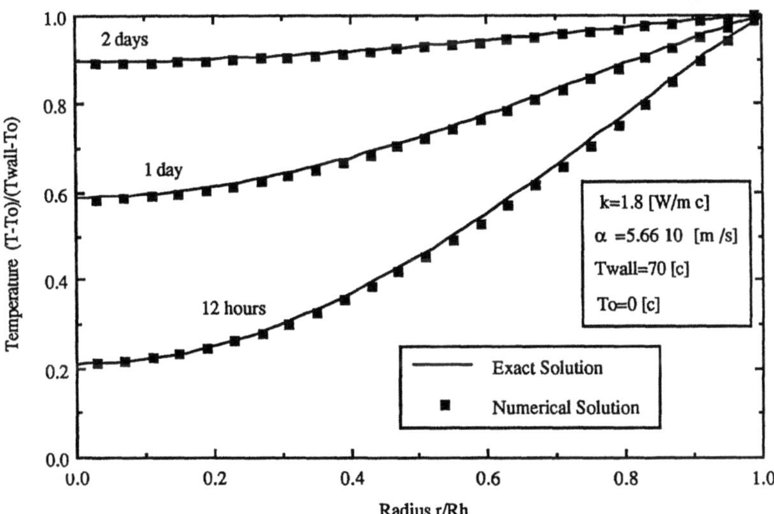

Fig. 3: Comparison between numerical results and the exact solution [7] for transient one-dimentional heat conduction in an infinite cylinder.

312

The initial condition is:

$$T(r,0) = T_0 \qquad (20a)$$

and the boundary conditions are:

$$\frac{\partial T}{\partial r}(0,t) = 0 \qquad (20b)$$

and

$$T(a,t) = T_{wall} \qquad (20c)$$

The exact solution to this problem as given in [7] is:

$$T(r,t) = T_0 - \frac{2 \cdot T_0}{a} \cdot \sum_{i=1}^{\infty} \exp\left(-\alpha \cdot \gamma_i^2 \cdot t\right) \cdot \frac{J_0(r \cdot \gamma_i)}{\gamma_i \cdot J_1(r \cdot a)} \qquad (21)$$

where γ_i is obtained from the solution of the Bessel equation $J_0(a \cdot \gamma_i) = 0$.

The test was carried out with the following numerical parameters: a very low thermal conductivity outside the well, thermal diffusivity of the soil in the well of $\alpha = 5.66 \cdot 10^{-7}$ m^2/s, a high water flow rate of m=10^5 kg/h and a very small pitch were taken in order to simulate a constant wall temperature. The results of the numerical model for this case are superimposed in Fig. 3. Very good agreement between the numerical and exact solution is clearly seen.

Experimental validation testing

The simplified theoretical model was tested vs. experimental data obtained from the experimental field system operating at the Institutes for Applied Research. In this system the helical heat exchanger made out of 0.03 m diameter polypropylene pipe with a 0.1 m pitch was 1 m in diameter and 6 m long. The heat exchanger was inserted into a 10-m deep well. The experimental system is described in detail in ref. 8. The thermophysical properties of the soil were estimated in previous experimental work: thermal conductivity $k = 1.3$ W/m·K and specific heat $Cp = 2.838 \cdot 10^6$ J/m^3·K. Solar radiation and dry and wet bulb air temperatures were provided by the meteorological station located in the neighborhood. The experiment was run for 30 days (2/2/90-4/3/90). Fig. 4 shows a comparison of the outlet temperature of the water from the heat exchanger as a function of time, as predicted by the simplified theoretical model, with that measured experimentally. Fig. 5 compares the temperature profiles in the soil at radius of 0.3 m from the center of the helical heat exchanger after 10, 20 and 30 days of the experiment. In both Figures it can be seen that the difference between measured and theoretical results is of the order of ±1°C, which is satisfactory agreement for engineering design purposes.

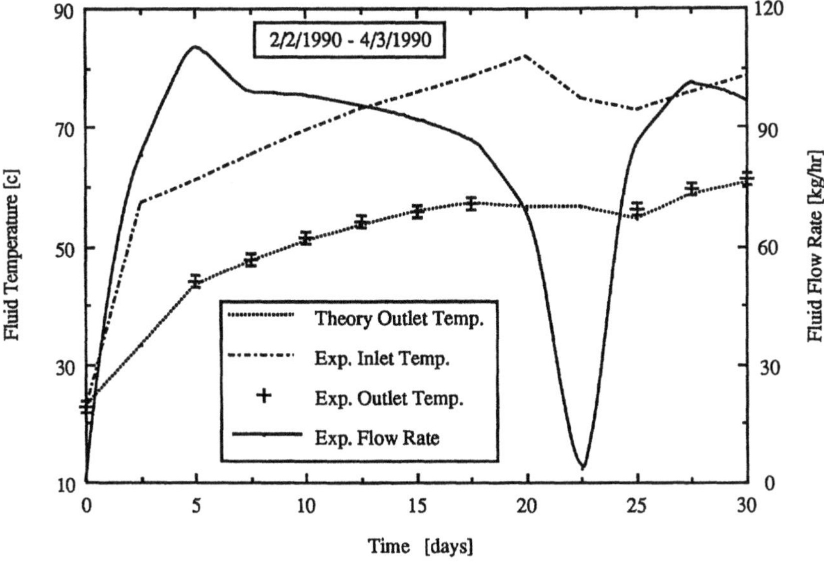

Fig. 4: Comparison between outlet fluid temperatures as predicted by the theoretical model and experimental results (measured from 2/2/90 to 4/3/90).

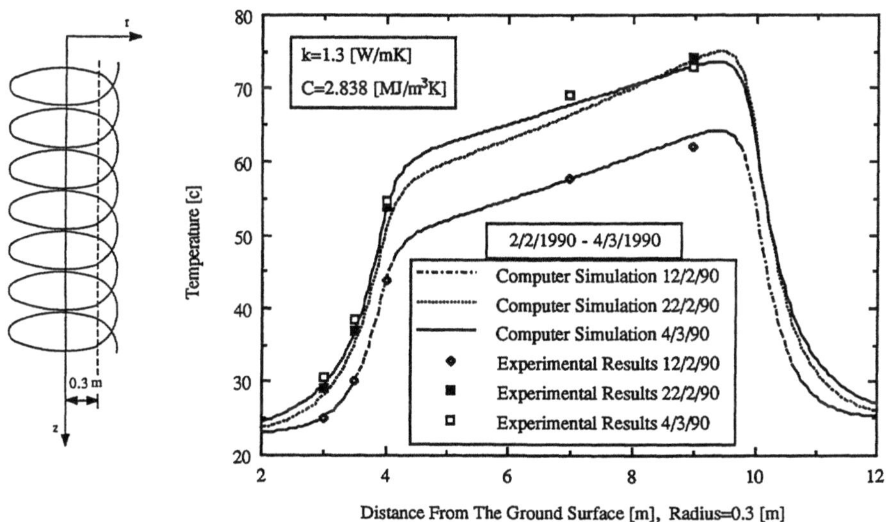

Fig. 5: Vertical temperature profiles in the soil, 0.3 m from the center of the well, as predicted by the theoretical model vs. experimental results.

Conclusions and recommendations for future work

A simplified numerical code for thermal analysis of a helical heat exchanger for thermal energy storage was developed. Theoretical verification testing and experimental validation testing showed that the model gives reliable results for parametric studies and engineering design purposes, for the type of soil with the properties tested in the present work, i.e., clay with a moisture content above 20%.

The numerical scheme was developed in a manner that can also predict the effect of incorporating a phase-change thermal energy storage element in the soil to improve the capacity of the thermal storage in the system. This effect is under study at present time and results will be published in the near future.

Acknowledgment

This work was partially supported by the U.S.-Israel Binational Science Foundation under Grant 10:85-00129.
The authors thanks Ms. I. Mureinik for editing the paper.

References

1. Reuss, M.; Schulz, H.; Wagner, B.; Solar Assisted Heat with Seasonal Storage. FAO Workshop at the IHTC 9, Jerusalem, Israel, (August 1990).
2. Nir, A.; Doughty, C.; Tsang, C.F.; Seasonal Heat Storage in Unsaturated Soils: Example of Design Study. Proc. 21 Intersol, Energy Conversion Engineering Conf., San Diego, Vol. 2 (1986) pp. 669-675.
3. Doughty, C.; Korin, E.; Nir, A.; Tsang, C.F.; Seasonal Storage of Thermal Energy in Unsaturated Soils: Simulation and Field Validation. Part I, STES, Vol. XII, No. 1, 1-5, 1990, Part II (in press)
4. McAdams, W.H.; Heat Transmission, 3rd edition, McGraw Hill, New York, 1954.
5. Pramelee, G.V.; Aubele, W.W.; Radiant Energy Emission of Atmosphere and Ground, Heating, Piping and Air conditioning. ASHRE Journal (1951) p. 123 .
6. Schmidt, F.W.; Szego, J.; Transient Response of Solid Sensible Heat Thermal Storage Units-Single Fluid. J. Heat Transfer, Aug. (1976) pp. 471-477.
7. Jager, J.C.; Carslaw, H.S.; Conduction of Heat in Solids, p. 199, p.397, Clarendon Press, Oxford, U.K.,1959.
8. Nir, A.; Korin, E.; Tsang, C.F.; Seasonal Heat Storage in Unsaturated Soils: Model Development and Field Validation, Internal Report, Ben-Gurion University of the Negev, Israel (1990).

Multiphase Systems

Pressure Drop During Condensation in Vertical Tubes

R. Numrich and N. Claus

Fachgruppe Verfahrenstechnik, FB 10 - Maschinentechnik
Universität-Gesamthochschule-Paderborn

Abstract

The calculation of required interfacial area during condensation process depends on the knowledge of the heat transfer coefficient in a falling condensate film. This transfer coefficient is a function of flow pattern, physical properties of the condensate film and also of shear stress at the condensate film surface due to the friction of gas flow. For the description of this influence on heat transfer the correct evaluation of shear stress will be necessary. Whereas many calculation methods exist in the range of atmospheric pressure and for systems without simultaneous phase change, data at higher pressures and in the presence of condensation are not available. At a testing plant friction pressure drop and consequently shear stress at the film surface were determined experimentally for downwards cocurrent flow. A method for computing the shear stress will be presented which provides good predictions of the experimental data at higher pressures. Applying this presented method for simultaneous condensation processes two phenomena have to be considered. Firstly, the condensing vapour increases the friction factor and consequently the shear stress at the condensate film surface following the film theory. Secondly, the momentum balance changes due to the decreasing velocity of the gas phase. Pressure drop measurements during partial condensation of vapour in the presence of air and heat transfer measurements during condensation with pure steam at increased pressure show good agreement with the presented method considering these cited phenomena.

1 Introduction

Vertical tube bundles with inside currents are normally used for condensation process under increased pressure. Knowledge of the heat transfer coefficient α_F in the arising condensate film is required for the correct dimensioning of the exchange area in the vertical tube for the condensation of pure steam or also for partial condensation. This heat transfer coefficient is normally determined by using the non-dimensional Nusselt number Nu_F.

$$Nu_F = \frac{\alpha_F \left(\frac{\nu_F^2}{g}\right)^{0,333}}{\lambda_F} \tag{1}$$

In addition to the Prandtl number Pr_F and the Reynolds number Re_F the Nusselt number also depends on the shear stress τ_O, which acts at the surface of the condensate film, due to the existing gas flow.

$$Nu_F = f(Re_F, Pr_F, \tau_O) \tag{2}$$

Figure 1 shows the results from Krebs [1] and Blangetti [2], who carried out measurements concerning the heat transfer in the condensate film using different Reynolds numbers Re_G from the gas phase and subsequently different shear stresses τ_O.

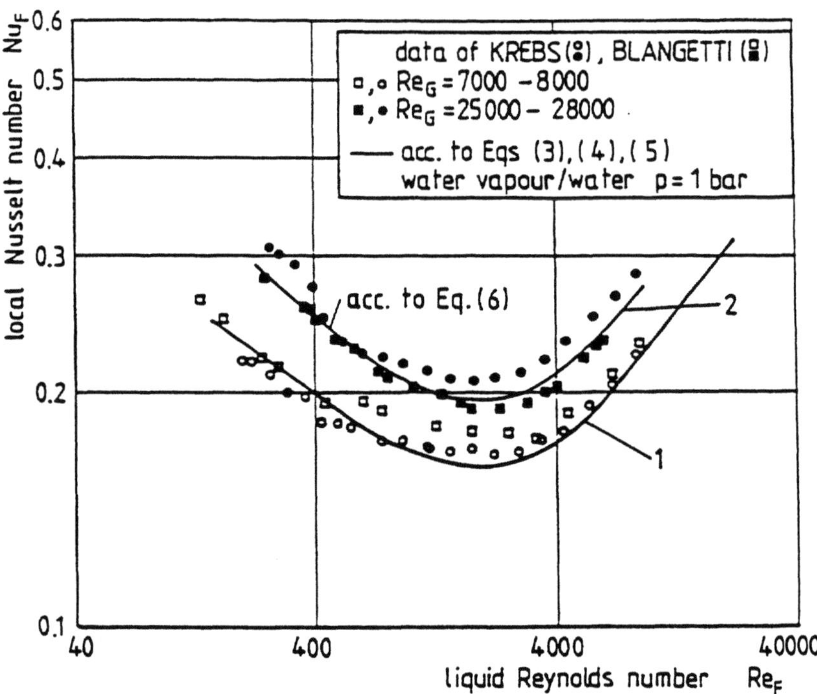

Fig. 1: Dependence of the Nusselt number on the condensate film Reynolds number for different Reynolds numbers from the gas phase.

It can clearly be seen that, with larger gas Reynolds numbers, the characteristic course of the curve moves towards larger Nusselt numbers. Based on experimentally proven Nusselt equations for a quasi static gas phase, laminar term (index *lam*) according to a suggestion by Zazoulja [3]

$$Nu_{F,lam} = 0.752\, Re_F^{-0.22} \tag{3}$$

and the turbulent term (index *tur*) from Blangetti

$$Nu_{F,tur} = 0.0051\, Re_F^{0.382}\, Pr_F^{0.569} \tag{4}$$

are combined with the function

$$Nu_F = (Nu_{F,lam}^4 + Nu_{F,tur}^4)^{0.25} \tag{5}$$

The influence of shear stress on the Nusselt number (shown by the high positioned +) can be represented, in accordance with [4], as follows:

$$\frac{Nu_F^+}{Nu_F} = \left(1 + 1.5 \frac{\tau_O}{\rho_F\, g\, \delta_F}\right)^{0.333} \tag{6}$$

In general, shear stress τ_O is defined by the following formulation:

$$\tau_O = \frac{f}{2}\, \rho_G\, \bar{u}_G^2 \tag{7}$$

f is generally the friction factor of the condensate film surface with respect to the flowing gas phase.

According to Andreussi [5], the thickness of the film results in the following:

$$\delta_F = \frac{6.\,59\, F}{(1 + 1400\, F)^{0.5}}\, d \tag{8}$$

Here, F is a modified Lockhart-Martinelli flow parameter expressed as follows:

$$F = \frac{[(0.5\, Re_F)^{1.25} + (0.02625\, Re_F)^{2.25}]^{0.4}}{Re_G^{0.9}} \left(\frac{\eta_F}{\eta_G}\right) \left(\frac{\rho_G}{\rho_F}\right)^{0.5} \tag{9}$$

Curve 2 in figure 1 represents the Nusselt numbers determined by equations (3-6). A high correspondence can be seen. This procedure has been verified at atmospheric pressure. It must also be checked whether this calculation is still valid for experiments concerning the condensation of steam at increased pressure. Therefore, pressure drop measurements from a gas/liquid flow were first performed on an existing test plant using air and water as the test media at increased pressure and at ambient temperature.

2 Test plant

A simplified flow sheet of the test plant is shown in figure 2. The principle item of the plant is the measuring section MS which consists of an upper part, a measuring tube (inside diameter d=30mm) and a lower part.

The volume flow of compressed air is controlled as it is fed into the measuring section and flows through this from top to bottom and escapes via a throttle valve into the atmosphere. Water is also conveyed out of the container B1 to the measuring section by means of the pump P1. A funnel shaped device, which distributes the water evenly onto the inside wall of the measuring tube to generate a homogeneous falling film, is situated in the upper part of the measuring section. The film of water, which is thereby produced, flows back into the container B1. It is possible that some droplets of water, which are entrained from the film, are conveyed by the air flow into the the lower part of the measuring section. There, they are separated from the air flow and are fed via a condensate separator to be weighed.

It should be noted that identically built measuring tubes with lenghts of $z_0 = 1\ m$ and $z_1 = 2\ m$ were used for the pressure drop measurements. Therefore, by subtracting the experimental pressure losses (measuring point 202) $\Delta p_{ex,z_1} - \Delta p_{ex,z_0}$ the friction pressure drop per metre of the tube length $(d\,p/d\,z)_{friction}$ and, subsequently, the shear stress can be determined.

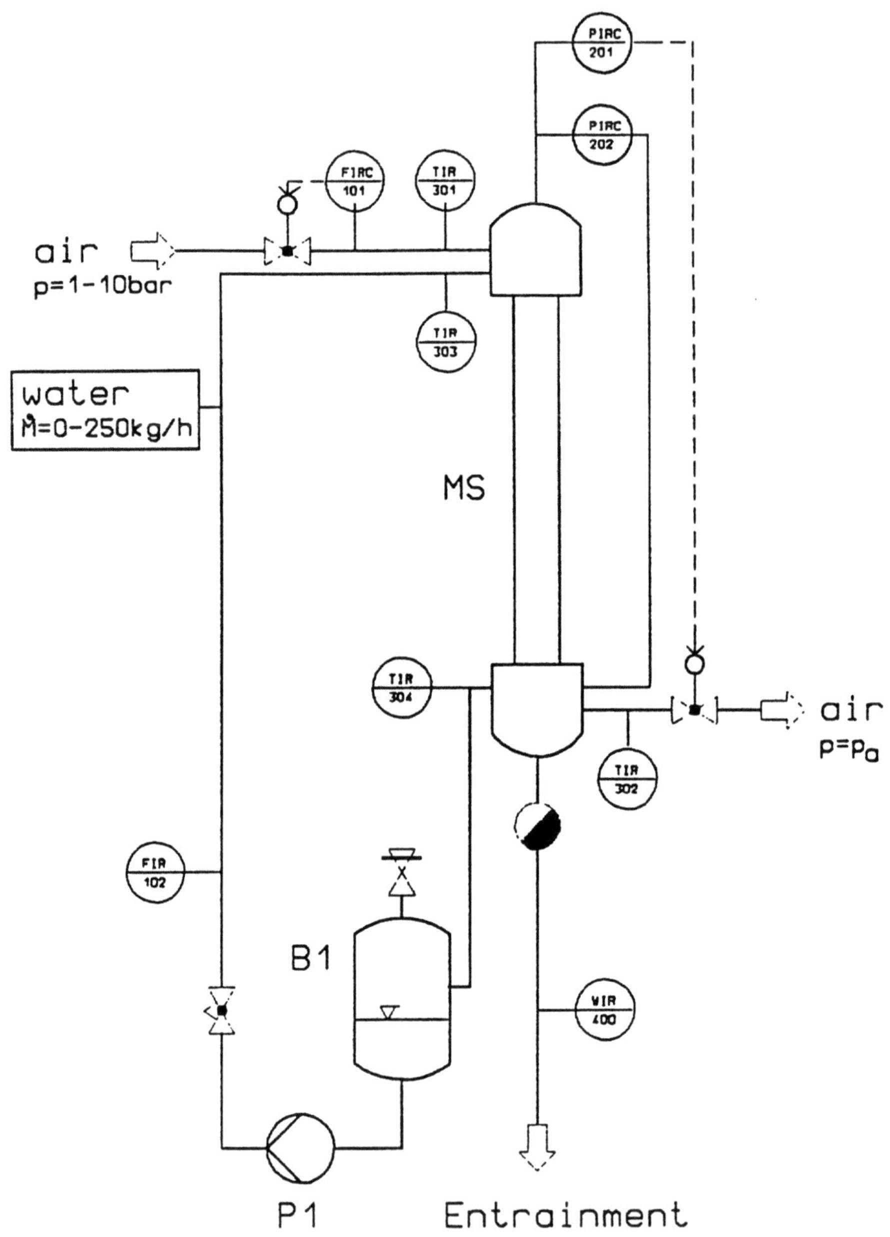

Fig. 2: Flow sheet

This correlation is shown by using the momentum balance for the gas phase (fig. 3). Under the condition that the pressure is evenly distributed over the cross section of the tube, the equation for the total pressure loss per unit of length (with $\delta_F \ll d$) is:

$$\frac{dp}{dz} = -\tau_O \frac{4}{d} + \rho_G\, g - \frac{d\left(\rho_G\, \bar{u}_G^2\right)}{dz} \tag{10}$$

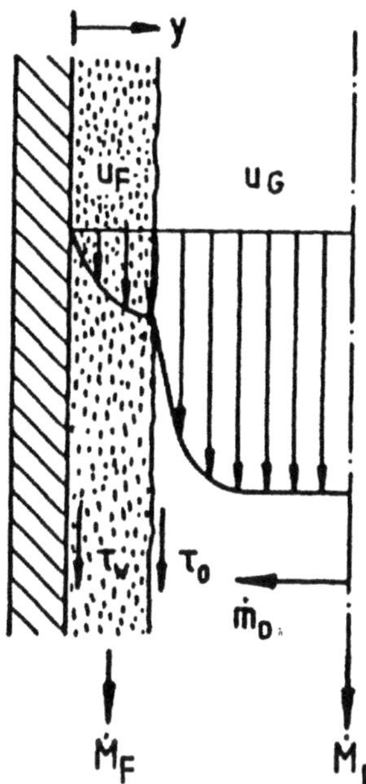

Fig. 3: Fig. 3: Momentum balance in the vertical tube

As the change in pressure in the measuring tube is small with regard to the absolute pressure and the temperatures of the test media in the measuring section remain constant, the density and consequently the gas velocity in the balance chamber do not change. The proportion of the change in momentum $d(\rho_G\, \bar{u}_G^2)/dz$ can, therefore, be ignored. Furthermore, the geodesic change in pressure is equalized by the arrangement of the pressure drop transmitter. Hence, the pressure drop in the measuring tube, which can be determined by experiments, is:

$$\Delta p_{ex,MR} = \Delta p_{friction} = \frac{4}{d} \int_0^z \tau_O\, dz \tag{11}$$

As the pressure measuring points are in the upper and the lower parts of the measuring section, there is an additional pressure loss in the inlets and outlets. Hence, the following equation is generally valid:

$$\Delta p_{ex,z} = \Delta p_E + \frac{4}{d} \int\limits_0^z \tau_O \, dz + \Delta p_A \tag{12}$$

Due to the identical construction of the measuring sections, which was also confirmed by testing the test plant with a pure gas flow, the following applies:

$$\Delta p_{ex,z_1} - \Delta p_{ex,z_0} = \frac{4}{d} \int\limits_{z_0}^{z_1} \tau_O \, dz \tag{13}$$

It should be noted that, by using this measuring process, all the inlet effects in the first metre of the measuring tube can be eliminated. After the inlet effects disappear for a stationary flow, the shear stress τ_O, acting at the surface of the film, is constant. For this reason, the pressure loss per unit of length is calculated by:

$$\Delta p_{ex} = \Delta p_{ex,z_1} - \Delta p_{ex,z_0} = \tau_O \frac{4\,(z_1 - z_0)}{d} \tag{14}$$

As mentioned before, the tests were carried out using water and air as the media. The temperature of the mass flows was approximately $20°C$. The measuring range for the pressure drop measurements is shown in table 1.

Table 1: Measuring range for the pressure drop measurement

Re_G	Re_F	$p\,/\,bar$
13000 - 96000	480 - 2880	1,04 - 10

3 Theoretical calculation

The pressure drop Δp_{th} of the adiabatic two phase flow is calculated according to different processes. Chawla [6] performed pressure drop measurements using evaporating refrigerant in horizontal tubes. He derived a generally applicable equation for the pressure drop from a dimension analysis. The calculation process from Theissing [7] is orientated on the process by Lockhart/Martinelli and requires a calculable one phase pressure drop. Reza [8] carried out tests using a water/air system on horizontal tubes. His equations are also valid for the entire flow range, for both horizontal and vertical tubes. In contrast to the above authors, Andreussi's approach is only valid for a downward adiabatic gas/liquid annular flow with and without entrainment. Under the conditions given for equation (14) and the definition of shear stress according to equation (7), the theoretical friction presssure drop Δp_{th} results in the following:

$$\Delta p_{th} = \frac{f}{2} \rho_G \, \bar{u}_G^2 \, \frac{4}{d}(z_1 - z_0) \tag{15}$$

Andreussi defines the following empirical equation for the friction factor f of the condensate film surface.

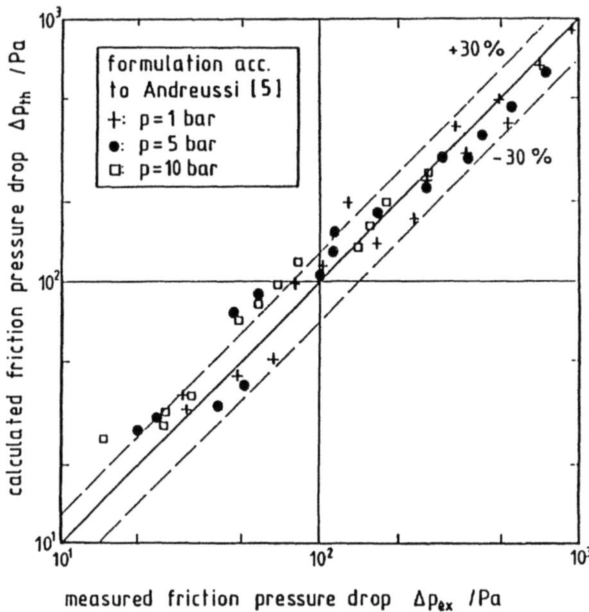

Fig. 4: Comparison between the measured friction pressure drop and the calculated values according to Andreussi [5]

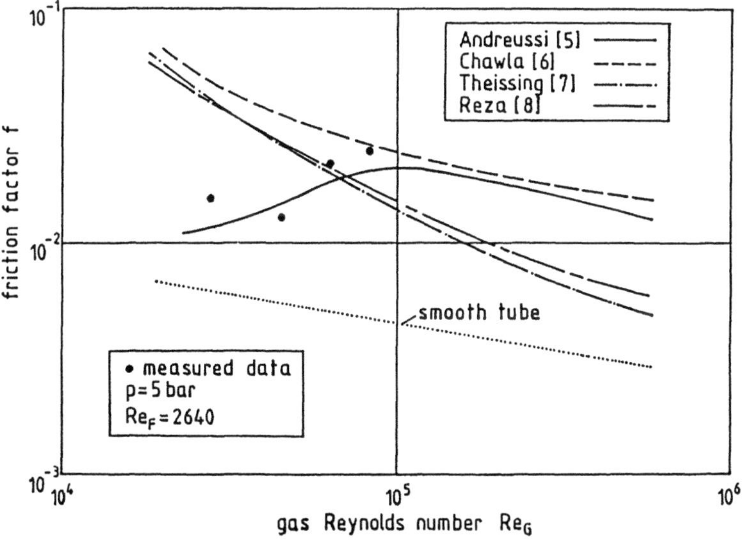

Fig. 5: Course of the friction factor f as a function of the gas Reynolds number

$$f = f_s \left(1 + 1400 \, \phi \, F\right) \tag{16}$$

In this formulation, f_s is the friction factor of a hydraulic smooth tube which is calculated, for example, according to Blasius. The factor ϕ stated in equation (16) reflects the influence of the relationship of the shear stress on the interface area to the weight of the falling film.

$$\phi = 0.27 \left(\frac{\tau_O}{\rho_F \, g \delta_F}\right)^{0.666} \quad \text{für} \quad \frac{\tau_O}{\rho_F \, g \delta_F} \leq 1.8 \tag{17}$$

$$\phi = 0.33 \left(\frac{\tau_O}{\rho_F \, g \delta_F}\right)^{0.333} \quad \text{für} \quad \frac{\tau_O}{\rho_F \, g \delta_F} > 1.8 \tag{18}$$

As the comparison between the variance of the measured values and the calculated values in Table 2 shows, however, the prediction accuracy of the generally applicable calculation approaches is not satisfactory.

Table 2: Comparison of the calculation approaches for pressure drop

	Andreussi	Chawla	Theissing	Reza
average deviation	21.7%	114.0%	89.2%	87.8%
deviation upwards	69.8%	525.3%	275.0%	287.0%
deviation downwards	-25.6%	-8.6%	-34.2%	-28.7%

Only the process developed by Andreussi can calculate the pressure drop satisfactorily which is shown in Fig. 4. This will be explicable because the other cited equations are valid for the whole two phase flow range and therefore they show greater deviations in the range of annular flow.

A more exact analysis also shows that Andreussi's process reflects the principle course of the friction factor f as a function of the Reynolds number Re_F and the Reynolds number Re_G (fig. 5). This is in accord with other known measurements in the literature [9,10] which, however, were made under ambient pressure.

4 Influence of mass flow on the friction factor

By using this process to determine the friction f on the condensation processes, an additional momentum exchange through the condensing mass flow $\dot{m}_D$ must be considered according to the film theory. According to [11], the following can be concluded for this correction:

$$C_f = \frac{f^\bullet}{f} = \frac{a_f}{1 - e^{-a_f}} \tag{19}$$

where

$$a_\tau = \frac{\dot{m}_D \, \bar{u}_G}{\tau_O} \tag{20}$$

In this formulation, $\dot{m}_D$ is the mass flow relating to the exchange surface and τ_O refers to equation (7). The actual shear stress, however, is now:

$$\tau_O^\bullet = \frac{f^\bullet}{2} \, \rho_G \, \bar{u}_G^2 \tag{21}$$

Whilst this correction can generally be ignored for condensation under atmospheric pressure, higher rates of condensation and subsequently changed friction factors, in accordance with equation (19), are obtained at increased pressure. This could be verified by tests for partial condensation. Values from C_f up to 1.5 are obtained. Using the same test technique, figure 6 shows measured friction pressure drops during tests for partial condensation of water steam in the presence of air at increased pressure and increased temperature.

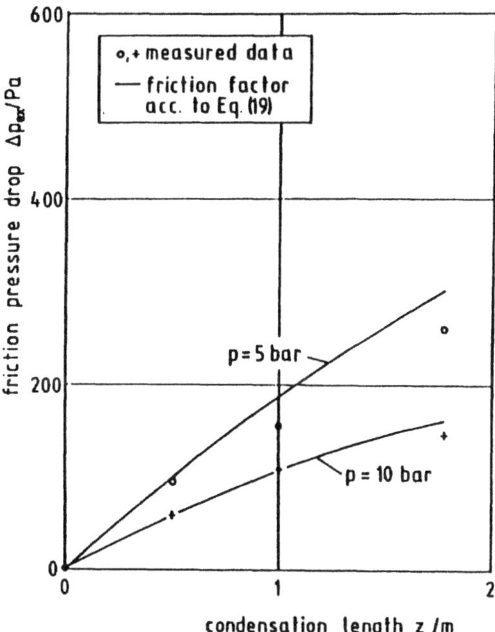

Fig. 6: Friction pressure drops for partial condensation along the condensation section

Corresponding to equation (10), it must be considered here that, on the one hand, the geodesic change in pressure must be calculated and, on the other hand, that the velocity gas phase is not constant, thereby producing a considerable axial change in momentum. In accordance with equation (13) the theoretical friction pressure drop Δp_{th} is calculated as follows:

$$\Delta p_{th} = \frac{4}{d} \int_{z_O}^{z} \tau_O^\bullet \, dz = \frac{4}{d} \int_{z_O}^{z} \frac{f^\bullet}{2} \, \rho_G \, \bar{u}_G^2 \, dz \tag{22}$$

The high correspondence between the measured values and the theoretical calculation, referring to equation (21), can be seen in figure 6. The described calculation process for the shear stress is, therefore, fundamentally valid for condensate film at a co-current flow from the gas and liquid phase also at increased pressure.

Condensation tests with pure water steam up to p=15 bar were then carried out using this test plant. Independent of the condensation rates, it was possible to vary the flow state in the falling film due to the existing dosing. Figure 7 shows measured condensate volumes along the condensation section, assuming $z_0 = 1 \, m$ for a dosed quantity of liquid at the inlet corresponding to film Reynolds number of $Re_F = 4640$.

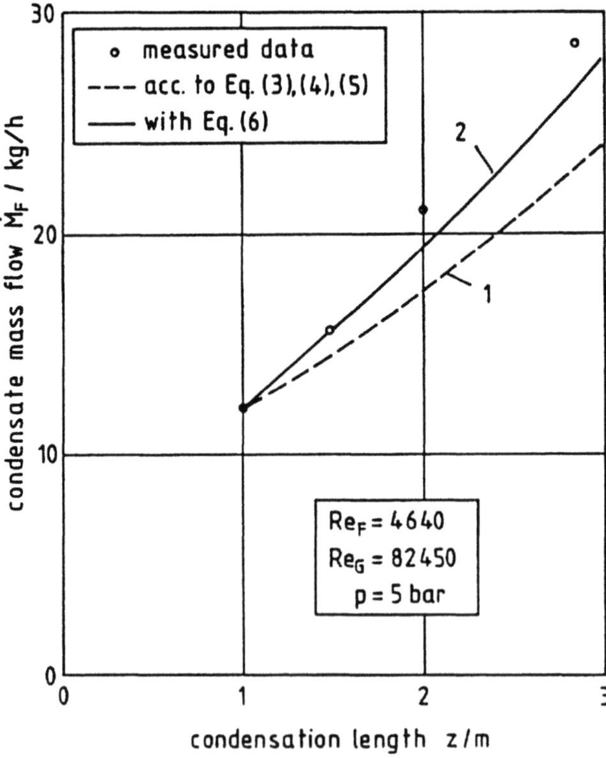

Fig. 7: Amount of condensate formed as a function of the condensation section for pure water steam

Ignoring the shear stress, the dashed curve (1) is obtained using equation (5). Smaller mass flows are produced. If equation (6) is applied, using the shear stress according to equation (21), there is a high correspondence between the experiment and the calculation (curve 2). Thus, the process, which had so far been verified under atmospheric conditions, can also be applied to determine the Nusselt number in a condensate film in the view of the higher condensation rates at increased pressure.

References

[1] *Krebs, R.:* Kondensation von Wasserdampf in Anwesenheit nichtkondensierbarer Gase in turbulent durchströmten senkrechten Kondensatoren. Disseration, Technische Universität Karlsruhe, 1984.

[2] *Blangetti, F.:* Lokaler Wärmeübergang bei der Kondensation mit überlagerter Konvektion im vertikalen Rohr. Dissertation, Technische Universität Karlsruhe, 1979.

[3] zitiert in: *Kutateladze, S. S.:* Fundamentals of Heat Transfer. Academic Press, New York, 1963.

[4] *Numrich, R.:* Influence of Gas Flow on Heat Transfer in Film Condensation. Chemical Engineering & Technology, 13 (1990), S.136 - 143.

[5] *Andreussi, P.; Zanelli, S.:* Downward Annular and Annular-Mist Flow of Air-Water Mixtures. Two-Phase Momentum, Heat and Mass Transfer in Chemical Process, S. 303-314.

[6] *Chawla, J. M.:* Reibungsdruckabfall bei der Strömung von Flüssigkeits/Gas-Gemischen in waagerechten Rohren. Chemie-Ingenieur-Technik 44 (1972), S. 58-63.

[7] *Theissing, P.:* Eine allgemeingültige Methode zur Berechnung des Reibungsdruckverlusts der Mehrphasenströmung. Chemie-Ingenieur-Technik 52 (1980), S. 344-345.

[8] *Reza, J. A.:* Reibungsdruckverlust bei der Gas-Flüssigkeits-Zweiphasenströmung in waagerechten Rohren mit kreisförmigem und ovalem Querschnitt. Dissertation, Universität Karlsruhe 1985.

[9] *Ueda, T.; Tanka, T.:* Studies of liquid film flow in twophase annular and annular-mist flow regions. Bulletin JSME 17 (1974), S. 603-613.

[10] *Konovalov, N. M.; Harin, V. F.; Nikolaev, N. A.:* Berechnung des Druckabfalls bei abwärtsgerichteter Gasströmung mit Flüssigkeitsfilm an der Rohrwand (russ.). Teoretičeskie Osnovi Himičeskoj Technologii 19 (1985), S. 48-52.

[11] *Bird, R. B.; Steward, W. E.; Lightfood, E. N.:* Transport Phenomena John Wiley & Sons, New York, 1960.

[12] *Numrich, R.:* Die partielle Kondensation eines Wasserdampf/Luftgemisches im senkrechten Rohr bei Drücken bis 21 bar. Dissertation, Universität Paderborn, 1988
VDI Fortschritt-Berichte, Reihe 3, Nr. 165, 1988.

Intensification of Heat Transfer in Horizontal-Tube Vapour Condensers

Y. E. TROKOZ, V. G. RIFERT

Kiev Politechnical Institute, Kiev

Summary

Improvement of existing horizontal-tube condensers involves
development of efficient types of heat exchange surface finning and rational arrangement of a tube bundle. The current
work is concerned with the search for optimal parameters of
tube surface shaping by means of the wall strain or fastening
to the fin surface. Results of investigations of heat exchange
in steam condensation at menisci and on horizontal tubes ribbed with a wire are given. In the investigation a numerical
simulation was employed. The optimal parameters of the spiral-
wire ribbing were obtained.

Introduction

Design of the efficient ribbing of the condenser tubes makes
use of physical and mathematical simulation. In this paper
numerical methods of mathematical simulation are applied to a
study of film condensation on a horizontal ribbed tube (Fig.1).

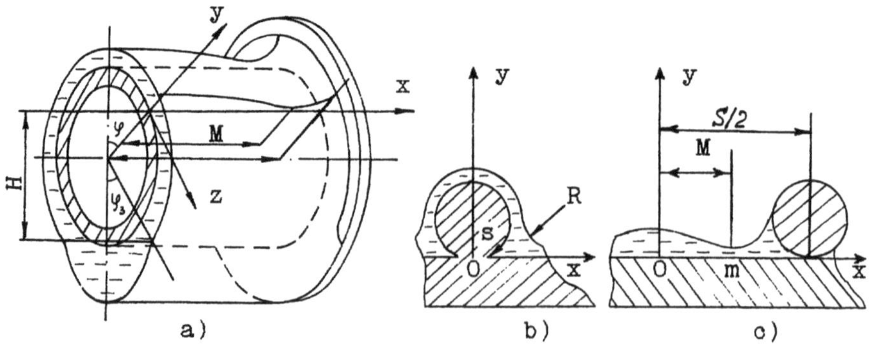

Fig.1. A tube with a cross ribbing

The proposed mathematical model incorporates the Navier-Stokes motion, energy and continuity equations with Nusselt assumptions for film condensation. It should be observed that the model takes into consideration a local gradient of the capillary pressure brought about by the varying film thickness and heat-exchange surface curvature. The film curvature change could be ocassioned by the meniscus at the rib base or by the changing surface curvature of the ribs themselves. Introduced into the boundary conditions are the condensate capillary confinement between ribs and in the under-bottom portion of tube. A detailed statement of the problem, description of algorithm for its numerical solution and some results in the dimensional form are presented in papers [1]-[3]. Taking into account the rib efficiency E influence on condesation, the rib parameters, corresponding to maximum heat transfer coefficient values were obtained in this work.

Condensation in the Film-Meniscus Transition Region

The transition region (shown as M or O-m in figure 1)is typical of the meniscus for all types of rib surfaces for condensers [2].

Using the results of works [1]-[4] for steam condensation on a flat surface, it is possible to obtain the following non-dimensional equation for the condensate film moving under the action of the surface tension forces:

$$\left(\eta^3 We\right)^{(1)} = \frac{3}{\eta} \tag{1}$$

The local Weber number, We, is given by

$$We = \frac{A\eta^{(3)}}{\left[1+(A\eta^{(1)})^2\right]^{1.5}} - \frac{3A^3\eta^{(1)}(\eta^{(2)})^2}{\left[1+(A\eta^{(1)})^2\right]^{2.5}} \; .$$

Where η is the dimensionless film thickness, δ/δ_0, where

δ is the dimensional film thickness and

$$\delta_0 = \left[\frac{\mu\lambda\Delta T}{\rho^2 h_{LG} g} \right]^{0.25}$$

The superscrips (1), (2) and (3) correspond to the first, second and third derivatives.

The Weber number shows the ratio of surface tension to gravity force which can be expand by the dimensional film thickness by

$$We = \frac{dp/dx}{\rho g}$$

where dp/dx is the local gradient of film capillary pressure ($\sigma dk/dx$).The film curvature, k, is given by

$$k = \delta^{(2)} / \left[1 + (\delta^{(1)})^2\right]^{1.5}$$

The parameter A is given by

$$A = (GaPrK)^{-0.25}$$

where the length term in the Galileo number is given by $L = (\sigma/\rho g)^{0.5}$ and K is the Kutateladze number: –

$$K = h_{LG} / c_p \Delta T$$

Equation (1) was solved numerically by the Runge–Kutta method subject to the following boundary conditions:

$$\tilde{x} = 0 \quad : \qquad \eta^{(1)} = 0 \; , \qquad\qquad \eta^{(3)} = 0 \qquad\qquad (2)$$

$$\tilde{x} = \tilde{m} \quad : \qquad \eta^{(1)} = 0 \; , \qquad\qquad \tilde{R} = \frac{\left[1 + (A\eta^{(1)})^2\right]^{1.5}}{A\eta^{(2)}} \qquad (3)$$

where $\tilde{x}=x/L$ and $\tilde{m}=m/L$. m is the coordinate of the joining point between the condensate film and the meniscus. $\tilde{R}=R/L$ is the non-dimensional radius of meniscus at the rib base.

To solve equation (1) numerically in for steam condensation on a curved surface with a profile s=f(x), representing a wire rib(Fig.1b), we changed the variable s to the variable x using the following conversion formula for increments:

$$ds=dx/\cos(arctg(f^{'}(x)))\tag{4}$$

Assuming, $\eta^{(1)}\ll 1$, it is possible to transform equation (1) into:

$$\eta^{(4)}=-3\eta^{(1)}\eta^{(3)}/\eta+3(GaPrK)^{0.25}/\eta^4\tag{5}$$

The equation (5) solution was obtained as an analytic function form, presented by a Taylor's series [5]

$$\eta=\sum_{i=1}^{n}\frac{\eta^{(n)}}{n!}x^n\tag{6}$$

The values of twelve derivatives at the point, where $\tilde{x}=0$ was obtained by consecutive differentiation of equation (5), where $\eta,\eta^{(2)}$ and initial conditions were given in this point.The Padet approximations [6] was used to improve the mathematical series gathering. This Taylor series solution with n=12 gave film thickness values only 2-2,5 % bigher than obtained by the Runge-Kutta method (1).This difference is decreasing with increasing the number of terms in the Taylor series.

In order to estimate quantitatively the surface forces, the lo-cal Weber number We was calculated at each integration step. The values obtained of the local were used to find the mean-integral value $\overline{We}$. The results of the numerical solution of equation (1) are presented in Figs.2-4.

The two computations shown in Fig.2 illustrate typical behaviour of the relative non-dimensional film thickness η/η_* (η_*- is the initial film thickness corresponding to dp/dx=0) in the condensation surface M region , where the film goes over into the meniscus. It is clearly seen that in this M region, the film tapers off towards the meniscus with respect to η_* and is a minimum at the joining point between the film and the meniscus.With the same initial film thick-ness (in the case under consideration $\eta_*=2$) and decreasing non-dimensional radius of the meniscus $\tilde{R}$,the mean-integral film thickness drops in the M region $(0-\tilde{m})$ and mean-integral $\overline{We}$ number increases. It is possible to divide the region into two zones: the M_g region$(0-\tilde{m}_g)$, where the values of the local We<1 and the M_σ region $(\tilde{m}_g-\tilde{m})$ where the We>1.

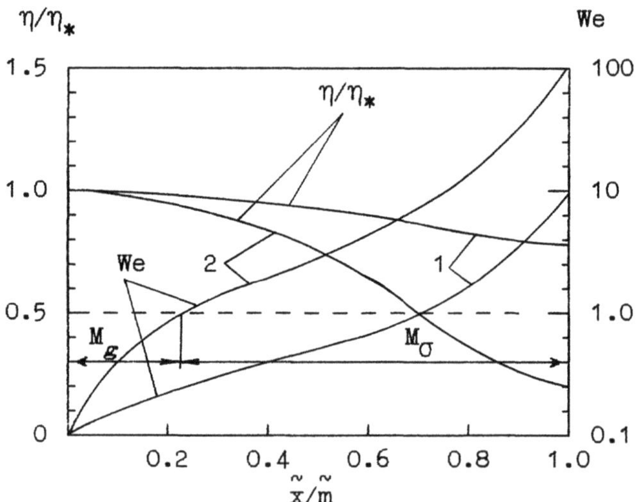

Fig.2. Distribution of the film thickness in the region where the film goes over into the meniscus:
GaPrK=2E9 ,$\eta_*=2$: 1 – $\tilde{R}=7.5$, $\overline{We}=0.8$; 2 – $\tilde{R}=0.5$, $\overline{We}=10$

It was determined that for films which have thickness $\eta_*<3$ where $\overline{We}>10$, the heat transfered from the steam to the wall in the

M_σ region equals 70-90% of all heat transfered in the M region.

Fig.3 demonstrates the relation of M region where the film goes over into the meniscus to the $\tilde{R}$ values of the initial film thickness η_*, and to the GaPrK group. The boundaries $(GaPrK)_1 =$ 1E7 and $(GaPrK)_2 =$2E9 confine all the working fluids now in use and, in particular, correspond to water and R-12 at T $=40°$C.

Relation of the $\overline{We}$ number in M region to the meniscus radius $\tilde{R}$, for various initial film thickness is presented in Fig.4. The region where the surface tension forces predominate over gravity lies above the line A-A in Fig.4 and to the left of the lines A $-$ A$_1$ and A $-$ A$_2$ corresponding to $(GaPrK)_1 =$1E7 and $(GaPrK)_2 =$2E9 respectively in Fig.3. For the same values of η_* and $\tilde{R}$, an increase in the GaPrK complex (Fig.3) involves a decreasing of the transition M region and an increasing the $\overline{We}$ value (Fig.4). For thin films $(\eta_* <0.5)$, $\overline{We} >1$ in the whole range of the analyzed values of $\tilde{R}$. With the increasing ini-

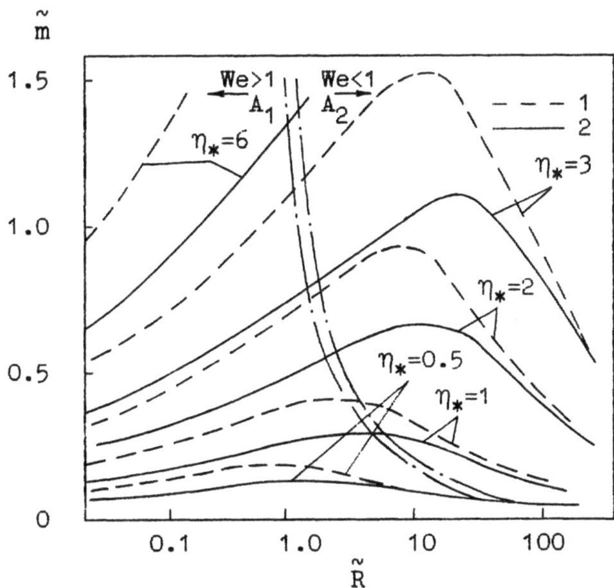

Fig.3. Relation of the extent of the film-meniscus transition region to the meniscus radius:1- GaPrK=1E7;2- 2E9

tial film thickness η_*, the transition M region extends, but the values of $\overline{We}$ >1 are observed for more reduced values of $\tilde{R}$.

It should be noted for comparison, that the value of η_*=2 corresponds to the thicknesses of water films and R-12 (at the temperature of saturation T_s=40°C) along the upper generatrices of smooth tubes 0.02 m and 0.01 m across respectively.

Proceeding from the results of numerical calculations,the following practical conclusions concerning the optimal design of the horizontal tube ribbing may be drawn:

1. If there exists a transition region where the film flow is determined by the surface tension forces ($\overline{We}$>10), then the location of its origin is primarily decided by the value of the meniscus radius and film thickness η_* influenced by the whole pre-history of the film undisturbed by the meniscus (the GaPrK value, the condensate infiltration, etc.).

2. In case when the transition M region occupies the whole

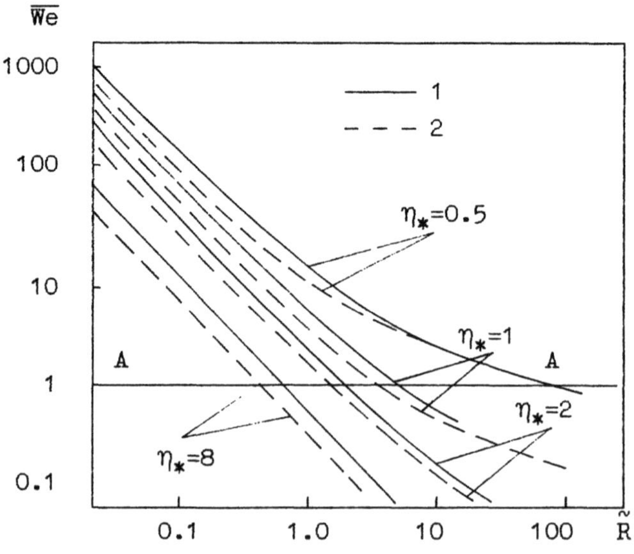

Fig.4. Relation of the $\overline{We}$ number in the film-meniscus transition region to the meniscus radius :1 − GaPrK=1E7; 2 − 2E9

element of the heat-exchange surface with a symmetrical profile
(a rib) the origin of the region lies on the axis of symmetry
(Fig.1b).

To estimate the value of surface tension, compared with the
gravity and the length of surface tension predominant influence ($\overline{We}>10$) and to calculate M region heat transfer, the results
of numerically calculations was approximated by the following
dependences $\tilde{m}=f(\tilde{R},\eta_*,GaPrK)$, $\tilde{m}=f(\tilde{R},\overline{We},GaPrK)$, $\overline{We}=f(\tilde{R},\tilde{m},GaPrK)$,
$\overline{Nu}=f(\tilde{R},\overline{We},GaPrK)$,which was used when solving the rib tube con-
densation problem.

Condensation on Tubes with The Spiral-Wire Ribbing

When designing the ribbed surface of a condensing tube, it
is necessary to try to extend the film-meniscus transition
region as much as possible. Moreover, to maximize the intensity, the extent of elements of the ribbed surface (faces of
straight ribs, radii of the curvilinear profiles of ribs)
should be kept at a minimum limited by a possible floading of
ribs by condensate. We have managed to find the solution to
this optimization problem in case of the spiral-wire ribbing.
The statement of the problem was much the same as in [1] and
[2],but the condensation on a wire rib was additionally included. The condensate was assumed to move over the wire surface by the action of the surface tension forces. In accordance with the previous part of paper, the wire diameter $\tilde{D}_w=D_w/L$
used in calculations did not exceed 1.2.

Numerical experiments were used to studied the effect of
GaPrK,tube diameter $\tilde{D}_t=D_t/L$, wire diameter $\tilde{D}_w=D_w/L$, ribbing
pitch $\tilde{S}=S/L$ and ribbing efficiency E on the variable $\tilde{Nu}=Nu/Nu_o$.
$\tilde{Nu}$ give the intensity of heat emission on a ribbed tube as compared with a smooth tube (the number Nu was computed according
to Nusselt for a smooth tube). In view of a large number of
factors (five), there was constructed a non-linear regression
model for condensation on the tubes furnished with the wire
ribbing. The optimal parameters of the spiral-wire ribbing were
chosen so as to meet possibilities of present-day technology.
Thus, for rib efficiency E=0.25 (varying with the type of rib

fastening) the optimal parameters are as follows: tube diameter $\tilde{D}_t$=15, wire diameter $\tilde{D}_w$=0.6 and ribbing pitch $\tilde{S}$ =2.1. Some typical calculations giving an insight into the mechanism of the process are presented in Fig.5-8.

CONCLUSION

The results obtained in the numerical experiments allow one to draw the following conclusions concerning ribbed tubes.

1. Within the analyzed range of the GaPrK values in all calculated variants, one can observe two different heat-emission GaPrK relations.As the GaPrK complex decreases (ΔT increases) to a certain level, the Nusselt heat-emission ΔT relation holds true. Further decrease in GaPrK causes a reduction of intensification due to the increased condensate flow rate at rib base. At that point the mean capillary radius $\tilde{R}_c$ governed by the condensate capillary confinement at the rib base would be less than the efficient value, $\tilde{R}_v$, necessary to afford the condensate transport. The process pattern is therefore set by the latter value (Fig.5).

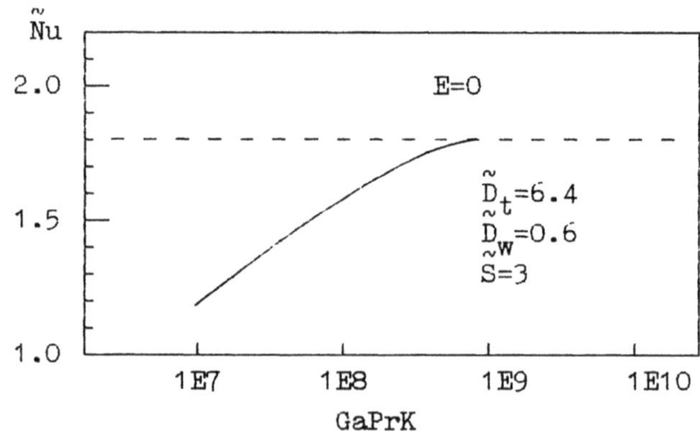

Fig.5. Relation of heat emission on a wire-ribbed tube to the GaPrK complex

2. The extremum of $\tilde{Nu}$ plotted against the tube diameter $\tilde{D}_t$ (Fig.6)is due to the fact that with a smaller tube diameter the

negative effect of the condensate capillary confinement in the
under portion of a tube is substantial, while with increasing
$\tilde{D}_t$, the value R_v increasing too (the total condensate expendi-
ture) and associated flooding of ribs become greater.

3. The maximum of $\tilde{Nu}$ plotted against the wire diameter $\tilde{D}_w$
(Fig.7) is caused by the fact that for a smaller wire diameter
$\tilde{D}_w$ there corresponds a greater $\overline{We}$ value on the wire surface,
but this effect is accompanied by a heavier flooding of the
wire by the condensate.

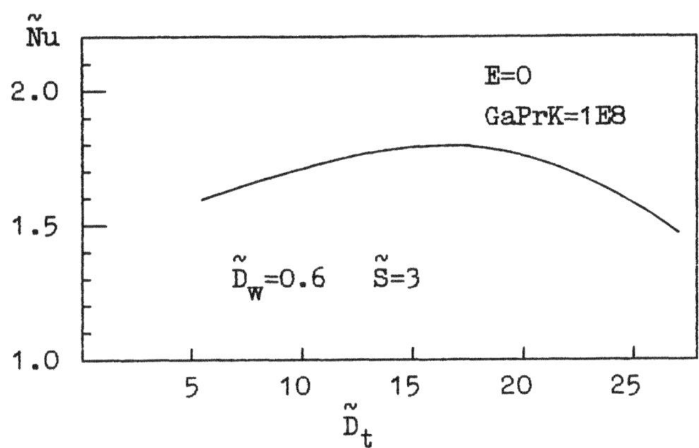

Fig.6. Relation of heat emission on a wire-ribbed
tube to its diameter

4. The extremum of $\tilde{Nu}$ plotted against the ribbing pitch $\tilde{S}$
(Fig.8)appears due to the fact that with decreasing the $\overline{We}$
value in the inter-rib clearance increases favouring thereby
the increase in the $\tilde{Nu}$ value, but at the same time the angle
of flooding of ribs and clearences between them by the
condensate also increases.

5. The increase in the rib efficiency, E, favours the increase in $\tilde{Nu}$ accompanied by the shift of the maximum of the latter
value against $\tilde{S}$ being shifted to the left (Fig.8).

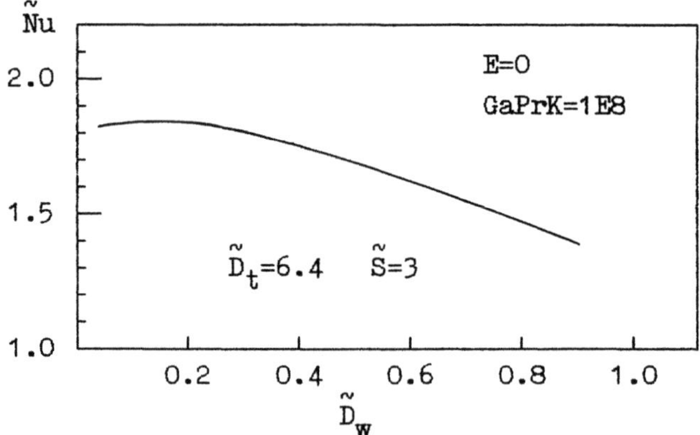

Fig.7. Relation of heat emission on a wire-ribbed tube to the wire diameter.

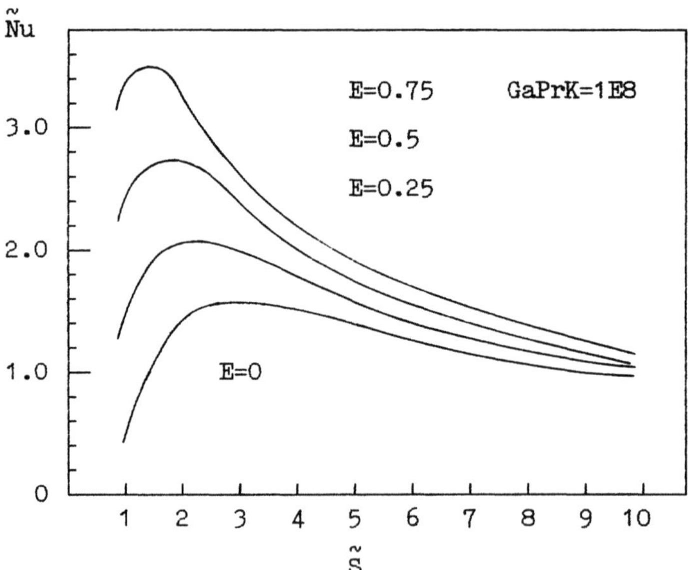

Fig.8. Relation of heat emission on a wire-ribbed tube to the spiral pitch: GaPrK=1E8, $\tilde{D}_t$=6.4, $\tilde{D}_w$=0.6

References

1. Rifert, V.G.; Trokoz, Ya.E.;.Barabash, P.A. : Theoretical study of effect of surface tension on heat exchange in condensation on profiled surface.The Proceedings of ISPCHT, May 20-23,1988, Choning, China, 290-295.

2. Rifert, V.G.; Trokoz, Ya.E.;.Barabash, P.A..Vizel, Ja.F.: Hydrodynamics anheat exchange in the region of menisci in condensation on a horizontaltube with transverse ribs. A book of scientific contributions 'Boilingand Condensation'.1988. pp 86-101 [In Russian].

3. Rifert, V.G.;.Barabash, P.A..Vizel, Ja.F.; Trokoz, Ya.E.: The effect of surface tension on hydrodynamics and heat exchange in steam condensa -tion on profiled surfaces. Industrial heat engineering. 1985, v.7,No 2. pp 20-25 [In Russian].

4. Mori, Hidzhicata, Hirasawa :'Heat Transfer'.1981 .t.103, N 1,.116-124 [In Russian].

5. Petrovsky, I.G.:Lecturers about the ordinary differential equations theory.M: MSU Publishing,1980 [In Russian].

6. Baker, J.; Graves-Morris, P.: The Padet Approximations. M: MIR Publishing,1986 [In Russian].

Measurements and Modelling:
A 350 MWe Power Plant Condenser

C. Zhang, C. Dutcher, W. Cooper,[+] K. Diab, A.C.M. Sousa and J.E.S. Venart
Fire Science Center and Department of Mechanical Engineering
University of New Brunswick
[+]New Brunswick Electric Power Commission
Fredericton, N.B. Canada

ABSTRACT

A measurement program to determine the tube-side and shell-side flows and heat transfers in a power plant condenser is described. Measurements included steam pressures on the tube bundle perimeter (96 points), steam temperatures (112 locations), inlet tubesheet water pressure distribution (26 measurements), outlet tubesheet flows and temperatures (26 points), hotwell flow and enthalpy, in addition to all makeup and extraction flowrates, as a function of load. The data were used to develop and test a numerical model for the shell-side flow, based on the governing differential equations in primitive form, and the constitutive relations for fluid flow and heat transfer. The tube bundle, baffles and additional steam flow obstacles are modeled using porous media concepts. The algorithm employs a segregated pressure correction linked formulation. A preliminary comparison of the experimental data to the model shows favorable agreement.

INTRODUCTION

EPRI [1] estimates that condenser problems cause approximately 3.8% loss of unit availability and 1.5% to 2% of performance loss of installed power in the U.S. utility industry. The costs of these losses during the period 1974-1984 was estimated to be more than US \$18 billion. An average 1 inch Hg increase of back pressure costs US \$1 million and US \$2 million annually for typical 600 MW_e oil-fired and nuclear units, respectively. Canada has a total installed generating capacity of approximately 85,000 MW_e, of which 10% is nuclear and 22%, conventional thermal sources. A conservative estimate indicates that improved Canadian condenser operation and design may lead to fuel savings in the range US \$150-200 million per year. Despite this economic incentive, few data have been obtained in condensers operating under realistic conditions in a form suitable for use with computer simulations [2, 3, 4]. To fill this gap the condenser of the 350 MW_e oil-fired Unit #1 of Coleson Cove Generating Station (New Brunswick Electric Power

Commission) was extensively instrumented as described in the following sections, and data obtained over the period October 1989 to mid July 1990. Over 2000 data records, representing a variety of loading and operational conditions, and resulting in over 5 MBytes of data, were obtained. Concurrent, with the measurement program, a sophisticated numerical model was developed so to provide for the performance evaluation of this and other units and design analysis of new units.

Early studies of fluid flow and heat transfer in power plant condensers have used the network method [5, 6]. This method, however, has the disadvantages of requiring "a priori" knowledge of the flow patterns, and of not taking into account the shape of the tube nest. In recent years, to overcome these shortcomings, sophisticated methods [7, 8, 9, 10] based on advanced numerical techniques have been suggested. They have the advantage of modelling the configuration of the tube nest and condenser internals. This permits investigation of different design alternatives, while providing detailed information of the pressure, temperature, velocity and non-condensable fields. Notwithstanding the progress made thus far, considerable work remains to be done. Previous development has invariably utilized two-dimensional flow assumptions for computational expediency in view of the limited understanding of and information about the detailed local thermal hydraulic phenomena occurring in condensers.

The shell-side flow within large power plant condensers is, in general, three-dimensional. This particular aspect has been neglected in the open literature, and is of primary concern in the present study. The numerical approach used here utilizes coupled heat transfer and fluid flow calculations with the three-dimensional effects, due to the temperature difference between inflow and outflow of the cooling water, being considered with a marching procedure along the tube length.

The governing equations describe the conservation of mass, momentum and non-condensable gas mass fraction, with the diffusive terms taken fully into account. Tube bundles and baffle plates are modelled using hydraulic resistances, and the discretization of the differential equations is carried out by a control-volume formulation over a staggered grid. The nodal equations are solved using a segregated, pressure-linked algorithm [11].

This work is an extension of ongoing research being conducted by the present authors [12, 13, 15, 16]. In [12], the algorithm is tested against steam flow and heat transfer data for an experimental condenser [14]. While in [13] preliminary computations were performed for the dual bundle condenser under study for a single load. The work presented in [15] is a preliminary report on the experimental data obtained over a variety of loads and inlet conditions. In [16] a description of the complete experimental program and its results is to be made. In the present study a variety of loads are to be considered.

FLOW CONFIGURATION

The geometrical and operating parameters for the Unit #1 steam condenser at Coleson Cove Generating Station (New Brunswick Electric Power Commission) are given in Table 1. Figure 1 depicts a side view of this dual bundle inline underslung condenser. The outside dimension of the condenser is 5.2 x 3.5 x 17 m^3 and consists of 6720 tubes in two identical bundles. There are 15 full partition plates (tube bundle supports) which divide the condenser into 16 sectors in the direction of the cooling water flow. The air extraction vent is located in the middle of the tube bundle with air being ducted out at the water inlet ends of the condenser. Steam enters the unit from top as shown in Figure 2. Cooling water flow enters each of the tube bundles from opposite ends of the condenser in order to equalize the steam flow from the LP turbines.

MEASUREMENT PROGRAM

The experimental data referred to in this study were obtained for three different operating conditions corresponding to two part loads (155 MW$_e$ and 245 MW$_e$), and a full load of 350 MW$_e$. The instrumentation layout is shown in Figure 1. Figure 2 indicates the steam pressure and temperature measurement locations. The measurements of steam pressure and steam temperature are made on 16 planes transverse to the tube bundle axis at points midspan between support plates. The steam total pressure and temperature measurements were obtained at six fixed locations in each plane. Cooling water flow rates, temperatures and pressures were obtained in 26 fully instrumented tubes located strategically in the bundle.

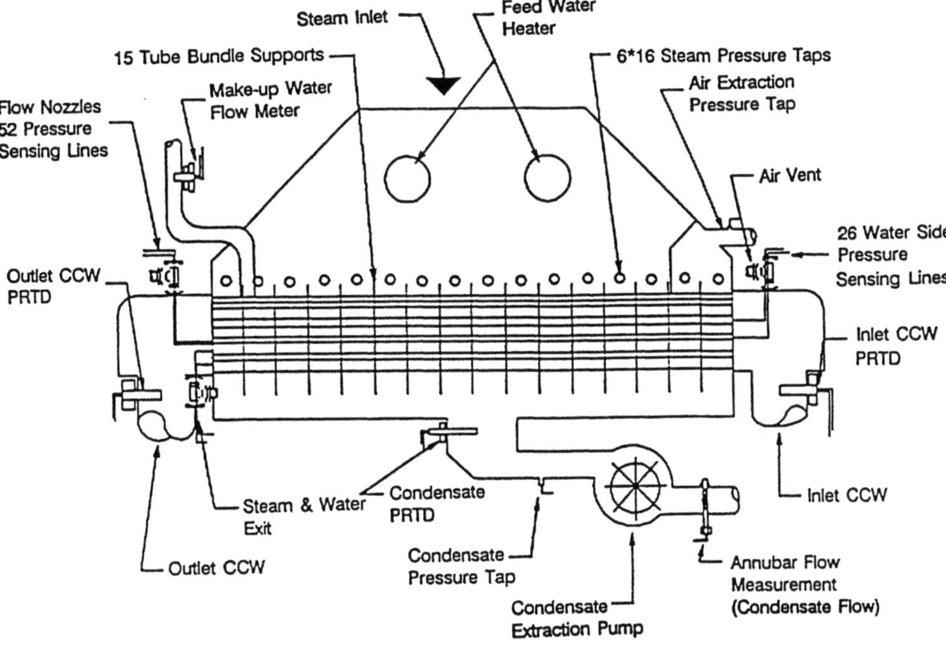

Figure 1: Longitudinal Section of Unit #1 Condenser at NBEPC Coleson Cove.

TABLE 1: Geometrical and Operating Parameters for a 350 MW Condenser

Geometrical Parameters	
Number of tube bundles	2
Number of tubes per bundle	6720
Condenser Length (m)	17
Tube Outer Diameter (mm)	25.4
Tube Inner Diameter (mm)	22.9
Tube Pitch (mm)	33.3
Operating Parameters	
Inlet Temperature of Cooling Water (C)	˜13
Inlet Velocity of Cooling Water (m/s)	2.2 (2 CCW Pumps) 1.1 (1 CCW Pump)

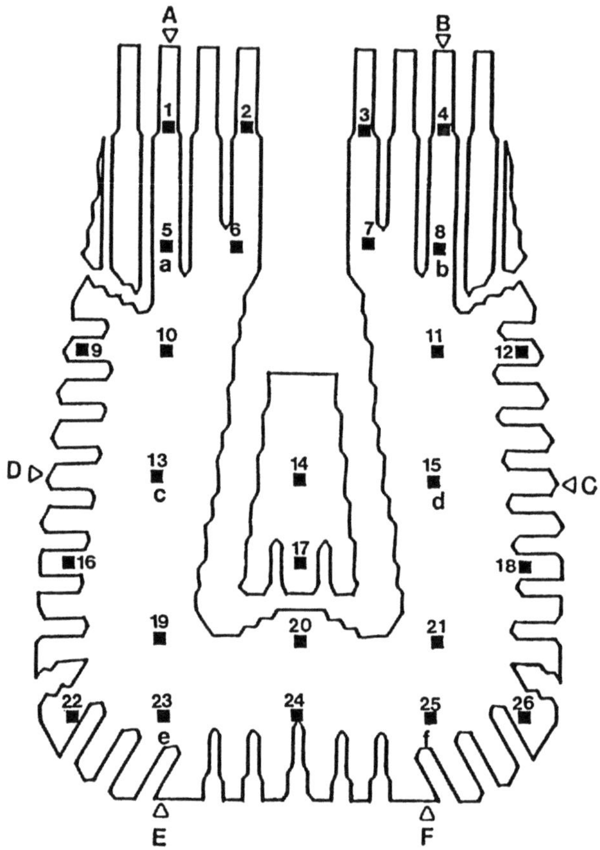

Figure 2: Locations of SteamPressure and Temperature Measurement
A-F: Pressures
1-26 CCW temperatures and flows
a-f (5, 8, 13, 14, 15, 23, 25) steam temperatures on 16 planes

Data acquisition is obtained with a network of pressure and temperature sensors, using a Multiple Scani-valve (MSV) System, and a custom built thermocouple scanner board connected to a Hewlett-Packard (HP3497A) controlled by a portable 8086 AT PC computer.

Steam and water side pressure measurements are made by four transducers of the MSV System, and the signals processed by an interface unit and read by the PC controller.

These measurements required the installation at the inlet and outlet tube sheets specially designed brass sensor heads. The steam and water side temperature measurements are made by T-type thermocouples. These thermocouples are connected to the switch board system, and their readings made by the HP 3497A DVM. The cooling water flow rates are measured by venturi-type flow nozzles, which have their high and low pressure points monitored by the MSV system. The two pressure taps for the air extraction flow measurement and one pressure tap for condensate pressure are also connected to the MSV system. The additional measurements linked to the switch board are: condensate flow rate (annubar), make-up flow rate (Tricon/E flow meter), inlet and outlet cooling water and condensate temperatures (plant PRTD's), unit load and back pressure (plant instrumentation). The data readings obtained from the MSV and switch board systems are transferred to the host micro computer, where they are manipulated and stored on hard disks. The data system and its stored data can be interogated from UNB (106 km) via a modem and data files downloaded or the program tested or restarted.

Steam Pressure Measurement

The steam total pressure measurements were obtained at each of the 16 midspan baffle plate stations at six fixed locations external to the tube bundle as shown in Figure 2. Each fixed pressure port on a given plane is connected to an absolute pressure transducer, externally located to the condenser shell through a manifold arrangement using the MSV stepping motor controlled "0" ring sealed valve. To obtain pressure measurements sequentially for each port, these valves are automatically operated by a data acquisition system. The lines are purged and the absolute reference pressure provided by a vacuum pump (Model DD50 Precision, 50 ℓ/m, 0.5 micron).

Steam Temperature Measurement

The steam temperature measurements were obtained using thermocouple probes in plugged tubes on the waterside. Since there is no water flow through these tubes, they are presumed to be in thermal equilibrium with the surrounding steam. The thermocouples are located at mid plane between support plates, and they are held tightly against the inside tube wall by an inflated bladder provide a good measure of the local total steam or adiabatic tube wall temperatures. The thermocouples are made of pure insulated strip copper wires and one constantan wire connected differentially, yielding both absolute and differential temperatures along the tube.

The estimated precision of the pressure, temperature and flow measurements is ±15 Pa, ±0.2 K and ±1-2%, respectively. These estimates were further corroborated by measurements taken at no load and shutdown when ambient conditions prevailed. An interim report [15] covers most aspects dealing with the measurement program, reliability, accuracy, and instrumentation. Complete details are to be made available elsewhere [16].

NUMERICAL METHOD

The numerical procedure developed has been described in detail in [12, 13] and only its main features will be described. The shell-side and tube-side flows are treated as steady-state. In the shell-side, the steam-air mixture is considered to be an ideal gas, and the steam is assumed saturated for computational convenience.

Due to the presence of partition plates and relatively small water temperature differences commonly experienced between adjacent partitions, a valid assumption is to consider that the shell-side flow has negligible velocity components parallel to the tube bundle. The condenser shell-side may thus be subdivided into a number of two-dimensional domains normal to the cooling water flow. Thus in each subdomain, the flow is assumed two-dimensional, with the domains interacting through the "thermal memory" of the tubeside cooling water. Calculations for each plane can then be made sequentially starting from the cooling water inlet end; the outlet cooling water temperature of the preceding subdomain being used as the inlet cooling water temperature for the subsequent subdomain. This marching procedure is used for each successive section of the condenser. The two-dimensional approach described below is thus extended to the third dimension.

Governing Equations

Two dimensional steady state porous medium volume-averaged conservation equations for mass, momentum and air mass fraction are used to describe the transport phenomena occurring in each two-dimensional subdomain. Heat and mass transfer resistances in these equations are employed to account for the wall effects. A locally isotropic porosity, β, defined as the ratio between volume occupied by the fluid and total volume, is used to represent the flow volume reduction due to the tube bundle and bafflers for each control volume. Consequently, β is a function of the local tube placement and grid selection. The resulting equations and the required constitutive relations have been given elsewhere [13].

Boundary Conditions

The boundary conditions for the inlet, outlet and solid walls are:

Inlet: The velocity and air mass fraction are specified at the inlet boundary.

Vent: A mass imbalance correction method based upon the evaluation of the total steam condensed.

Walls: The shell walls of the condenser are assumed to be non-slip, impervious to flow, and adiabatic. Thus, the normal velocity components are equal to zero and air mass fraction gradients normal to the walls are set to zero.

Plane of symmetry: Along the center line the derivatives with respect to the cross stream direction of all field variables are set to zero.

Solution Procedure

The discretization of the differential equations is carried out by integrating over small control volumes in a staggered grid. Since these equations are coupled together and are highly non-linear, an iterative approach is used for their solution. A cyclic outer iteration is employed comprising the following sequence of operations:

(i)   The momentum equations are solved based on a pressure field taken from the previous iteration.

(ii) A Poisson equation for the pressure correction, derived from the continuity equation is solved, and at the end of each outer iteration loop, pressures and velocities are corrected.

(iii) The air mass fraction, ϕ, is obtained from the discretized form of its transport equation.

(iv)   The temperatures of mixture and cooling water, density, mass source term, and momentum source term are then updated.

(v)   A new cycle is started unless the prescribed accuracy has been reached.

(vi)   The overall energy balance for cooling water, steam, and condensate is checked.

PREDICTIONS

The calculations are performed in a mesh of 31 by 26 in the main and cross flow directions, respectively. Previous studies [12, 13, 20] indicate that a grid of this size adequately reflects the geometry, flow and heat transfer. An average of 90 cycles provides adequate convergence for each plane.

In order to solve the governing equations, the inlet boundary conditions for the steam flow must be known. The steam inlet conditions are complicated due to the rotational effects of the turbine exhaust flow, the hood configuration, and internal structures. It is assumed here, as a first approximation, that the inlet velocity profile is uniform for each sector. The mass flow rate for each sector is assumed to be equal to the maximum possible condensation rate at this sector, and the magnitude of inlet velocity in each sector is determined by its mass flow rate. The pressure at B (Figure 2) is chosen as a reference pressure. Since the inlet air mass fraction was not available, a tentative value of 0.15% of inlet air mass fraction was made based upon the maximum capacity of the air extraction system. To illustrate the capability and flexibility of the developed numerical model typical predictions are depicted for full load (350 MW$_e$) at the first partition (Figures 4-6).

The velocity vector plot is shown in Figure 3. It can be seen from this figure that the velocity distribution in the vicinity of the tube bundle is nearly "parallel" to the tube bundle edge except for the "hot well" region. This particular flow pattern can be inferred from the experimental observations since a large proportion of the steam flow goes through the steam lanes.

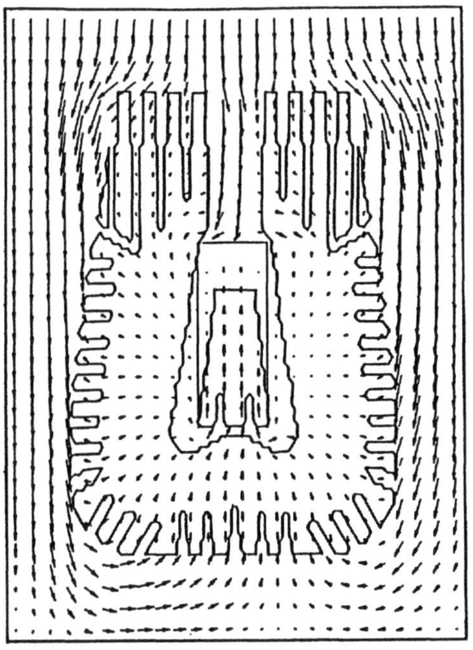

Figure 3. Velocity Vector Plot (Plane No. 1) for Full Load.

Figure 4 provides a contour map for the pressure distribution in the first computational plane of the condenser. The pressure distribution around the tube bundle, as corroborated by the measurements, is not uniform. It should be noted, however, that design codes based on the network method [e.g., 21] routinely make the assumption of pressure uniformity around the tube bundle, which may yield seriously inaccurate predictions.

The air mass fraction contour map is given in Figure 5 and, as expected, a large air bubble can be easily identified in the venting region.

Figures 6 and 7 depict, for the three power loads considered, the distribution of the condensation rate and measured outlet cooling water temperature, respectively, for each computational domain. The high outlet temperature at 155 MW_e is due to the fact that only one cooling water pump is running.

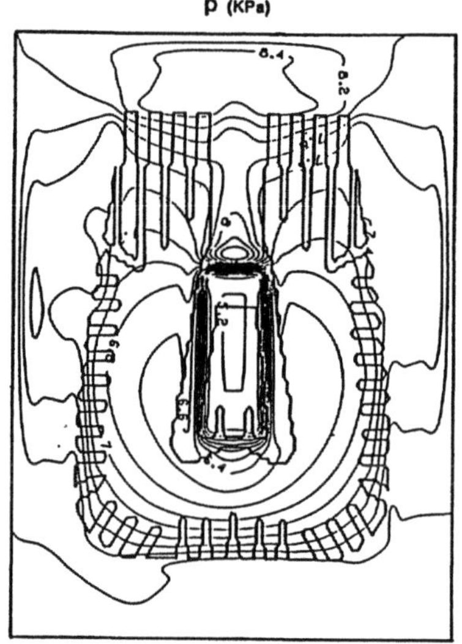

Figure 4. Pressure Distribution (Plane No. 1) for Full Load.

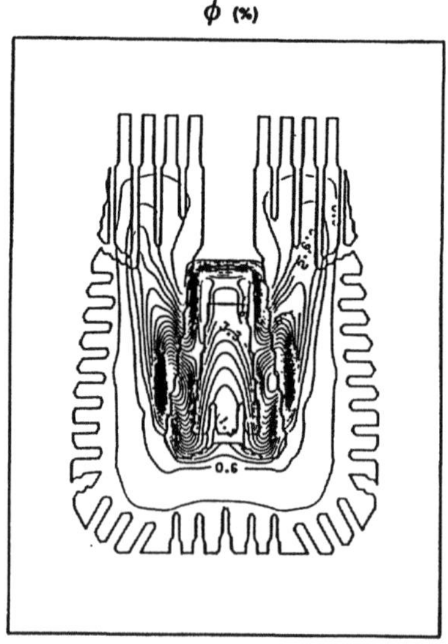

Figure 5. Air Concentration Distribution (Plane No. 1) for Full Load.

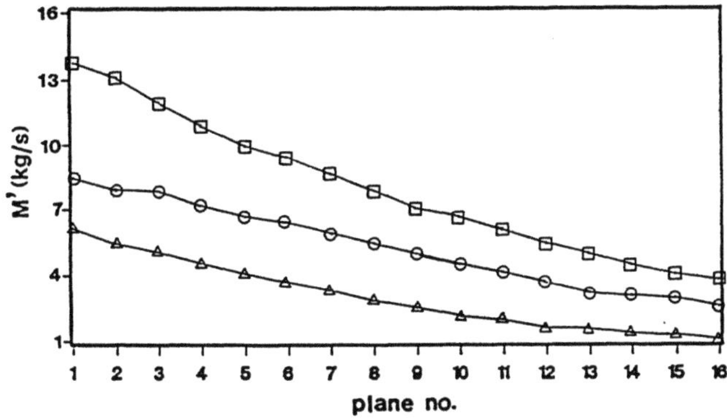

Figure 6. Distribution of Condensation Rate
($\Box$: 350 MW$_e$; $\bigcirc$: 245 MW$_e$; $\triangle$: 155 MW$_e$)

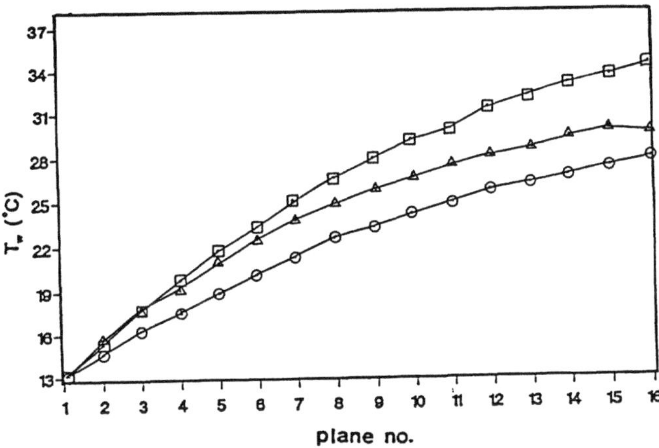

Figure 7. Distribution of Averaged Cooling Water Temperature
($\Box$: 350 MW$_e$; $\bigcirc$: 245 MW$_e$; $\triangle$: 155 MW$_e$)

COMPARISON BETWEEN PREDICTIONS AND EXPERIMENTS

Table 2, for the three cases studied, provides a first comparison between predicted and experimental results for the total condensation rate, M'_{total}, averaged condensate temperature, T_{cond}, and mean outlet cooling water temperature, $(T_w)_{out}$. Despite the uncertainties of the numerical model, particularly the imposition of symmetry conditions along the condenser centre line, the comparison is encouraging. Of some concern is the relatively large discrepancy between prediction and experiment for mass flow condensed (M_{total}) at the lowest load. This may indicate departures from the assumed inlet conditions, and, in particular, the 0.15% air mass fraction utilized.

Table 2. Comparison between Prediction and Experiments

Case (MW$_e$)	M'$_{total}$ (Kg/s)			T$_{cond}$ (°C)			(T$_w$)$_{out}$ (°C)		
	Pred.	Exp.	%	Pred.	Exp.	%	Pred.	Exp.	%
155	52.15	61.23	14.8	33.16	29.21	13.5	31.11	31.0	0.3
245	89.03	91.01	2.2	32.20	29.34	9.7	28.54	26.87	6.2
355	131.03	135.9	3.6	41.20	37.66	9.4	35.24	33.05	6.6

Comparisons for steam pressures and temperatures for each load are listed in Tables 3a to 5b for each of the 16 planes. Tables 3a, 4a, 5a relate the computed and experimental pressures while 3b, 4b, and 5b indicate the same for the interior bundle steam temperatures. The overall agreement for all cases is very good, despite the variety of loads considered and the simplifying assumptions made. In general the predicted pressures at location A are higher than the experimental results for the first four planes, while the predicted temperatures at location b are lower than the experimental data. For the first four planes for all loads the difference of experimental pressures at A and B is large and pressure A is much lower than pressure B, while the differences of the predicted pressures at A and B is small. After the fourth plane, the experimental pressure differences between A and B is small, and the predicted pressures agree well with the experimental data in most cases. A possible explanation for the large pressure difference between A and B may be attributed to the inlet flow distributions which unfortunately, at this stage of the experimental program, there is as yet no available information. This is a prime source of uncertainty between the predictions and the experimental data. The assumption of flow symmetry between bundles for this type of condenser is also questionable, since the cooling water enters each tube bundle from opposite ends of the condenser.

Table 3a: Predicted and Experimental Steam Pressures, 155 MW$_c$

Pressure (Pa)

Location		A	B	C	D	E	F
Plane #1	Pred.	5088.8	5053.1	4728.1	4836.8	4885.4	4849.5
	Exp.	4835.4	5053.1	4996.28	4873.21	5075.2	5100.5
Plane #2	Pred.	5268.1	5236.1	4954.4	5050.0	5089.6	5058.8
	Exp.	5163.5	5236.1	5255.1	5034.2	5217.3	5261.5
Plane #3	Pred.	5348.4	5221.4	5099.7	5177.6	5205.1	5180.5
	Exp.	5239.3	5221.4	5324.5	5113.1	5264.6	5305.6
Plane #4	Pred.	5382.2	5359.3	5183.1	5247.2	5266.1	5246.4
	Exp.	5321.4	5359.3	5356.1	4983.7	5280.3	5362.4
Plane #5	Pred.	5366.3	5346.7	5203.6	5257.4	5270.6	5254.6
	Exp.	5321.4	5346.7	5387.7	5308.7	5359.3	5371.9
Plane #6	Pred.	5468.0	5450.8	5329.3	5376.0	5385.9	5372.3
	Exp.	5460.3	5450.8	5397.1	5245.7	5340.3	5387.7
Plane #7	Pred.	5456.0	5441.3	5341.9	5381.2	5387.9	5376.8
	Exp.	5390.8	5441.3	5394.0	5267.7	5365.6	5425.5
Plane #8	Pred.	5450.7	5438.2	5356.0	5389.2	5393.8	5384.7
	Exp.	5403.5	5438.2	5406.6	5299.3	5398.2	5416.1
Plane #9	Pred.	5437.2	5426.7	5361.1	5388.3	5391.1	5383.9
	Exp.	5150.9	5226.7	5226.7	5173.06	5270.9	5340.3
Plane #10	Pred.	5415.4	5406.6	5353.5	5375.9	5377.6	5371.8
	Exp.	5409.8	5406.6	5378.2	5242.5	5349.8	5409.8
Plane #11	Pred.	5445.3	5438.2	5397.2	5414.7	5415.7	5411.3
	Exp.	5444.5	5438.2	5428.7	5349.8	5412.9	5441.3
Plane #12	Pred.	5444.0	5438.2	5405.8	5419.7	5420.3	5416.9
	Exp.	5507.6	5438.2	5463.4	5428.7	5431.9	5447.6
Plane #13	Pred.	5478.0	5472.9	5444.5	5456.7	5457.1	5454.1
	Exp.	5558.1	5472.9	5501.3	5375.1	5397.1	5450.8
Plane #14	Pred.	5476.8	5472.9	5452.3	54611.1	5461.3	5459.3
	Exp.	5494.9	5472.9	5453.9	5384.5	5513.9	5513.9
Plane #15	Pred.	5587.1	5583.4	5563.3	5571.9	5572.0	5570.0
	Exp.	5548.6	5583.4	5523.4	5412.9	5482.4	5435.0
Plane #16	Pred.	5545.2	5542.3	5528.0	5534.0	5534.1	5532.7
	Exp.	5586.5	5542.3	5435.0	544.3	5428.7	5435.0

Table 3b: Predicted and Experimental Steam Temperatures, 155 MW$_e$

Temperature (°C)

Location		5	8	13	14	15	23	25
Plane #1	Pred.	31.45	31.75	30.66	28.54	30.98	31.35	31.54
	Exp.	32.16	33.46	30.44	30.23	*	35.46	32.86
Plane #2	Pred.	32.36	32.62	31.68	29.39	31.95	32.25	32.40
	Exp.	32.40	33.86	30.82	*	*	35.39	33.05
Plane #3	Pred.	32.93	33.14	32.42	31.02	32.63	32.85	32.97
	Exp.	*	34.03	31.09	*	*	35.46	33.14
Plane #4	Pred.	33.27	33.44	32.87	31.90	33.03	33.20	33.30
	Exp.	33.60	34.29	31.31	32.04	*	35.46	*
Plane #5	Pred.	33.38	33.52	33.05	33.20	33.19	33.32	33.40
	Exp.	33.19	34.70	*	32.4	*	35.41	33.50
Plane #6	Pred.	33.83	33.96	33.55	32.76	33.67	33.78	33.84
	Exp.	33.41	34.82	32.13	32.20	*	35.39	33.55
Plane #7	Pred.	33.90	34.00	33.67	32.94	33.77	33.85	33.90
	Exp.	33.67	35.04	*	32.73	*	35.36	33.84
Plane #8	Pred.	33.97	34.06	33.77	33.09	33.86	33.92	33.96
	Exp.	33.86	35.13	32.89	33.19	*	35.63	33.94
Plane #9	Pred.	34.00	34.07	33.84	33.30	33.91	33.97	34.00
	Exp.	34.08	35.18	33.30	33.55	*	35.94	34.20
Plane #10	Pred.	33.99	34.05	33.86	33.38	33.92	33.95	33.98
	Exp.	34.44	*	33.64	33.81	*	36.15	34.32
Plane #11	Pred.	34.14	34.19	34.05	33.77	34.09	34.12	34.14
	Exp.	34.29	35.44	34.00	34.05	*	36.34	34.54
Plane #12	Pred.	34.18	34.22	34.10	33.90	34.14	34.16	34.18
	Exp.	34.65	35.66	34.45	34.32	*	36.87	34.77
Plane #13	Pred.	34.31	34.35	34.24	34.01	34.27	34.29	34.31
	Exp.	34.77	35.73	34.81	34.41	*	36.51	34.8
Plane #14	Pred.	34.34	34.37	34.29	34.19	34.32	34.33	34.34
	Exp.	34.82	35.68	*	34.68	*	36.27	35.11
Plane #15	Pred.	34.71	34.73	34.66	34.52	34.68	34.69	34.70
	Exp.	35.01	35.48	*	34.56	*	36.18	35.44
Plane #16	Pred.	34.60	34.61	34.56	34.49	34.58	34.59	34.59
	Exp.	35.04	35.36	*	35.23	*	35.75	35.71

* Data not available.

Table 4a: Predicted and Experimental Pressures, 245 MW$_e$

Pressure (Pa)

Location		A	B	C	D	E	F
Plane #1	Pred.	5045.3	4990.0	4349.9	4549.0	4677.2	4605.8
	Exp.	4614.4	4990.0	4892.2	4668.0	4926.9	4958.4
Plane #2	Pred.	5462.1	5409.8	4809.5	4990.0	5111.8	5047.6
	Exp.	5027.9	5409.8	5150.9	4876.4	5116.2	5141.5
Plane #3	Pred.	5422.7	5375.1	4859.8	5017.6	5116.0	5061.2
	Exp.	5160.4	5375.1	5223.5	4917.4	5119.4	5242.5
Plane #4	Pred.	5344.8	5302.5	4876.6	5010.8	5085.6	5040.0
	Exp.	5267.7	5302.5	5283.5	4677.5	5191.9	5302.5
Plane #5	Pred.	5296.0	5258.3	4902.8	5018.3	5075.5	5037.3
	Exp.	5207.7	5258.3	5311.9	5176.2	5296.1	5315.1
Plane #6	Pred.	5529.8	5495.0	5174.6	5279.0	5329.2	5294.9
	Exp.	5482.4	5495.0	5302.5	5062.6	5270.9	5337.2
Plane #7	Pred.	5510.4	5479.2	5206.5	5297.9	5337.1	5307.9
	Exp.	5311.9	5479.2	5337.2	5106.7	5337.2	5406.6
Plane #8	Pred.	5485.2	5457.1	5222.9	5303.6	5334.5	5309.4
	Exp.	5321.4	5457.1	5324.5	5289.9	5384.5	5400.3
Plane #9	Pred.	5452.1	5426.7	5223.1	5295.2	5319.7	5397.9
	Exp.	5040.5	5226.7	5106.7	5037.3	5169.9	5311.9
Plane #10	Pred.	5441.8	5419.2	5245.2	5308.4	5327.3	5308.7
	Exp.	5356.1	5419.2	5220.4	4914.2	5210.9	5340.3
Plane #11	Pred.	5401.3	5381.3	5235.3	5289.9	5303.9	5288.2
	Exp.	5311.9	5381.3	5261.5	5207.7	5324.5	5343.5
Plane #12	Pred.	5364.3	5346.7	5222.3	5270.0	5280.5	5267.2
	Exp.	5400.3	5346.7	5330.9	5318.3	5308.7	5330.9
Plane #13	Pred.	5308.3	5293.0	5189.1	5230.0	5237.4	5226.3
	Exp.	5428.7	5293.0	5334.0	5097.3	5239.3	5346.7
Plane #14	Pred.	5388.9	5375.1	5284.4	5320.6	5326.5	5316.8
	Exp.	5334.0	5375.1	5293.0	5160.4	5324.5	5362.4
Plane #15	Pred.	5457.2	5444.5	5362.9	5395.8	5400.7	5392.0
	Exp.	5375.1	5444.5	5346.7	5201.5	5318.3	5274.1
Plane #16	Pred.	5455.7	5444.5	5374.6	5403.3	5406.8	5399.4
	Exp.	5394.0	5444.5	5233.0	5223.5	5270.9	5270.9

Table 4b: Predicted and Experimental Steam Temperatures, 245 MW$_e$

Temperature (°C)

Location		5	8	13	14	15	23	25
Plane #1	Pred.	26.95	30.18	28.00	23.44	28.65	29.52	29.91
	Exp.	31.22	32.72	31.76	32.04	31.52	35.05	33.75
Plane #2	Pred.	31.50	31.96	30.07	25.81	30.61	31.37	31.37
	Exp.	31.41	33.08	32.03	*	32.12	35.05	34.07
Plane #3	Pred.	31.77	32.18	30.54	26.70	31.01	31.64	31.92
	Exp.	*	33.30	32.37	*	31.93	35.13	34.02
Plane #4	Pred.	31.91	32.26	30.90	28.20	31.29	31.80	32.03
	Exp.	32.06	33.54	32.61	33.79	32.07	35.25	*
Plane #5	Pred.	32.07	32.38	31.24	29.24	31.57	31.98	32.17
	Exp.	32.37	34.14	*	33.12	32.89	35.37	34.35
Plane #6	Pred.	33.08	33.35	32.36	30.81	32.65	33.00	33.17
	Exp.	32.71	34.31	33.23	33.46	32.86	35.32	34.31
Plane #7	Pred.	33.24	33.48	32.63	31.32	32.88	33.17	33.31
	Exp.	33.07	34.62	*	33.0	33.18	35.44	34.64
Plane #8	Pred.	33.34	33.55	32.81	31.59	33.03	33.27	33.40
	Exp.	33.43	34.93	33.88	33.10	33.85	36.11	34.88
Plane #9	Pred.	33.38	33.57	32.91	31.64	33.11	33.31	33.42
	Exp.	33.65	34.93	34.05	33.07	33.87	36.80	34.90
Plane #10	Pred.	33.49	33.66	33.09	31.99	33.26	33.42	33.52
	Exp.	34.06	*	34.05	33.00	33.90	37.37	34.64
Plane #11	Pred.	33.49	33.63	33.14	32.23	33.29	33.42	33.51
	Exp.	33.86	35.26	34.31	33.07	34.42	37.75	35.10
Plane #12	Pred.	33.47	33.60	33.17	32.33	33.30	33.41	33.48
	Exp.	34.32	35.53	34.50	33.14	34.95	38.80	35.29
Plane #13	Pred.	33.37	33.49	33.12	32.38	33.23	33.32	33.38
	Exp.	34.42	35.55	34.60	33.00	34.88	38.01	34.59
Plane #14	Pred.	33.71	33.81	33.50	32.96	33.59	33.67	33.72
	Exp.	34.46	35.29	36.34	33.10	35.31	37.75	35.07
Plane #15	Pred.	33.99	34.08	33.80	33.31	33.88	33.95	33.99
	Exp.	34.66	35.01	36.19	33.07	35.76	37.78	35.36
Plane #16	Pred.	34.04	34.12	33.87	33.45	33.95	34.00	34.04
	Exp.	34.70	34.86	36.24	33.61	36.41	37.25	35.50

* Data not available.

Table 5a: Predicted and Experimental Steam Pressures, 350 MW$_e$

Pressure (Pa)

Location		A	B	C	D	E	F
Plane #1	Pred.	8213.9	8155.7	7298.4	5702.9	7727.2	7643.6
	Exp.	7615.8	8155.7	7959.9	7628.2	8007.5	8013.7
Plane #2	Pred.	8696.1	8641.8	7848.9	8034.8	8240.9	8167.4
	Exp.	8130.2	8641.8	8303.9	7890.2	8253.6	8288.1
Plane #3	Pred.	8649.8	8600.4	7920.0	8084.0	8253.1	8189.7
	Exp.	8338.4	8600.4	8398.4	7966.1	8266.0	8408.3
Plane #4	Pred.	8578.7	3534.2	7959.9	8103.5	8238.6	8184.4
	Exp.	8487.3	8534.2	8512.1	7641.3	8361.1	8537.6
Plane #5	Pred.	8479.4	8439.7	7964.4	8088.7	8193.4	8147.4
	Exp.	8385.0	8439.7	8515.6	8351.5	8509.4	8515.6
Plane #6	Pred.	8791.7	8755.5	8335.5	8447.0	8536.7	8496.0
	Exp.	8771.4	8755.5	5825.2	8124.0	8490.1	8581.8
Plane #7	Pred.	8765.8	8733.5	8379.4	8477.4	8548.2	8513.3
	Exp.	8543.9	8733.5	8565.9	8301.2	8581.8	8670.0
Plane #8	Pred.	8749.0	8720.4	8427.9	8512.6	8566.8	8537.2
	Exp.	8572.1	8720.4	8550.1	8525.2	8632.1	8651.4
Plane #9	Pred.	8765.7	8740.4	8497.1	8570.5	8612.2	8587.1
	Exp.	8297.7	8540.4	8329.4	8209.5	8452.2	8607.3
Plane #10	Pred.	8781.0	8758.3	8550.0	8615.1	8648.3	8626.6
	Exp.	8641.8	8758.3	8461.8	8045.4	8505.9	8645.2
Plane #11	Pred.	8705.9	8685.9	8513.0	8569.5	8594.4	8576.2
	Exp.	8585.2	8685.9	8565.9	8398.4	8622.5	8654.2
Plane #12	Pred.	8663.1	8645.2	8498.4	8548.2	8567.4	8551.8
	Exp.	8739.7	8645.2	8641.8	8648.0	8636.3	8648.0
Plane #13	Pred.	8673.4	8657.6	8533.5	8577.0	8591.7	8578.3
	Exp.	8739.7	8657.6	8597.6	8316.3	8537.6	8654.2
Plane #14	Pred.	8684.0	8670.0	8565.6	8603.3	8614.4	8603.1
	Exp.	8585.2	8670.0	8559.7	8389.4	8625.9	8657.6
Plane #15	Pred.	8723.9	8711.4	8621.2	8654.6	8663.3	8653.5
	Exp.	8619.7	8711.4	8603.8	8477.7	8588.0	8547.3
Plane #16	Pred.	8725.2	8714.2	8636.9	8666.3	8672.8	8664.5
	Exp.	8670.0	8714.2	8483.9	8480.4	8550.1	8528.0

358

Table 5b: Predicted and Experimental Steam Temperatures, 350 MW$_e$

Temperature (°C)

Location		5	8	13	14	15	23	25
Plane #1	Pred.	39.00	39.36	37.72	35.91	38.13	39.01	39.29
	Exp.	38.7	42.38	36.53	35.91	*	43.25	40.13
Plane #2	Pred.	40.50	40.81	39.31	36.38	39.68	40.45	40.69
	Exp.	38.96	42.19	36.89	*	*	43.13	40.58
Plane #3	Pred.	40.76	41.04	39.73	36.83	40.05	40.70	40.90
	Exp.	*	42.59	32.27	*	*	43.11	40.61
Plane #4	Pred.	40.94	41.18	40.06	37.54	40.34	40.86	41.04
	Exp.	39.62	43.8	37.44	37.89	*	42.99	*
Plane #5	Pred.	41.01	41.23	40.28	38.34	40.53	40.95	41.10
	Exp.	40.07	43.23	*	38.48	*	42.82	41.15
Plane #6	Pred.	41.93	42.12	41.31	39.54	41.52	41.87	42.00
	Exp.	40.38	42.95	38.77	37.58	*	42.59	41.18
Plane #7	Pred.	42.08	42.25	41.56	40.08	41.74	42.02	42.13
	Exp.	40.83	43.70	*	38.63	*	42.35	41.70
Plane #8	Pred.	42.23	42.38	41.80	40.73	41.96	42.18	42.28
	Exp.	41.16	43.61	40.03	39.03	*	42.35	42.07
Plane #9	Pred.	42.42	42.55	42.07	41.28	42.20	42.38	42.46
	Exp.	41.52	43.75	40.53	39.88	*	42.28	42.17
Plane #10	Pred.	42.57	42.68	42.27	41.56	42.38	42.53	42.60
	Exp.	42.09	*	40.96	40.07	*	42.16	42.19
Plane #11	Pred.	42.51	42.61	42.26	41.69	42.36	42.48	42.54
	Exp.	41.90	44.10	41.57	40.41	*	41.93	42.62
Plane #12	Pred.	42.50	42.59	42.28	41.77	42.37	42.47	42.52
	Exp.	42.44	44.06	42.16	40.52	*	41.83	43.00
Plane #13	Pred.	42.60	42.67	42.41	41.99	42.49	42.57	42.61
	Exp.	42.56	43.96	42.73	40.86	*	41.29	42.69
Plane #14	Pred.	42.68	42.75	42.53	42.20	42.59	42.66	42.70
	Exp.	42.68	43.80	*	41.31	*	41.24	43.14
Plane #15	Pred.	42.82	42.88	42.68	42.39	42.74	42.80	42.83
	Exp.	42.96	43.06		41.12	*	41.34	43.61
Plane #16	Pred.	42.86	42.92	42.75	42.49	42.80	42.84	42.87
	Exp.	43.03	43.01	*	41.29	*	40.86	43.87

* Data not available.

The experimental data show that the temperatures at a and b nearly correspond to the saturation temperatures for the pressures at A and B, respectively. In the simulation, however, a significant pressure drop occurs between A and a, and B and b. Since the pressure B is set equal to the experimental value, it results in predicted temperatures at b lower than experimental values.

The predicted condensing rate, the mean outlet cooling water and mean condensate temperatures (Table 2) also compare well with the experimental values.

CONCLUSION

An extensive experimental program on a 350 MW_e thermal condenser has been conducted over a variety of loads ranging from full to 50% of load. Detailed pressure and temperature field measurements indicate that the steam flows are highly three dimensional and that condensing rate per support plate section varies by a factor of 3 from the cooling water inlet to exit. A simulation of the shell-side steam flow and heat transfer for the condenser has also been carried out. The flow was assumed to be quasi-three-dimensional and incompressible, however, density was allowed to vary with temperature and air concentration. The numerical method proposed has shown the capability of predicting the performance of condensers including three dimensional effects due to the increase of cooling water temperature. The predictions have produced physically meaningful results, and when consideration is given to the uncertainties of the experimental data, the assumptions undertaken, and the limitations of the computational procedure, the predictive capability of the model appears very encouraging. Overall the algorithm developed shows good potential as an analytical and design tool.

Modification of the code to evaluate the clearly non-symmetric boundary between bundles is necessary before final assessment of its capabilities can be made.

ACKNOWLEDGEMENT

The participation of the New Brunswick Electric Power Commission and the staff at the Coleson Cove Generating Station, and in particular, Mr. W. Brown and Mr. L. Webb,

were essential to the successful completion of the measurement program. The work has be
financially supported by the New Brunswick Electric Power Commission, the Natural Sciences a
Engineering Research Council through an NSERC University-Industry Collaborative Resear
Grant (NSERC Grant No. CRD-0039112) and the NSERC Operating Grant of one of the auth
(ACMS) (NSERC A1398).

REFERENCES

[1]   Diaz-Tous, I.A., "Keynote Address", Proc. Symposium on State-of-the-Art Condens
        Technology, Eds. I.A. Diaz-Tous and R.J. Bell, Orlando, Florida, June 7-9, 1983, pp. 1:
        1:22.

[2]   Rowe, M. and Ferrison, S.A., "Air Cooling and Venting Arrangements in Stea
        Condensers", ASME 84 - JPGC - PWR - 12.

[3]   Davidson, B.J., "Steam Condenser Thermal Design Theories", Lecture Series 1983-06, V
        2, Von Karman Inst. for Fluid Dynamics, Belgium.

[4]   Mussalli, Y.G., Bell, R.J., Impugliazzo, A.M., "High-Reliability Condenser Design Stud
        CS-3200, Research Project 1689-10, EPRI, 1983.

[5]   Barsness, E.J., "Calculation of the Performance of Surface Condenser by Digi
        Computer", ASME Paper No. 63 - PWR-2, 1963.

[6]   Chisholm, D., Osment, B.D.J., McFarlane, Mrs. M.W. and Choudhury, M.H., "T
        Performance of an Experimental Condenser", Proc. of 3rd Int. Heat Transfer Conferenc
        Chicago, IL, Vol. 1, 1966, pp. 179-185.

[7]   Davidson, B.J. and Rowe, M. "Simulation of Power Plant Condenser Performance
        Computational Method: An Overview", Power Condenser Heat Transfer Technology, Ed
        P. Marto and R. Nunn, Hemisphere, Washington, 1981, pp. 17-49.

[8]   Caremoli, C., "Numerical Computation of Steam Flows in Power Plant Condenser
        Numerical Methods in Thermal Problems, Eds., R.W. Lewis and K. Morgan, Pinerid
        Press, Swansea, U.K., Volume IV, 1985, pp. 315-325.

[9]   Al-Sanea, S., Rhodes, N., Tatchell, D.G. and Wilkinson, T.S., "A Computer Model f
        Detailed Calculation of the Flow in Power Station Condensers", Condensers: Theory ar
        Practice. Inst. Chem. E. Symposium Series, No. 75, Pergamon Press, 1983, pp. 70-88.

[10]   Shida, H., Kuragasaki, M. and Adachi, T., "On the Numerical Analysis Method of Flo
        and Heat Transfer in Condensers", Proc. 7th Int. Heat Transfer Conference, Munche
        Fed. Rep. of Germany, Vol. 6, 1982, pp. 347-352.

[1] Van Doormaal, J.P. and Raithby, G.D., Enhancements of the SIMPLE Method for Predicting Incompressible Fluid Flow", Numerical Heat Transfer, Vol. 7, 1984, pp. 147-163.

[2] Zhang, C. and Sousa, A.C.M., "Numerical Predictions of Steam Flow and Heat Transfer in a Condenser", Numerical Methods in Thermal Problems, Eds. R.W. Lewis and K. Morgan, Pineridge Press, Swansea, U.K., Volume VI, Part 2, 1989, pp. 1368-1378.

[3] Zhang, C., Sousa, A.C.M. and Venart, J.E.S., "The Numerical and Experimental Study of a Power Plant Condenser" in Heat Transfer in Advanced Energy Systems, Eds. R.F. Boehm and G. Vliet, ASME, HTD, Vol. 151, AES, Vol. 18, 1990, pp. 1-7.

[4] Fujii, T., Uehara, H., Hirata, K. and Oda, K., "Heat Transfer and Flow Resistance in Condensation of Low Pressure Steam Flowing Through Tube Banks", Int. J. Heat Mass Transfer, Vol. 15, 1972, pp. 247-260.

[5] Diab, K.A. and Venart, J.E.S., "Monitoring and Analysis of an On-Line 350 MW_e Condenser for Coleson Cove Generating Station", Report No. CRD 0039112 (for The New Brunswick Electric Power Commission), Department of Mechanical Engineering, UNB, Fredericton, N.B., Canada E3B 5A3, May, 1990, 55 pages.

[6] Cooper, W., Diab, K., Sollows, K., Venart, J.E.S. and Sousa, A.C.M., "Experimental Study of a 350 MW Surface Condenser", (to be published).

[7] Naviglio, A., Sala, M., Socrate, S., Stefani, A. and Vigevano, L., "Distribution of Non-Condensable Gases Within the Tube Bundle of Surface Condensers", TEC 88-Conference, Recent Advances in Heat Exchangers, Grenoble, France, Dec. 10-13, 1988-.

The Computer Aided Design of Steam Surface Condensers

J. Y. Jang and J. S. Leu

Department of Mechanical Engineering
National Cheng-kung University
Tainan, Taiwan 70101
            R. O. C.

Summary
    Steam surface condenser is an important component in a
power plant. It condenses the latent heat of the turbine exhaust steam. Therefore,the turbine efficiency is directly influenced by the performance of the steam surface condenser .
This work is to develop a computer-aided condenser design
software, written in Quick Basica language. This software can
be run in a personal computer(PC/XT,PC/AT). The design methodology is based a method proposed by Heat Exchange Institute
for Steam Surface Condensers[1]. This program has the following four functions:(1)Rating the condenser (2)Sizing the con-
denser(3) Calculating the cleanliness factor and (4) Determining the condenser absolute pressure. The features of this
program are powerful interactive ability and high accuracy.
Calculation results are available within seconds. The program
has been checked against controlled performance test and operating data on installed steam surface condensers.

1. Introduction

    The primary purpose of the surface condenser is to con-
dense the exhaust steam from the turbine. Thus, the surface
condenser performance directly influences the efficiency of
the power plant. Accurate and detailed knowledge of the thermal and hydraulic design for the surface condenser is essential. Although the computer codes for the simulation of two
dimensional or three dimensional power plant condenser are
available [2-4]. They are time consuming and are difficult to
run for a practice engineer. In general,manufactures have usually based their design in general accordance with a method
proposed by the Heat Exchange Institute for Steam Surface

Condensers[1].However,the engineers still need to use many ta
-bles and charts to do the thermal and hydraulic calculations
.The purpose of this work is to develop a highly interactive
computer-aided design software. This program has the four
functions:(1)Rating the condenser(2) Sizing the condenser (3)
Calculating the cleanliness factor and(4)Determining the condenser absolute pressure. This program has been checked accu-
rately and wildly used in the steam power plants of Taiwan
Power Company.

2. The Theory

2.1 Physical Property of steam and cooling water

This software used the mathematical correlations as described in Thomas and Peter [5] to evaluate the physical properties of the steam, such as saturated temperature,saturated
pressure, density, viscosity and latent heat. The errors of
results from these equations are within 0.15% in saturated
region. As for the physical properties of the cooling water,
a polynomial regression method is used to correlate the data
taken from the Steam Table by Keenan et al.[6]. The correlations for the thermal conductivity(k), dynamic viscosity(μ)
,density(ρ) and specific heat(C_P),which are accurate to 0.5%
for the temperature range from 10°C to 90°C, are given below:

$$K(T) = -3.991479 \times 10^{-12}T^5 + 7.374313 \times 10^{-10}T^4 - 8.466236 \times 10^{-9}$$
$$T^3 - 1.219797 \times 10^{-5}T^2 + 2.124257 \times 10^{-3}T + .5169788$$
$$\mu(T) = -6.037238 \times 10^{-13}T^5 + 1.76243 \times 10^{-10}T^4 - 2.109807 \times 10^{-8}$$
$$T^3 + 1.40047 \times 10^{-6}T^2 - 6.042551 \times 10^{-5}T + 1.791604 \times 10^{-3}$$
$$\rho(T) = 6.137309 \times 10^{-9}T^5 - 1.560358 \times 10^{-6}T^4 + 1.599051 \times 10^{-4}T^3$$
$$- 1.161664 \times 10^{-2}T^2 + .1017765 \times T + 999.799$$
$$C_P(T) = -6.3845 \times 10^{-11}T^5 + 1.804488 \times 10^{-8}T^4 - 1.98933 \times 10^{-6}T^3$$
$$+ 1.154485 \times 10^{-4}T^2 - 3.376351 \times 10^{-3}T + 4.217034$$

where T : cooling water inlet temperature (°C)

2.2 The heat Load Q and the overall heat transfer coefficient U

The heat transferred in a condenser can be written as:
$$Q = UA\triangle T_{lm}$$

where A is the total outside surface area and $\triangle T_{lm}$ is the log mean temperature difference in the condenser. The overall heat transfer coefficient U is given empirically by[1]

$$U = C_c\ C_t\ C_m\ C\ \sqrt{V}\qquad (\ Btu/(h \cdot ft^2 \cdot {}^\circ F)\ )$$

where V =cooling water velocity (ft/s)
C_t =correction factor for cooling water inlet temperature
C_m =correction factor for tube material and gauge
C_c =correction factor (0.85 for clean tubes)
C =factor depending upon tube outer diameter

Based upon the information for C_t and C_m as presented in the Heat Exchange Institute Standards[1],the following correlations, which are accurate to 0.5% , are obtained from the polynomial regression method.

C_t = $3.42405 \times 10^{-8} \times T^4$ - $8.87872 \times 10^{-6} \times T^3$ + $7.112049 \times 10^{-4} \times T^2$ - $.0097041 \times T$ + $.4130287$

(a) For Admiralty Metal,Arsenical Copper and Aluminum
C_m = $2.77778 \times 10^{-5} \times BWG^4$ - $.002014 \times BWG^3$ + $.05278 \times BWG^2$ - $.57417 \times BWG$ + 3.064

(b) For Aluminum Brass, Aluminum Bronze and Muntz Metal
C_m = $-1.909711 \times 10^{-5} \times BWG^4$ + $1.423603 \times 10^{-3} \times BWG^3$ - $.0400345 \times BWG^2$ + $.51708 \times BWG$ - 1.663989

(c) For 70-30 Cu-Ni
C_m = $-3.47 \times 10^{-6} \times BWG^4$ + $2.7777 \times 10^{-4} \times BWG^3$ - $9.096991 \times 10^{-3} \times BWG^2$ + $.161665 \times BWG$ - $.397993$

(d) For 90-10 Cu-Ni
C_m = $8.680322 \times 10^{-6} \times BWG^4$ - $6.770663 \times 10^{-4} \times BWG^3$ + $1.840232 \times 10^{-2} \times BWG^2$ - $.1841614 \times BWG$ + 1.289978

(e) For Carbon Steels
C_m = $3.472238 \times 10^{-6} \times BWG^4$ - $2.7778 \times 10^{-4} \times BWG^3$ + $7.01392 \times 10^{-3} \times BWG^2$ - $4.083376 \times 10^{-2} \times BWG$ + $.628002$

(f) For Stainless Steels (410/430)
C_m = $1.38888 \times 10^{-5} \times BWG^4$ - $.0011111 \times BWG^3$ + $3.180543 \times 10^{-2} \times BWG^2$ - $.3608319 \times BWG$ + 1.971994

(g) For Titanium

$$C_m = 1.283068 \times 10^{-4} \times BWG^4 - 1.036111 \times 10^{-2} \times BWG^3 + 0.3107168 \times BWG^2 - 4.075777 \times BWG + 20.35857$$

where

      T   : Cooling Water Inlet Temperature (°F)

      BWG : Tube Gauge (Birmingham Wire Gauge)

2.3 The pressure drop and pumping power

The pressure drop of a condenser is the sum of these four losses: tube loss, tube end loss, water box inlet loss , and water box outlet loss. The Heat Exchange Institute has published a series of curves for an estimate of condenser pressure drop . The mathematical correlations for the four losses developed by Kam and Priddy[7] are used in this program. The total hydraulic loss is the sum of the condenser pressure drop and the pipe external friction loss. Then the condenser pumping power is given by:

$$\text{pumping power} = \frac{(\text{water mass flow rate})(\text{total hydraulic loss})}{\text{Pump Efficiency}}$$

2.4. Cleanliness Factor(C.F.)

The definition of cleanliness factor is $C.F. = U_a/U_c$, where U_c is the overall heat transfer coefficient for new and clean tubes , and U_a is the measured overall heat transfer coefficient. The cleanliness factor will show a trend indicating an accumulation of tube fouling matter.

2.5. Condenser absolute pressure

It can be shown that the steam saturated temperature (T_s) of the condenser is a function of condenser heat load(Q) and circulating water inlet temperature(T_1) [7].

$$T_s = C_1 \, Q + T_1$$

   where

$$C_1 = \frac{1}{m_\omega Cp[1 - EXP(-NTU)]}$$

   and m_ω : cooling water mass flow rates

In practice, the condenser absolute pressure is frequently used instead of condenser saturated temperature. Once T_s

is known, the value of condenser absolute pressure can be determined from the mathematical correlations as described in Thomas and Peter[5].

3. The Program

This program is written in Quick Basica language in order to have powerful interactive ability. It can be run in a personal computer(PC/XT,PC/AT). It has four main functions:

3.1. Rating the condenser:

Rating means to predict the performance of a surface condenser , when the condenser dimensions are given. The effectiveness—number of transfer unit (ε - NTU) formula [8] for the condenser is used to calculate the outlet temperature of cooling water. The input and output of rating problem are presented below:

Input:
  (a) Operating conditions
    1. cooling water mass flow rates, inlet temperature
    2. steam absolute pressure, quality
  (b) Geometrical data
    1. tube outside diameter, tube thickness(BWG), material
    2. tube number, tube pass, tube length
Output:
    1. cooling water outlet temperature, water velocity
    2. pressure drop and pumping power
    3. overall heat transfer coefficient (U)

3.2. Sizing the condenser:

Sizing means to determine the necessary size of a surface condenser, when given the required condenser performance . The Log-Mean-Temperature Difference(LMTD) method is used to solve heat transfer area .

Input:
  (a) Operating conditions
  1. cooling water mass flow rates
  2. cooling water inlet and outlet temperature

3. steam absolute pressure, quality, design heat load
(b) Geometrical data
1. tube outside diameter, BWG, tube material, tube pass
Output:
1. tube length, heat transfer area, tube number
2. pressure drop and pumping power

3.3. Calculating the cleanliness factor:

This program is to calculate the present cleanliness factor of a surface condenser, when given the actual condenser performance.

Input   : Actual operating conditions and Geometrical data
Output : Cleanliness factor

3.4. Calculating the condenser absolute pressure:

This program is to determine the condenser absolute pressure, when given the heat load and cooling water inlet temperature.

Input:
(a) Operating conditions
1. cooling water mass flow rates; inlet and outlet temperature
2. design heat load
(b) Geometrical data
Output: Steam absolute pressure

4. Results and Discussion

This program has been checked against several controlled performance tests and operating data on installed steam surface condenser of Taiwan Power Company(TPC). Calculation results can be obtained within 20 seconds. One of the several tests is given as follows:

```
condenser pressure    = 1.97 in-Hg
heat load             = 6.4719x10⁹ BTU/hr
water mass flow rate = 3.1774x10⁸ lb/hr; velocity=6.24 ft/s
water inlet temp.    = 72.4°F ; outlet temp. = 92.7°F
tube O.D. = 1 in ; BWG = 18 ; tube length = 60 ft
tube number = 51104 ; tube pass = 1 ; material:90-10 Cu-Ni
```

The complete input and output on the screen for the sizing problem are shown in the Appendix A.

4.1. Rating Program

The computer output for the rating problem is shown in the Appendix B. The following table shows the comparison between the present results and those taken from the TPC. It is seen that the data predicted by the present program are very close to those from TPC.

	program	TPC data	% error
water outlet temp.(°F)	92.38	92.77	-0.42
heat load (BTU/hr)	6.342×10^9	6.471×10^9	-2.00
water velocity (ft/s)	6.249	6.24	+0.14
U (BTU/(hr·ft²·F))	507.07	508.79	-0.34
pressure drop(ft-H₂O)	12.40	12.51	-0.88

4.2. Sizing Program

The comparison between the present result for heat transfer area and that taken from TPC is shown below.

	program	TPC data	% error
heat transfer area (ft²)	802739.1	800480	-0.28

4.3. Cleanliness Factor Calculation

The computer output of the cleanliness factor calculation is shown in the Appendix C. The value of cleanliness factor is 0.885. In general, 0.85 is the most common value to account for the fouling situation. Thus, in this example, the cleanliness factor 0.885 means that the surface condenser is in good operating condition.

4.4. Determining the condenser absolute pressure

The condenser absolute pressure often represents the condenser performance. Low absolute pressure indicates good condenser performance and large turbine output work.The computer output is presented in the Appendix D. The following table shows the comparison between the present result and that from TPC.

	program	TPC data	% error
absolute pressure (in-Hg)	2.016	1.97	2.3

5. References

[1] Heat Exchange Institute,"Standards for Steam Surface Condensers ",eighth edition ,1984 .

[2] Davidson, B. J.,"Computational methods for evaluating the performance of condensers ", Proceedings of a meeting on Steam Turbine Condensers, National engineering Lab. Glasgow, Report 619, Sept. , 1974.

[3] Butterworth, D. ," The development of a model for three-dimensional flow in tube bundles", Int. J. of Heat Mass Transfer Vol.21, PP253-256 , 1978.

[4] Sha, W. T., Yang, C. I., Kao, T. T., Cho S. M., " Multi-dimensional numerical modeling of heat exchangers", J. of Heat Transfer, Vol. 104, Aug. , 1982.

[5] Thomas F. I. and Peter E. L. ,"Steam and Gas Tables with Computer Equations", first edition, Academic Press, 1984.

[6] Keenan J. H. et. al."Steam Tables-Thermodynamics Properties of Water Including Vapor, Liquid and Solid Phases" , Wiley , New York, 1978.

[7] Kam W. L. and Priddy A. P. , " Power Plant System design" .1st edition ,Chap. 9, 1985 .

[8] kays W. M. and London A. L. ,"Compact Heat Exchangers" , third edition , McGraw-Hill, 1984.

Acknowledgment

Financial support of this work was provided by the Taiwan Power Company.

Appendix A
The complete input and output for the sizing problem.

PLEASE SELECT THE FUNCTIONS

```
****************************************************************
```

1. Rating the condenser --- Prediction of the performance of a
                condenser,when the condenser's dimensions
                are given .

2. Sizing the condenser --- Determination of the necessary size of
                a condenser,when given the condenser's required
                performance .

3. Calculating the cleanliness factor --- Determination of the
                cleanliness factor a condenser,when given the
                actual condenser's performance .

4. Calculating the operating pressure --- Determination of the
                operating pressure a condenser,when given the
                condenser's dimensions and heat duty.

5. Exit this program

PLEASE INPUT THE NUMBER ? 2

```
*****************************************************************
```

Do you want to correct the data , Type (Y/N) ? n

PLEASE SELECT THE UNITS

```
****************************************************************
```

1.   METRIC   UNIT

2.   ENGLISH   UNIT

PLEASE INPUT THE NUMBER ? 2

```
****************************************************************
```

Do you want to correct the data , Type (Y/N) ? n

```
Shell Side (Steam)
======================
      Heat Load            (Btu/hr)     ----->    ? 6.4719e9
      Inlet Pressure       (in-Hg )     ----->    ? 1.97
      Steam Quality      (0.80-1.00)    ----->    ? 1

Tube Side (Sea Water)
=====================

      Inlet Temperature             ('F ) ----->   ? 72.4
      Max. Required Outlet Temp.('F ) ----->   ? 92.77
      Water Pump Efficiency           ----->   ? 0.86
      Cleanliness Factor    (0.7-1.0)  ----->   ? 0.85
External loss from coast to condenser (ft)    ? 0


==================================================================

Do you want to correct the data , Type (Y/N) ---> ? n




         Physical properties of the Shell side (Steam)
      --------------------------------------------------------

      Density          (Lb/ft²3)     ----->     2.899469E-03
      Latent Heat      (Btu/lbm)     ----->     1036.346
      Saturated Temperature ('F) ----->     100.0574


         Physical propertise of the Tube side (Sea Water)
      --------------------------------------------------------

      Conductivity  (Btu/hr.ft'f)----->     .3228229
      Viscosity     (Lb/ft.hr)   ----->     2.282583
      Specific heat (Btu/lb'F)   ----->     .998673
      Density       (Lb/ft²3)    ----->     62.28094
      --------------------------------------------------------

         Press  ''RETURN'' to continue   ?
```

```
********** INPUT THE GEOMETRICAL DATA ************

A. Input Tube  O.D.   (in)
      Choices are : 1. 5/8      2. 3/4      3. 7/8
                    4. 1        5. 1 1/8  6. 1 1/4
   Please input Number 1,2,3 .... ----->    ? 4
B. Input Tube Gage (BWG) (12-24)  ----->    ? 18

C. Input Tube Material:
      1. Aluminum    2. Admiralty Metal  3. Arsenical Copper
      4. Aluminum Brass  5. Aluminum Bronze  6. Muntz Metal
      7. 70-30 Cu-Ni        8. Carbon Steels
      9. Stainless Steels   10. Titanium
         (Type 410/430)     11. 90-10 Cu-Ni
      Please input the appropriate number:  ? 11

D. Tube Passes Number (1 OR 2)              ? 1


****************************************************
Do you want to correct the data , Type (Y/N)       ? n




E. Input Water Velocity         (ft/s) (3-10)   ----> ? 6.24
F. Input Nozzle Water Velocity (ft/s) (3-10)    ----> ? 7
G. Calculated Heat Transfer Area        (ft²2) :   836333.3
   Calculated Tube Number                     :   51244
   Calculated Tube Length               (ft)  :   62.33957

    -------------------------------------------------------
   |  If the length is not acceptable , you must  change   |
   |        the water velocity or tube number !!           |
   |-------------------------------------------------------|

E. Input Water Velocity         (ft/s) (3-10)   ----> ? 6.24
F. Input Nozzle Water Velocity (ft/s) (3-10)    ----> ? 7
G. Input Tube Number                            ----> ? 51104
H. New Calculated Mass Flow Rate of Sea Water(Lb/hr):   3.172672E+0
I. New Calculated Heat Transfer Area (ft²2)        : 838864.6
J. New Calculated Tube Length          (ft)        : 62.70017
    =====================================================

   Do you want to change the velocity or tube number (Y/N)? n
```

```
            *******  GEOMETRICAL    DATA  ********
TUBE SIDE  :
O.D. (in)= 1              I.D. (in)= .9019874           BWG= 18
Length (ft)= 62.70017     Number= 51104    Material is: 90-10 Cu-Ni
-----------------------------------------------------------------
                             Tube Side (Sea Water)Shell Side (Steam)
                             ------------------------------------
Flow Rate                (Lbm/hr)  3.172672E+08      6244921  ( 1 )
Inlet Temperature        ('F)      72.4                100.0574
Max. Required Outlet Temp.  ('F)   92.77
Calculated Outlet Temperature ('F) 92.826            ( 1.97 in-Hg)
Water Temperature Rise   ('F)      20.426
Heat Transfer Area       (ft²2)    838864.6
Water Velocity           (ft/s)    6.24
Nozzle Water Velocity    (ft/s)    7
Pressure Drop            (in-Hg)   11.34304
Reynolds Number                    46071.78
           **************************************************
Total Conductance (Btu/hr.'F.ft²2)=  506.6818   Cleanliness Facter= .85
Heat Load  (Btu/hr)=  6.4719E+09        Pump horsepower (H.P.)=  2394.431
ITD= 27.65738 ('F)           TTD= 7.231377 ('F)      LMTD= 15.22666 ('F)
-----------------------------------------------------------------
        Do you want to change the dimension , Type (Y/N)?
```

Appendix B
The computer output for the rating problem.

```
            *******  GEOMETRICAL    DATA  ********
TUBE SIDE  :
O.D. (in)= 1              I.D. (in)= .9019874           BWG= 18
Length (ft)= 60           Number= 51104    Material is: 90-10 Cu-Ni
-----------------------------------------------------------------
                             Tube Side (Sea Water)Shell Side (Steam)
                             ------------------------------------
Flow Rate                (Lbm/hr)  3.1775E+08       6120489  ( 1 )
Inlet Temperature        ('F)      72.4                100.0574
Calculated Outlet Temperature ('F) 92.38859         ( 1.97 in-Hg)
Water Temperature Rise   ('F)      19.98859
Heat Transfer Area       (ft²2)    802739.1
Water Velocity           (ft/s)    6.249496
Nozzle Water Velocity    (ft/s)    7
Pressure Drop            (in-Hg)   10.94625
Reynolds Number                    46141.89
           **************************************************
Total Conductance (Btu/hr.'F.ft²2)=  507.0736   Cleanliness Facter= .85
Heat Load  (Btu/hr)=  6.342945E+09      Pump horsepower (H.P.)=  2314.188
ITD= 27.65738 ('F)           TTD= 7.668793 ('F)      LMTD= 15.58281 ('F)
-----------------------------------------------------------------
        Do you still want to try again , Type (Y/N) ?
```

Appendix C

The computer output of the cleanliness factor calculation.

```
******** GEOMETRICAL    DATA  ********
TUBE SIDE  :
O.D. (in)= 1              I.D. (in)= .9019874          BWG= 18
Length (ft)= 60          Number= 51104    Material is: 90-10 Cu-Ni
------------------------------------------------------------------
                                  Tube Side (Sea Water)Shell Side (Steam)
                                  ------------------------------------
Flow Rate                (Lbm/hr)  3.1775E+08        6244921  ( 1 )
Inlet Temperature        ('F)      72.4               100.0574
Calculated Outlet Temperature ('F) 92.7             ( 1.97 in-Hg)
Water Temperature Rise   ('F)      20.3
Heat Transfer Area       (ft²2)    802739.1
Water Velocity           (ft/s)    6.249496
Nozzle Water Velocity    (ft/s)    7
Pressure Drop            (in-Hg)   10.94625
Reynolds Number                    46141.89
                         ************************************************
Total Conductance (Btu/hr.'F.ft²2)= 525.9097    Cleanliness Facter= .8815747
Heat Load  (Btu/hr)= 6.4719E+09        Pump horsepower (H.P.)=  2314.188
ITD= 27.65738 ('F)        TTD= 7.357384 ('F)      LMTD= 15.33014 ('F)
------------------------------------------------------------------
```

Do you still want to try again , Type (Y/N) ?

Appendix D

The computer output of the absolute pressure determination

```
******** GEOMETRICAL    DATA  ********
TUBE SIDE  :
O.D. (in)= 1              I.D. (in)= .9019874          BWG= 18
Length (ft)= 60          Number= 51104    Material is: 90-10 Cu-Ni
------------------------------------------------------------------
                                  Tube Side (Sea Water)Shell Side (Steam)
                                  ------------------------------------
Flow Rate                (Lbm/hr)  3.1775E+08        6243584  ( 1 )
Inlet Temperature        ('F)      72.4               100.6197
Calculated Outlet Temperature ('F) 92.79497        ( 2.016 in-Hg)
Water Temperature Rise   ('F)      20.39497
Heat Transfer Area       (ft²2)    802739.1
Water Velocity           (ft/s)    6.249496
Nozzle Water Velocity    (ft/s)    7
Pressure Drop            (in-Hg)   10.94625
Reynolds Number                    46141.89
                         ************************************************
Total Conductance (Btu/hr.'F.ft²2)= 507.0736    Cleanliness Facter= .85
Heat Load  (Btu/hr)= 6.4719E+09        Pump horsepower (H.P.)=  2314.188
ITD= 28.21966 ('F)        TTD= 7.824692 ('F)      LMTD= 15.8996 ('F)
------------------------------------------------------------------
```

Do you still want to try again , Type (Y/N) ?

Some Comments on the Use of Mixed Bundles
of Smooth and Enhanced Tubes in Reboilers

A. MACIVER, B. M. BURNSIDE

Department of Mechanical Engineering,
Heriot-Watt University, Edinburgh, Scotland.

Summary

A theoretical study of mixed HIGHFLUX/smooth and LOWFIN/smooth tube bundles
boiling R113 at atmospheric pressure is described. A circulation model
adapted from that developed by Palen and Yang and based on data obtained
from an experimental smooth tubed boiler in the laboratory was used in the
study. Curves of bundle tubing cost versus proportion of enhanced tube rows
are drawn for different enhanced:smooth tube unit cost ratios. It is
concluded that there is no cost advantage to be gained from employing mixed
bundles of smooth and enhanced tubes.

Introduction

For many years it has been evident that there are potential rewards to be
gained by designers of kettle reboilers and cooler evaporators in employing
enhanced boiling surfaces. In recent years porous sintered surfaced tubes
and a variety of worked enhanced surfaced tubes have become available
commercially. Further, increased energy costs and other financial pressures
have dictated reduced heat exchanger temperature driving forces for which
enhanced boiling surfaces have been designed. It has now become quite
common to use bundles of enhanced tubes in reboilers and evaporators in
non-fouling environments either in newly designed plant or in uprating
existing equipment.

Another possible commercial approach is to sacrifice some of the decrease in
Δt afforded by enhanced tubing for reduction in the number of tubes used,
thus reducing capital cost. To an extent this approach is taken currently,
since when using enhanced surface tubes with order of magnitude higher
performance than smooth tubes, heat transfer coefficients are so much
higher, even at the low Δt used, than at the much higher Δt used in smooth
tube designs. Use of less expensive tubes with less enhancement, such as
low-finned tubes, should allow more flexibility in trading-off Δt against
tube numbers.

376

Another design possibility which has been suggested is the use of mixed bundles of smooth and enhanced tubing. It is well known [1,2,3] that two phase flow convective heat transfer augments considerably nucleate boiling in the upper rows of smooth tubed bundles. Thus, the questions arise - is it necessary to use enhanced surfaces there? Could expense be reduced by using enhanced surfaces only in the lower rows of the bundle? If the object of the exercise is the maximum reduction in Δt, then clearly the only option is the use of a full bundle of tubes with the highest enhancement possible. An order of magnitude enhancement cannot be replaced by a smooth tube surface with convection, even allowing for any suppresssion of performance of the enhanced surface by two-phase flow over it [4].

However, if the main objective is reduction in capital cost, with reduction in Δt a secondary consideration, there may be a trade-off between cost and performance of smooth and enhanced tubes, of the type shown in figure 1.

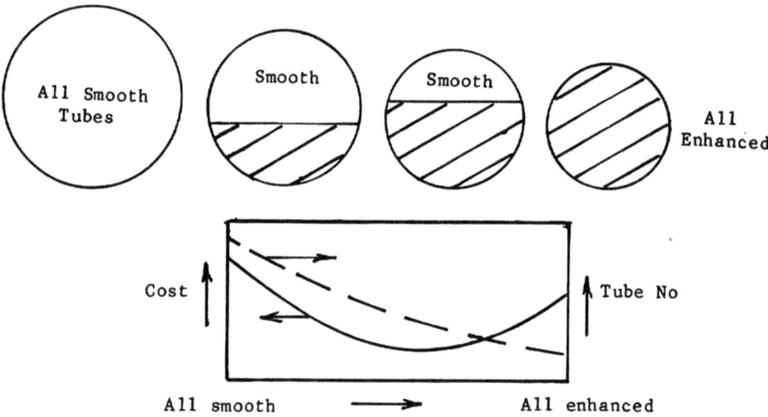

Figure 1.   Does this situation exist?

Increase in the proportion of enhanced tubes in the bundle reduces the bundle size. However, the bundle cost does not fall in proportion because the unit cost of an enhanced tube is greater than that of a smooth tube. It is possible that there will be a minimum bundle tube cost which will depend evidently on the relative degree of enhancement and unit cost of the enhanced and smooth tubes.

Obviously there are other related design considerations. One of these is that the size of the bundle affects the shell and support size and cost. This analysis, however, has the more limited objective of establishing whether there is an optimum enhanced:smooth tube ratio leading to minimum bundle cost. The analysis is based on the one-dimensional circulation model described by Palen and Yang [2] applied to boiling R113 at 1 atm, for which experimental data is available to the authors. An earlier analysis of boiling heptane [5] assumed uniform wall-to-boiling liquid Δt. It indicated that a bundle cost v. proportion of enhanced tubes characteristic of the type shown in figure 1, existed for highly enhanced tubes.

The present analysis includes the effect of steam condensing side heat transfer resistance. The effect of the use of enhanced tubes with high and low degrees of enhancement on the results of the study is considered.

Method of Analysis

Calculations were based on a one-dimensional circulation model of the boiler, assuming a square bundle. The theory of the model has been described in a number of papers [2,3,4]. However, for commercial reasons, many of the details relating to 2-phase pressure drop, flow pattern determination, mixture and enhanced tubing effects have not been disclosed. For this work a model developed by Ahmad [6], based on that of Palen and Yang [2], was adapted. The model fluid was R113 boiling at a pressure of 1 atm. Two versions of this model were developed for (a) uniform heat flux and (b) uniform condensing-to boiling-side temperature difference. The heating medium was assumed to be steam in case (b).

The flow-pattern map of Grant and Murray [7] was included in the model computer program in an attempt to predict the flow regime and therefore the appropriate 2-phase flow relations to use at different levels in the bundle – bubbly, slug or spray flow. However, at all heat flux levels considered, this flow map predicted bubbly flow throughout the boiler. Although this is not likely, the 2-phase flow calculations were based or. the same separated two phase flow model throughout the bundle [5,6].

The Palen and Yang model [2] contained four undetermined constants r, r_v, a and b which appear in the expressions for the 2-phase convection correction factor $F_{tp} = (\phi^2_\ell)^r$, for the vapour correction factor to nucleate boiling

$F_v = (\varepsilon/0.6)^{r_v}$ and, in the case of a and b, in the expression for the thin film conduction heat transfer coefficient [2]. An investigation was carried out by Chua [8] to determine the best fit of the uniform heat flux model to data obtained boiling R113 at atmospheric pressure over a 241 tube, 25.4mm square pitch, thin slice model reboiler bundle. This data did not show the very large heat transfer coefficients reported by Leong and Cornwell [1] at the top of the boiler. Chua found that it was not possible to obtain a set of values of r, r_v, a and b which would give low errors in the prediction of measured tube boiling heat transfer coefficients and at the same time predict reasonable vertical profiles of thin film conductive heat transfer coefficients. Very good fits were obtained if only nucleate boiling and two phase convective components of heat transfer were assumed to occur [8], when r and r_v only are determined from the model reboiler data.

Pool boiling heat transfer coefficients for R113 on smooth tubes were based on the Mostinski [9] correlation. The influence of HIGHFLUX and LOW-FIN (19FPI) enhanced tubes in the bottom rows of the boiler was based on isolated tube pool boiling data measured in this laboratory. For R113 at atmospheric pressure the correlations were $\alpha_{HF} = 6.77\Delta T^{0.661}$ and $\alpha_{LF} = 0.501\Delta T^{0.892}$. For the HIGHFLUX surface this corresponds to $\alpha_{HF} = 15$ kW/m^{2}K at a heat flux density of 50 kW/m^2. The corresponding value for the LOW-FIN surface is 4.4 kW/m^{2}K. Two-phase flow convection corrections were ignored in the case of the HIGHFLUX tubes and the LOW-FIN tubes were treated as smooth tubes in the analysis, apart from their enhanced nucleate boiling heat transfer coefficient.

On the condensing side the method described by Butterworth [10] was used to calculate the average coefficient of heat transfer over a metre length of tube. Steam was assumed to enter saturated and at a velocity sufficient to ensure that condensation occurred over the whole length of the tube. Tube wall heat transfer resistance was ignored.

The calculations were handled by computer. For uniform heat flux, a single iterative loop on the total mass velocity to balance the pressure drops sufficed. The resulting tube wall temperatures were compared with measured values at heat fluxes between 10 and 50 kW/m^2 and the calculation repeated with different values of r and r_v until the best fit was obtained. The uniform ΔT program incorporated iteration loops on tubewall temperature, heat flux and mass velocity through the bundle.

HIGHFLUX/Smooth Tube Bundles

For this configuration steam to R113 temperature differences Δt = 5, 7.5 and 10 degC. Steam velocities at entry to the HIGHFLUX and smooth tubes was taken to be C_S = 120 and 60 m/s respectively. Calculations were based on a bundle heat rating of 700 kW/m length. The numbers of tubes in the bundle were calculated for several proportions of enhanced rows at the bottom of the bundle, $0 \leq f \leq 1$. The first and last of these cases correspond to an all smooth tube bundle and an all HIGHFLUX bundle.

Assuming that the unit costs of smooth and HIGHFLUX tubes are c_S and c_H respectively, then the cost of tubes, C_T, for a square bundle of N tubes with a fraction f of its rows HIGHFLUX tubes would be given by

$$C_T = [(1 - f) \, c_S + f c_H] \, N$$

The ratio of the cost of the bundle to the unit cost of a smooth tube, r_c, is therefore

$$r_c = N \left[1 + f \left\{ \frac{c_H}{c_S} - 1 \right\} \right]$$

Values of r_c were calculated for $\dfrac{c_H}{c_S}$ = 2, 4 and 8 and r_c and N plotted in figures 2, 3 and 4 versus f for a fixed temperature driving force Δt. Also shown in the figures are the average heat flux densities in the enhanced and smooth tubes, q_H and q_S and the ratio of condensing side to boiling side heat transfer resistance for the enhanced tubes, R_H.

There is a steady rise in tube number as the proportion of HIGHFLUX tubes falls, up to about f = 0.25. When the proportion, f, falls further there is then a steep rise in the number of tubes. This is caused by the very low heat flux densities on the smooth tubes, q_S, which reaches a maximum of only 11 kW/m^2 at Δt = 10.2K. At the lowest temperature driving force, Δt = 5, figure 2 the tube cost rises also steadily as the proportion of enhanced tubes falls, when c_H/c_S = 2 and 4. The all enhanced bundle proves the cheapest if the unit cost of the enhanced tubes is less than about 5-6 times the plain tube cost. For c_H/c_S = 8 the tube cost is about the same for bundles with down to 75% HIGHFLUX tubes as for the all HIGHFLUX bundle. It rises as the proportion of smooth tubes rises further.

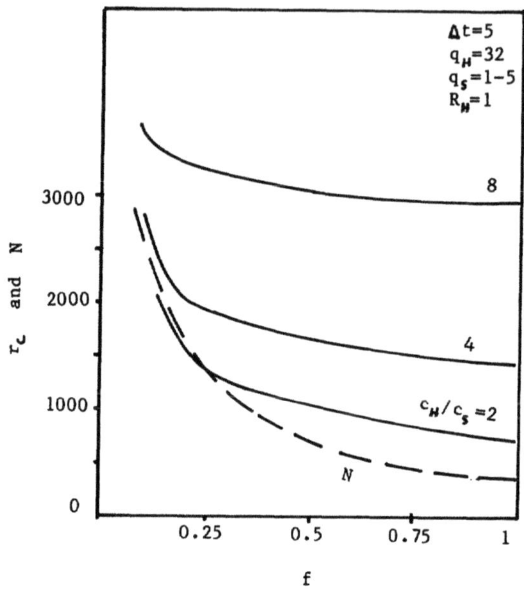

Figure 2. Bundle cost r_c and tube number N v.
HIGHFLUX row proportion (Δt = 5K)

For higher temperature driving forces, Δt = 7.7 and 10.2K, there is a shallow minimum in tube cost at c_H/c_S = 8 but it is not sufficient to justify a mixed bundle configuration, figures 3 and 4. The enhanced tube heat flux densities vary from 32 to 70 kW/m^2 and the ratio of boiling to condensing side heat transfer resistance, R_H, from 1 to 0.66 as Δt is increased from 5 to 10.2K.

LOWFIN/Smooth Tube Bundles

For this configuration Δt values of 10.2, 16.5 and 19.9 were considered. The steam velocities and bundle heat rating were the same as used in the HIGHFLUX/smooth bundle study. Again the numbers of tubes in the bundle were calculated for a range of proportions of LOWFIN rows at the bottom of the

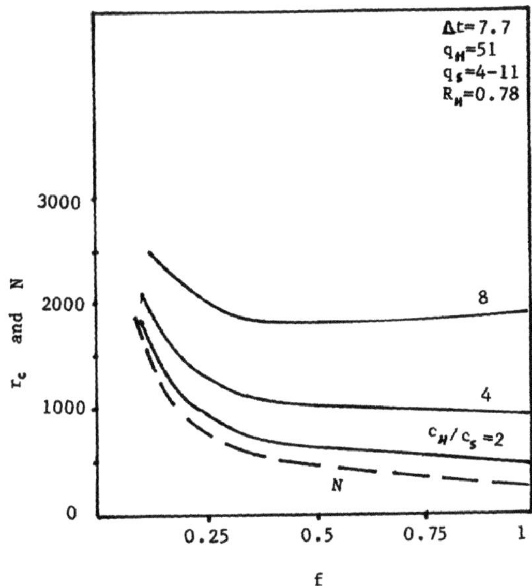

Figure 3. Bundle cost r_c and tube number N v.
HIGHFLUX row proportion (Δt = 7.7K)

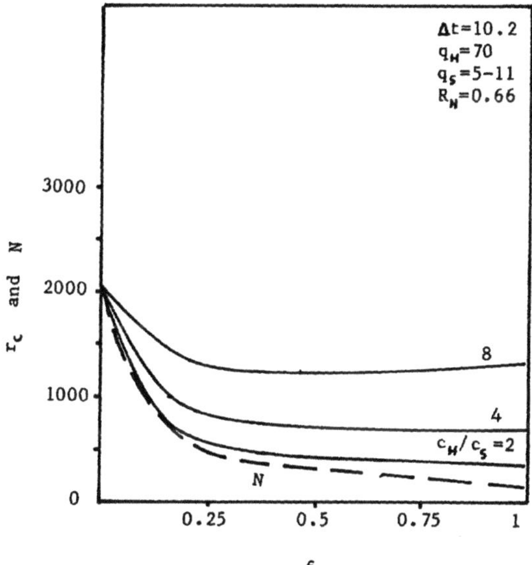

Figure 4. Bundle cost r_c and tube number N v.
HIGHFLUX row proportion (Δt = 5K)

bundle. Again, the ratio of the tube cost of the of the bundle to the unit cost of a smooth tube is

$$r_c = N \left[1 + f \left\{ \frac{C_{LF}}{C_s} - 1 \right\} \right]$$

where c_{LF} is the unit cost of a low finned tube. Values of r_c were calculated for $\frac{C_{LF}}{C_s} = 1.5$, 2.5 and 4 and r_c and N plotted in figures 5, 6 and 7 for fixed temperature driving force Δt.

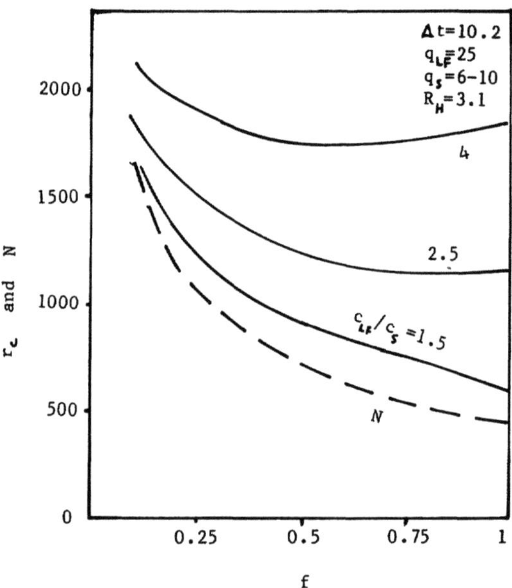

Figure 5. Bundle cost r_c and tube number N v. LOWFIN row proportion (Δt = 10.2K)

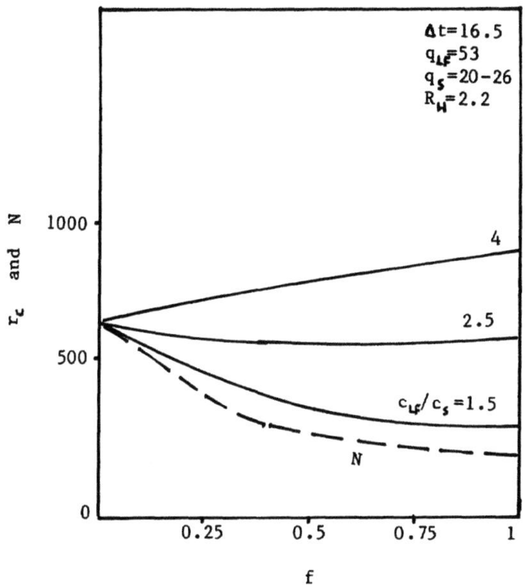

Figure 6. Bundle cost r_c and tube number N v.
LOWFIN row proportion (Δt = 16.5K)

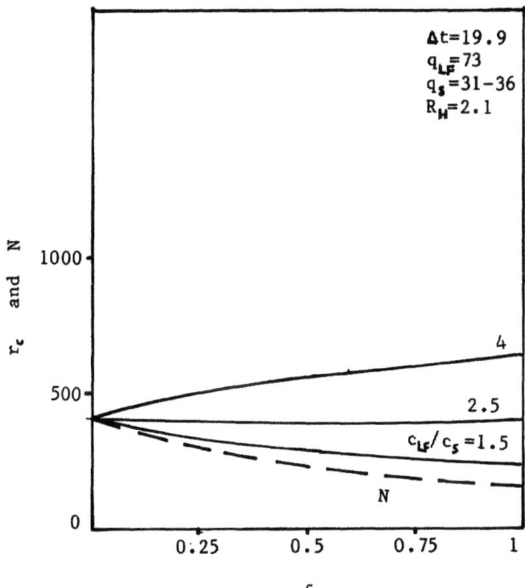

Figure 7. Bundle cost r_c and tube number N v.
LOWFIN row proportion (Δt = 19.9K)

These figures show the same characteristic behaviour as described above for the HIGHFLUX/smooth bundles. Tube number increases more gradually with decrease in f, however, due to the less significant difference between smooth and enhanced tube performance. At the lowest Δt = 10.2K studied, figure 5, a shallow minimum in tube cost occurs at about f = 0.80 if the unit cost of an enhanced tube is about $2\frac{1}{2}$ times that of a smooth one. For cheaper enhanced tubes the minimum cost is decidedly that of the all-LOWFIN bundle. At c_{LF}/C_s = 4, the minimum cost shifts to 0.45<f<0.7 but is only about 3% less than the cost of the all LOWFIN bundle. The average heat flux density of the LOWFIN tubes is q_{LF} = 25kW/m^2 and of the smooth tubes q_S = 8kW/m^2, at this temperature driving force. Boiling side resistance on the LOWFIN tubes, R_H, is three times that on the condensing side. When Δt is raised to 16.5 and 19.9K, figures 6 and 7, the cost v. proportion of enhanced tube characteristics change. The rise in N with fall in f is less pronounced, due to the improved performance of the smooth tubes. Again there is a shallow minimum in cost at Δt = 16.5K and at Δt = 19.9K, indeed, the tube cost is unaffected by the proportion of enhanced tubes used. At higher LOWFIN tube cost, c_{LF}/c_S = 4, the lowest cost tubing is in the all smooth tube bundle at both temperature driving forces. The average q_{LF} = 53kW/m^2 for Δt = 16.5K and 73 kW/m^2 at Δt = 19.9K. Corresponding smooth tube heat flux densities are 23 and 34 kW/m^2 respectively. In both cases the LOWFIN tube boiling side heat transfer resistance is about twice that on the condensing side, figures 6 and 7.

<u>Conclusion</u>

The results of this investigation show emphatically that there is no significant advantage in cost to be gained in the use of mixed bundles of smooth and enhanced tubes. Even where there is a slight tube cost advantage, when the unit cost of the enhanced tubes is high, the increase in bundle diameter surely nullifies it. Initial enthusiasm for the concept of mixed tube bundles was fuelled by the very high heat transfer coefficients measured by Leong and Cornwell [1] in the upper rows of their experimental smooth tubed boiler. Work carried out in this laboratory on a similar boiler with low and known heat losses, using the same fluid, R113, showed much less dramatic rise in heat transfer coefficients with level in the boiler. Basing the circulation model calculations on this data, even allowing for the increased flux of vapour in the upper smooth tube rows due to the enhanced tubes at the bottom, the performance of the smooth tubes is still very much poorer than that of the enhanced tubes. As a result, too many more of them are required to achieve the same output as an all enhanced tube bundle to allow any cost saving. This is in addition to problems of assuring positive Δt throughout which might arise in a reboiler with multipass condensing side configuration and both vertical and longitudinal boiling side temperature variation. Such problems are accentuated by requirements and possibility to use very low temperature driving forces in modern reboilers.

Acknowledgements

The authors are indebted to Ali Tarrad for isolated tube R113 boiling data for HIGHFLUX and LOWFIN tubes and to HTFS and HTRI for support to carry out the multitube boiling work.

References

1. Leong, L.S.; Cornwell, K. Heat transfer coefficients in a reboiler tube bundle, The Chemical Engineer no.343 (1979) 219-221.

2. Palen, J.S.; Yang, C.C. Circulation boiling model for analysis of kettle and internal reboiler performance, Heat Exchangers for 2-Phase Applications, ASME Publication, HTD, 27, (1981) 55-61.

3. Brisbane, T.W.C.; Grant, I.D.R.; Whalley, P.B. A prediction method for kettle reboiler performance, ASME Paper no.80-HT-42 (1980) 19th National Heat Transfer Conf. Orlando.

4. Palen, J.W.; Taborek, J.; Yilmaz, S. Comments to the application of enhanced boiling surfaces in tube bundles, Chapter 13, Heat Exchanger Source Book, ed. J.W. Palen, Hemisphere, 1986.

5. Ahmad, H.H.; Maciver, A.; Burnside, B.M. Economic assessment of replacing plain by enhanced tubes in reboiler tube bundles, Eurotherm Seminar No.8, Advances in Pool Boiling Heat Transfer, Paper No.3, Session 6, May 1989.

6. Ahmad, H.H.; Boiling of immiscible systems over tube bundles, Ph.D. Thesis, Heriot-Watt University, 1989.

7. Grant, I.D.R.; Chisholm, D. Two-phase flow on the shell side of a segmentally baffled shell and tube heat exchanger, J. Heat Transfer, 101 (1979) 38-42.

8. Chuah, W.T. A recirculation model for kettle reboilers, B.Eng. Disseration, Dept. of Mech. Eng., Heriot-Watt University, 1990.

9. Mostinski, I.L. Application of the rule of corresponding states for calculation of heat transfer and critical heat flux to boiling mixtures, Chem. Eng. Abstracts, Folio no.150 (1963) 580.

10. Butterworth, D. Film condensation of pure vapour, Chapter 2.6.2, Heat Exchanger Design Handbook, Hemisphere 1983.

The Heat Pipe Heat Exchangers:
Design, Technology and Applications

S. CHAUDOURNE

GRETh
CEA / CENG
BP 85X, 38041 Grenoble Cedex, FRANCE

ABSTRACT

In this paper the Heat Pipe Heat Exchanger (HPHE) is presented as a new and very attractive
Heat Exchanger type. In the first part of the paper a presentation of the modelization of the
HPHE is given, in the second part some technologocal considerations are developped and in
the third part two exemples of industrial applications are described.

1. INTRODUCTION

A Heat Pipe Heat Exchanger (HPHE) is a special type of Heat Exchanger using the excellent
thermal properties of the heat pipes.

A Heat Pipe is a tube containing a liquid in equilibrium with its vapour (the working fluid). It
does not contain air nor any other gas and it is hermetically closed. The thermal transfer in a
heat pipe uses the evaporation of the liquid working fluid at the hot end (the evaporator) and
the condensation of the vapor at the cold end (the condensor). The condensed liquid returns to
the evaporator by means of a capillary structure or by the gravity effect (fig. 1). This two-
phase thermal transfer inside the heat pipe gives it a very high thermal conductance (several
hundred times the conductance of a copper rod having the same dimensions).

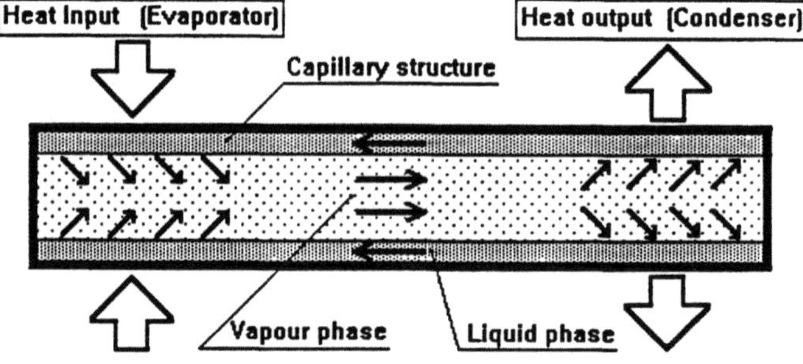

Fig. 1: Principle of a heat pipe

In an HPHE the heat pipes are arranged in bundles in such a way that the evaporators are plunged into the hot fluid and the condensors are plunged into the cold fluid (fig. 2). The thermal transfer is obtained through the heat pipes which constitute a thermal bridge between the two fluids.

The HPHE has several advantages over the classical heat exchangers. The most attractive features of these heat exchangers are the following:

- great design flexibility: the heat pipes are independant componants

- high thermal effectiveness: heat pipes have a very high conductance and it is easy to increase the exchange surface between these heat pipes and the fluid as when using finned tubes for heat pipes for exemple

- excellent mechanical isolation between the two fluids: the heat pipes make a double wall between the two fluids

- small pressure loss on the two fluids: they flow only along the outside of the heat pipe tubes.

- possibility to adjust the heat exchange surface temperature according to the choice of the heat exchange areas on the hot side and the cold side (this is useful to avoid corrosion due to acid condensation). This characteristic results from the independance of the hot and cold exchange surfaces.

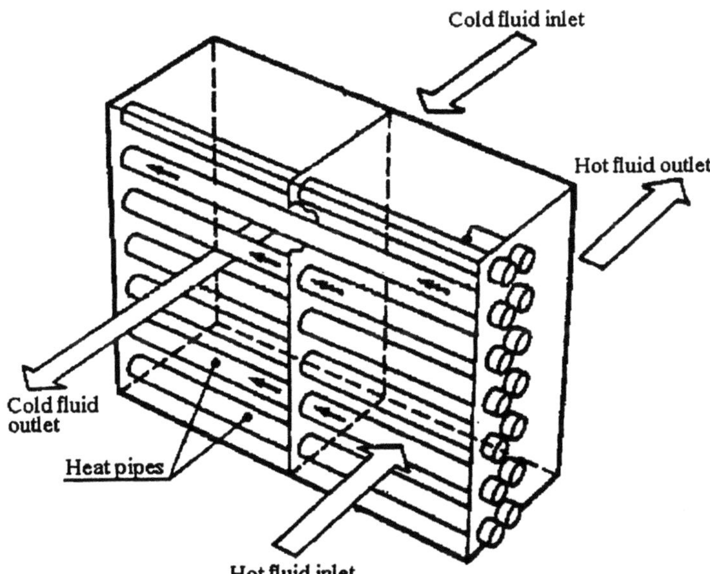

Fig. 2: Principle of a Heat Pipe Heat Exchanger

2. MODELIZATION OF THE HPHE

2.1. Principle of the modelization

The principle of the modelization consists in dividing the HPHE into elementary heat exchangers. The thermal effictiveness of these elementary heat exchangers is calculated from their Number of Transfer Unit (NTU). The HPHE is then considered as a network of elementary heat exchangers and its overall effectiveness is calculated using certain rules of heat exchanger associations.

The modelization can be divided into three steps: the heat pipe modelization, the modelization of one row of heat pipes, and the modelization of all the HPHE.

2.2. Modelization of a heat pipe

A heat pipe is considered as an evaporator (heated by the hot fluid) coupled by a coupling fluid (the heat pipe working fluid) with a condenser (cooled by the cold fluid). The vapour pressure inside the heat pipe being almost constant, the internal temperature (which is the saturation temperature) is also constant, so that the coupling fluid can be considered as isothermal (fig. 3).

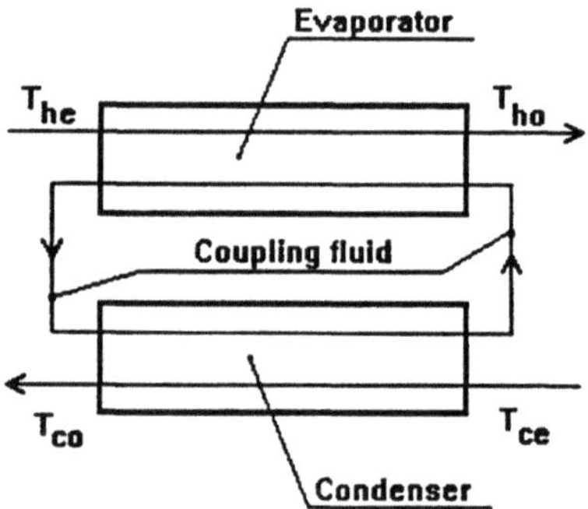

Fig.3: Modelization of a heat pipe

2.3. Modelization of a one-row HPHE

It can be assumed that all the evaporators of the heat pipes situated on the same row of the heat exchanger are at the same temperature. The same remark is valid for the heat pipe condensers. A heat pipe row can thus be considered as a unique evaporator coupled with a unique condenser. Obviously the heat exchange area $A_{h,i}$ (for the i row hot side) and $A_{c,i}$ (for the i row cold side) are the sum of the respective heat exchange areas of all the heat pipes of the i row.

The Number of Transfer Unit (NTU) can be expressed as following:

$$NUT_{h,i} = \frac{U_{h,i} \cdot A_{h,i}}{C_{h,i}} \tag{1}$$

$$NUT_{c,i} = \frac{U_{c,i} \cdot A_{c,i}}{C_{c,i}} \tag{2}$$

where $U_{h,i}$, $U_{c,i}$ are the overall heat transfer coefficient and $C_{h,i}$, $C_{c,i}$ the capacity rate for the hot side and cold side respectively.

The thermal effectiveness $E_{h,i}$ and $E_{c,i}$ corresponding to an evaporator and a condenser are:

$$E_{h,i} = 1 - \exp(-NUT_{h,i}) \tag{3}$$

$$E_{c,i} = 1 - \exp(-NUT_{c,i}) \tag{4}$$

It can be shown [6] that the effectiveness of the coupled evaporator condenser is given by:

$$E_i = \frac{1}{\dfrac{1}{E_{min,i}} + \dfrac{C_{min,i}}{C_{max,i} \cdot E_{max,i}}} \tag{5}$$

with:

$$C_{min,i} = MIN(C_{h,i}, C_{c,i})$$

$$C_{max,i} = MAX(C_{h,i}, C_{c,i})$$

$$E_{min,i} = E_{h,i} \text{ si } C_{h,i} = C_{min,i}, \; E_{c,i} \text{ si } C_{c,i} = C_{min,i}$$

$$E_{max,i} = E_{h,i} \text{ si } C_{h,i} = C_{max,i}, \; E_{c,i} \text{ si } C_{c,i} = C_{max,i}$$

2.4. Modelization of a P row HPHE

The successive rows of the HPHE constitute an overall counter-flow heat exchanger association (fig.4). For each one-row heat exchanger, two equations can be written: one expressing the effectiveness definition and the other expressing the enthalpy balance:

$$E_i = \frac{C_h \cdot (T_{h,i-1} - T_{h,i})}{C_{min} \cdot (T_{h,i-1} - T_{c,i})} \tag{6}$$

$$C_h \cdot (T_{h,i-1} - T_{h,i}) = C_c \cdot (T_{c,i-1} - T_{c,i}) \tag{7}$$

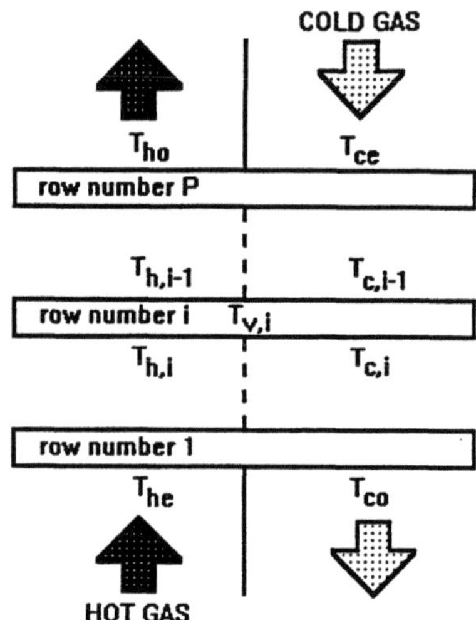

Fig. 4: Counter flow series heat exchangers association

$T_{h,i}$ and $T_{c,i}$ are the hot and cold fluid temperature between the i row and the i+1 row.

These equations can be written in a matrix form:

$$[T_i] = [N_i] \cdot [T_{i-1}] \qquad \text{for i=1 to P} \tag{8}$$

with:

$$[T_i] = \begin{pmatrix} T_{h,i} \\ T_{c,i} \end{pmatrix} \qquad\qquad [N_i] = \begin{pmatrix} N_{11,i} & N_{12,i} \\ N_{21,i} & N_{22,i} \end{pmatrix}$$

$$N_{11,i} = \frac{q_i - (1+r_i).E_i}{d_i} \qquad\qquad N_{12,i} = \frac{E_i}{d_i}$$

$$N_{21,i} = \frac{- r_i.E_i}{d_i} \qquad\qquad N_{22,i} = \frac{q_i}{d_i}$$

$$q_i = \frac{C_{h,i}}{C_{min,i}} \qquad\qquad r_i = \frac{C_{h,i}}{C_{c,i}} \qquad\qquad d_i = q_i - r_i \cdot E_i$$

By resolving the system of equations (8),

we obtain the outlet fluid temperatures: $T_{ho} = T_{h,P}$ and $T_{co} = T_{c,0}$

from the inlet fluid temperatures: $T_{he} = T_{h,0}$ and $T_{ce} = T_{c,P}$:

$$\begin{pmatrix} T_{cs} \\ T_{fs} \end{pmatrix} = \begin{pmatrix} \dfrac{N_{11}.N_{22} - N_{12}.N_{21}}{N_{22}} & \dfrac{N_{12}}{N_{22}} \\ \dfrac{-N_{21}}{N_{22}} & \dfrac{1}{N_{22}} \end{pmatrix} \begin{pmatrix} T_{ce} \\ T_{fe} \end{pmatrix} \qquad (9)$$

with:

$$[N] = [N_P].[N_{P-1}]...[N_1] = \begin{pmatrix} N_{11} & N_{12} \\ N_{21} & N_{22} \end{pmatrix}$$

The intermediate fluid temperatures $T_{h,i}$ and $T_{c,i}$ can be obtained then from equations (6) and (7). The heat flux carried by each heat pipe is given by:

$$\Phi_i = \frac{C_{h,i}.(T_{h,i-1} - T_{h,i})}{N_t} \qquad (N_t \text{ number of heat}$$

pipes in one row)

and the vapour temperature can be calculated by the following relation:

$$T_{v,i} = \frac{C_{h,i}.E_{h,i}.T_{h,i-1} + C_{c,i}.E_{c,i}.T_{c,i}}{C_{h,i}.E_{h,i} + C_{c,i}.E_{c,i}}$$

When these results are obtained, it is necessary to verify that each heat flux Φ_i at the vapour temperature $T_{v,i}$ is lower than the maximum heat flux of the heat pipe (cf. §3).

2.5. The computer code ECCO

The modelization of a HPHE presented above has been used to develope a computation code named ECCO. ECCO has been validated with experimental data obtained on the GRETh's heat exchanger test facility ESTHER [9]. ECCO is a powerful design tool for HPHE.

3. HEAT PIPE TECHNOLOGICAL CONSIDERATIONS

3.1. General considerations

The choice of a specific type of heat pipe for a given HPHE depends on several characteristics:

The working temperature level of the heat pipe which determines the working fluid.

. The position of the heat pipe and the heat flux which determines the capillary structure. Whenever possible, a gravity assisted heat pipe or thermosiphon will be used.

. The corrosion risk from the external fluid which determines the heat pipe tube material. This material must be compatible with the heat pipe working fluid to avoid internal corrosion that might give off incondensable gas and cause the heat pipe to break down.

3.2. Working fluids

The pressure inside the heat pipe is the saturated vapour pressure of the fluid at the heat pipe working temperature; this temperature determines the fluid to choose in order to have an internal pressure that is neither too low nor too high. The internal pressure would typically be from 0.01MPa to 2MPa. The fluid chosen must be compatible with the envelop and capillary structure material. The following table gives the working temperature and compatibility of some usual heat pipe fluids often used for HPHE.

Fluid	Temperature field (°C)	Compatible materials	Comments
Freon 11[1]	-20 to 110	Steel, Stainless steel, Aluminium, Copper	Cheap and useful for ambiant temperature
Freon 113[1]	-10 to 90	Steel, Stainless steel, Aluminium, Copper	Cheap and useful for ambiant temperature
Water	70 to 220	Copper, Steel[2]	Very good thermal performances
Toluene	80 to 270	Aluminium, Copper, Steel, Stainless steel	Thermal performances lower than water but lower pressure and compatibility with steel without treatment
Gilotherm DO	150 to 300	Copper, Steel, Stainless steel	Risk of decomposition above 300°C
Naphtaline	200 to 420	Steel, Stainless steel	Solid at ambiant temperature
Mercury	250 to 550	Steel, Stainless steel	Very good thermal performances but high toxicity

3.3. Capillary structures and working limit

A porous structure developing capillary forces is used to drive the liquid phase from the condenser to the evaporator. The most usual capillary structures are the grooves (straight or helicoïd) and the screen wick. The maximum liquid flow rate in the capillary structure is the

---

[1] These fluids will have soon to be replaced by substitution fluids in order to respect environmental legislations

[2] Only with a special treatment for the steel (passivation) and eventually for the water (corrosion inhibitor)

result of the balance between the capillary forces and the pressure losses. It determines the maximum heat flux through the heat pipe which is named the capillary limit.

Whenever possible, gravity is used to replace capillary forces. In this case heat pipes must be tilted at least 5 degrees from the horizontal position with the evaporator above the condenser. The capillary structure is no longer necessary, meanwhile, grooves can be useful to increase inside heat transfer coefficients. In these gravity-assisted heat pipes grooves are often used in aluminium and copper tubes but not in steel tubes (because their manufacturing is too difficult). For gravity-assisted heat pipes the working limit is the entrainment limit which is due to liquid entrainment by the vapour counter flow.

For each heat pipe type the working limit must be calculated in order to verify that it cannot be reached in the heat exchanger working conditions (cf. §2).

3.4. Tubes materials

Aluminium is the most common material for moderate temperature (up to 150°C). It is a good thermal conductor, it is easy to work by extrusion to obtain grooves capillary structures and it is light. Aluminium is often used with refrigerant (such as Freon 11, Freon 113 or their substitute) as working fluid for air conditioning applications. Aluminium is not compatible with water.

Copper is often used with water as working fluid because it has an excellent compatibility. Copper has a very high thermal conductivity and copper tubes with grooves capillary structures are commercialized. Copper can be used up to 200°C.

Steel is the best suited material to high temperatures and corrosive external fluids. It is compatible with all organic fluids. A special treatment is necessary with water. Stainless Steel is also used, especially for food industry.

4. APPLICATIONS OF HPHE

4.1. Using domains of HPHE

The main areas of application of HPHE are the following:

Air conditioning: The HPHE advantages for these applications are: compactness, very good mechanical isolation between the the two fluids, very low susceptibility to freezing (absence of any cold point).

Heat recovery: The HPHE advantages for these applications are: great design flexibility, high thermal effectiveness, low pressure losses, very good mechanical isolation between the two fluids, possibility to avoid corrosion by acid condensation (absence of any cold point).

Agro industry: The HPHE advantages for these applications are: possibility to design a heat exchanger that can easily be dismantled and cleaned because of the independance of the heat pipes. Very good mechanical isolation between the the two fluids.

4.2. Exemple of heat recovery on flue gas for preheating air

In 1986 a 1300kW HPHE was designed to recover heat on flue gas coming from a distillation furnace in a petroleum refinery. The heat recovered is used to heat the combustion air going to the furnace gas burner (fig. 5).

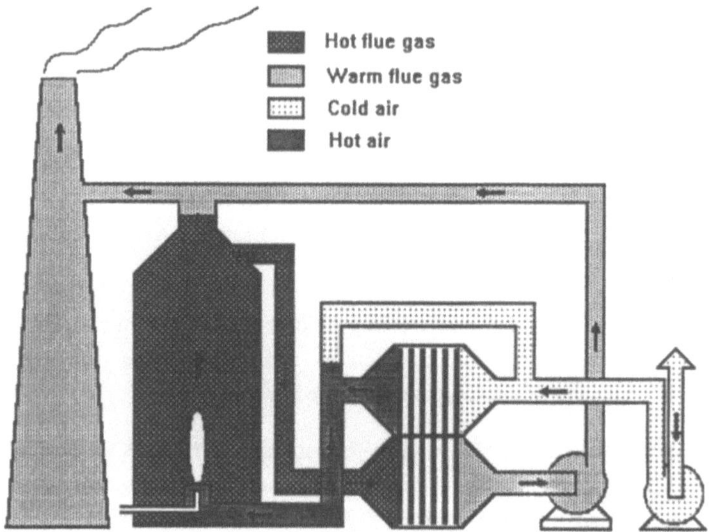

Fig. 5: Diagram of the heat recovery installation on a refinery furnace

The HPHE is composed of 345 heat pipes. The heat pipes are made of 5m long steel finned tubes (diameter=44.5mm) with toluene as the working fluid.

The working conditions of the HPHE are summarized in the following table:

Flue gas input temperature (°C)	325
Flue gas output temperature (°C)	158
Flue gas mass flow rate (kg/s)	7.081
Air input temperature (°C)	50
Air output temperature (°C)	242
Air mass flow rate (kg/s)	6.686
Thermal power recovered (kW)	1309
Thermal effectiveness (%)	70

result of the balance between the capillary forces and the pressure losses. It determines the maximum heat flux through the heat pipe which is named the capillary limit.

Whenever possible, gravity is used to replace capillary forces. In this case heat pipes must be tilted at least 5 degrees from the horizontal position with the evaporator above the condenser. The capillary structure is no longer necessary, meanwhile, grooves can be useful to increase inside heat transfer coefficients. In these gravity-assisted heat pipes grooves are often used in aluminium and copper tubes but not in steel tubes (because their manufacturing is too difficult). For gravity-assisted heat pipes the working limit is the entrainment limit which is due to liquid entrainment by the vapour counter flow.

For each heat pipe type the working limit must be calculated in order to verify that it cannot be reached in the heat exchanger working conditions (cf. §2).

3.4. Tubes materials

Aluminium is the most common material for moderate temperature (up to 150°C). It is a good thermal conductor, it is easy to work by extrusion to obtain grooves capillary structures and it is light. Aluminium is often used with refrigerant (such as Freon 11, Freon 113 or their substitute) as working fluid for air conditioning applications. Aluminium is not compatible with water.

Copper is often used with water as working fluid because it has an excellent compatibility. Copper has a very high thermal conductivity and copper tubes with grooves capillary structures are commercialized. Copper can be used up to 200°C.

Steel is the best suited material to high temperatures and corrosive external fluids. It is compatible with all organic fluids. A special treatment is necessary with water. Stainless Steel is also used, especially for food industry.

4. APPLICATIONS OF HPHE

4.1. Using domains of HPHE

The main areas of application of HPHE are the following:

Air conditioning: The HPHE advantages for these applications are: compactness, very good mechanical isolation between the the two fluids, very low susceptibility to freezing (absence of any cold point).

Heat recovery: The HPHE advantages for these applications are: great design flexibility, high thermal effectiveness, low pressure losses, very good mechanical isolation between the two fluids, possibility to avoid corrosion by acid condensation (absence of any cold point).

Flue gas input temperature (°C)	240.0
Flue gas output temperature (°C)	232.6
Flue gas mass flow rate (kg/s)	1.222
Flue gas pressure loss (Pa)	34
Water input temperature (°C)	15.0
Water output temperature (°C)	35.3
Water mass flow rate (kg/s)	0.11
Water pressure loss (Pa)	116
Thermal power recovered (W)	9338

5. CONCLUSION

This paper has tried to show that Heat Pipe Heat Exchangers have many advantages for numerous applications such as air conditioning, heat recovery or agro industry. Design tools are existant and qualified, real size industrial applications have been successfully carried out. It is now up to European industry to take advantage of this new heat exchanger type.

6. REFERENCES

[1] DUNN P.D., REAY D.A., Heat Pipes, Pergamon Press, Oxford, 1982

[2] CHAUDOURNE S., Les Echangeurs à Caloducs, Technique et Documentation, Paris, 1987

[3] CHAUDOURNE S., GRUSS A., Theoritical and experimental study of a high temperature heat pipe heat exchanger, application to a 1300 kW recuperator, 6th International Heat pipe Conference, Grenoble (France), 1987

[6] CHAUDOURNE S., Présentation du logiciel de calcul d'échangeurs à caloducs: ECCO, Note Technique GRETh 91/244, Février 1991

[7] CHAUDOURNE S., Validation expérimentale du programme de calcul d'échangeurs à caloducs ECCO, Note Technique GRETh/86/92, Août 1986

[8] KAYS W.M., LONDON A.L., Compact Heat Exchangers, Mc Graw Hill, New York, 1984

[9] DOMINGOS J.D., Analysis of complex assemblies of heat exchangers, Int. J. Heat and Mass Transfer, Vol.12, pp.537-548, 1969

Performance Analysis and Test of a Two-Phase Closed Thermosyphon Heat Exchanger

C. S. Chang, C. Tao, R. J. Shyu

Energy & Resources Laboratories
Industrial Technology Research Institute
Chutung, 31015, Hsinchu, Taiwan R.O.C.

Abstract

In this study, a thermal conductance model was used to predict the performance characteristics of an air to air two-phase closed thermosyphon heat exchanger for waste heat recovery. To validate the analysis, a heat exchanger composed of 124 copper /water thermosyphons was manufactured and tested under various operating conditions. The experimental performance of the heat exchanger showed good agreement with the theoretical prediction from the aspect of engineering applications.

Introduction

Heat exchangers made of heat pipes or two-phase closed thermosyphons are one of the most effective equipments for low temperature waste heat recovery. In an early study, Lee and Bedrossian [1] developed an analytic model to predict the characteristics of counter-flow heat exchangers and compared performance with experimental results. Huang and Tsuie [2] analyzed the thermal performance of heat pipe heat exchangers by using a conductance model. The heat conductance of the heat pipe used in the model was obtained from a performance test of a single heat pipe. Stulc et al. [3] reported an intensive study of various heat pipe heat exchangers for use in heat recovery. Azad and Goola [4] developed a design procedure for gravity-assisted heat pipe heat exchangers. The variations of overall effectiveness with various design parameters were presented and discussed. In the present study, a two-phase closed thermosyphon heat exchanger was designed and analyzed by a thermal conductance model. Its thermal performance was measured experimentally and compared with analytical prediction.

Analysis

1.Thermal Conductance Model

Assume that a two-phase closed thermosyphon heat exchanger can be divided into n sections along the air flow direction, and each section consists a row of two-phase closed thermosyphons of the same kind, as shown in Fig.1. The overall thermal resistances across the hot flow to the cold flow for each thermosyphon in the j-th section can be expressed by the following equation for the ideal model shown in Fig.1.

$$(R)_j = (\sum_{i=1}^{7} R_i)_j \quad , \quad j = 1 \cdots n \tag{1}$$

where R_1 and R_7 are the thermal resistances between the heat source and the evaporator external surface, and between the condenser external surface and the heat sink, respectively. R_1 and R_7 are defined as

$$R_1 = 1/(h_e A_{oe} \eta_{oe}) \tag{2}$$
$$R_7 = 1/(h_c A_{oc} \eta_{oc}) \tag{3}$$

In Eqs.(2) and (3), h_e and h_c are convective heat transfer coefficients across the thermosyphon tube wall in the hot and the cold flows.They can be obtained by using Briggs & Yang correlation[5].

$$hD_0/k = 0.134 \, Re^{0.681} Pr^{1/3} (y/H)^{0.2} (y/\delta)^{0.1134} \tag{4}$$

The total surface temperature effectiveness η_0 in Eqs.(2) and (3) is defined as

$$\eta_0 = 1 - A_f (1 - \eta_f)/A_0 \tag{5}$$

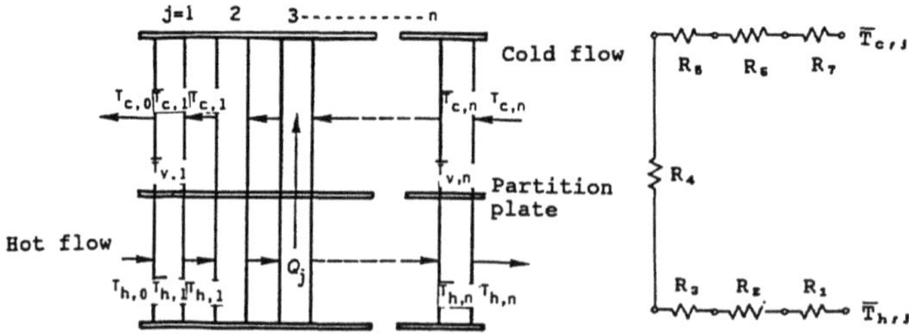

Fig.1 Thermal conductance model

where the approximation of the fin temperature effectiveness for annular low-finned tubes, η_f, is recommended [6] as

$$\eta_f = \tanh(mz)/(mz) \tag{6}$$
$$z = H[1+(\delta/2H)][1+0.35\ln(D_f/D_o)] \tag{7}$$
$$m = [2h/(k_f\delta)]^{0.5} \tag{8}$$

R_2 and R_6 represent the thermal resistances across the thickness of the thermosyphon wall in the evaporator and the condenser respectively. For a tube geometry, R_2 and R_6 are

$$R_2 = \ln(D_o/D_i)/(2\pi k_w l_e) \tag{9}$$
$$R_6 = \ln(D_o/D_i)/(2\pi k_w l_c) \tag{10}$$

R_3 and R_5 are the internal thermal resistances of working fluid in the boiling and condensing phase inside the thermosyphon .For a vertical tube, the condensation thermal resistance R_5 can be calculated from Nusselt's theory of filmwise condensation, thus

$$R_5 = 0.235 \ q^{1/3}/(D_i^{4/3} g^{1/3} l_c \emptyset_2^{4/3}) \tag{11}$$

For the evaporation thermal resistance, ESDU [7] recommands a method for the calculation of an approximate value. It examines the thermal resistance a combination of the nucleate boiling in the pool and the evaporating liquid film in the evaporator section.The result can be described by the following equations,

$$R_{3P} = 1/[\emptyset_3 g^{0.2} q^{0.4} (\pi D_i l_e)^{0.6}] \tag{12}$$
$$R_{3f} = 0.235 \ q^{1/3}/(D_i^{4/3} g^{1/3} l_e \emptyset_2^{4/3}) \tag{13}$$
$$\text{if } R_{3P} < R_{3f}, \quad R_3 = R_{3P} \tag{14}$$
$$\text{otherwise,} \quad R_3 = R_{3P}F + R_{3f}(1-F) \tag{15}$$

In Eq.(15), F is the fill ratio of the thermosyphon, which is defined as the ratio of the fill volume to the evaporator volume.

R_4 is the thermal resistance related to the saturation temperature drop between evaporator and condenser. Under normal operating conditions, R_4 can be neglected. Therefore, the overall thermal resistance across a two-phase closed thermosyphon can be determined. The heat transfer rate in the j-th section can be expressed as

$$Q_j = N_j (\overline{T}_{h,j} - \overline{T}_{c,j})/(R)_j \qquad (16)$$

where N_j is the number of the thermosyphons in the j-th section. $\overline{T}_{h,j}$ and $\overline{T}_{c,j}$ are the mean temperatures of the hot and the cold flows which are defined as

$$\overline{T}_{h,j} = (T_{h,j} + T_{h,j-1})/2 \qquad (17)$$
$$\overline{T}_{c,j} = (T_{c,j} + T_{c,j-1})/2 \qquad (18)$$

The total heat transfer rate delivered by the heat exchanger can be calculated by an summation over the N sections.

$$Q_t = \sum_{j=1}^{n} Q_j \qquad (19)$$

2.Temperature Distributions and Effectiveness
In order to calculate the total heat transfer rate, the temperature distributions in the hot and the cold flows must be calculated first. By applying the energy balance to the j-th section, the temperature difference of the flow across each section can be written as

$$T_{h,j-1} - T_{h,j} = Q_j/C_h \qquad (20)$$
$$T_{c,j-1} - T_{c,j} = Q_j/C_c \qquad (21)$$

Where C_h and C_c are capacity rates of the hot and the cold flows.

$$C_h = \dot{M}_h c_{ph} \qquad (22)$$
$$C_c = \dot{M}_c c_{pc} \qquad (23)$$

By substituting Eqs.(17), (18) and (20)-(23) into Eqs.(16), the heat transfer rate in j-th section can be derived as

$$Q_j = N_j (T_{h,j-1} - T_{c,j-1})/[(R)_j + N_j (1-C_r)/(2C_h)] \qquad (24)$$

$$\text{where } C_r = C_h/C_c \qquad (25)$$

The temperature distributions of the hot and the cold flows are obtained by substituting Eq.(24) into Eqs.(20) and (21).

$$T_{h,j} = T_{h,j-1} - N_j (T_{h,j-1} - T_{c,j-1})/[R_j C_h + N_j (1-C_r)/2] \qquad (26)$$
$$T_{c,j} = T_{c,j-1} - C_r N_j (T_{h,j-1} - T_{c,j-1})/[R_j C_h + N_j (1-C_r)/2] \qquad (27)$$

The vapor temperature $\overline{T}_{v,j}$ can be expressed as

$$\overline{T}_{v,j} = \overline{T}_{h,j} - Q_j \left(\sum_{i=1}^{3} R_i \right)_j / N_j = \overline{T}_{c,j} + Q_j \left(\sum_{i=5}^{7} R_i \right)_j / N_j \qquad (28)$$

Because the thermal resistances R_3 and R_7 are functions of the heat transfer rate and the working temperature of the two-phase closed thermosyphon, an iteration method was employed to solve the temperature distributions.

The effectiveness of the heat exchanger is defined as

$$\varepsilon = (T_{h,o} - T_{h,n}) / (T_{h,o} - T_{c,n}) \qquad \text{if } C_h \leq C_c \qquad (29)$$
$$\varepsilon = (T_{c,o} - T_{c,n}) / (T_{h,o} - T_{c,n}) \qquad \text{if } C_h > C_c \qquad (30)$$

and the overall heat transfer coefficient of the heat exchanger is defined as

$$U = Q_t / (A_t \Delta T_m) \qquad (31)$$

Where ΔT_m is the log mean temperature difference across the heat exchanger

$$\Delta T_m = [(T_{h,o} - T_{c,o}) - (T_{h,n} - T_{c,n})] / \ln[(T_{h,o} - T_{c,o}) / (T_{h,n} - T_{c,n})] \qquad (32)$$

The pressure drop across the tube boundle is given by [6] as follow,

$$\Delta P = 2fnG^2_{max} / \rho \qquad (33)$$

where f is defined as

$$f = 3.805\ Re^{-0.234} (y/D_f)^{0.251} (H/y)^{0.759} (D_o/D_f)^{0.729}$$
$$(D_o/S_t)^{0.709} (S_t/S_l)^{0.379} \qquad (34)$$

Experiment

A heat exchanger using two-phase closed thermosyphons was constructed and tested. The specifications of the two-phase closed thermosyphon and the heat exchanger are listed in Table 1. The thermosyphon is 1 meter long with both evaporator and condenser are 0.495 m. They are 25.4 mm outside diameter copper tube with spiral copper fin. The working medium is distilled

Table 1 Design specifications of the heat exchanger

Item	Descriptions	
1.Case	1100 mm(high)x 900 mm(wide)x 400 mm(deep)	
2.Thermosyphon	overall length	1000 mm
	evaporator length	495 mm
	condenser length	495 mm
	adiabatic length	10 mm
	outside diameter	25.4 mm
	inside diameter	23.4 mm
	total number	124
	number of rows	8
3.Fin	density	393/m
	thickness	0.4 mm
	height	7 mm
4.Tube arrangement	staggered	
	transverse pitch	49.2 mm
	longitudinal pitch	42.7 mm

water and its fill ratio is 50% of the evaporator volume. The total number of thermosyphons is 124 in 8 rows with staggered tube arrangement.The frontal area of the heat exchanger is 0.8 m². Figure 2 shows the schematic diagram of the experimental setup. The heat exchanger was designed to be operated in counter-flow condition with air as heat transfer fluid. A 50 KW steam boiler and a 30 KW electric heater provided a hot air flow up to 150 °C. The inlet and outlet temperatures of the two air flows through the heat exchanger were measured by T-type thermocouples at the center of the cross section of each channel. The operating temperature of the central thermosyphon in each row was measured. To do this, the foil junction of a cement-on thermocouple was bonded to the adiabatic section of the thermosyphon. The velocity distribution of each flow was measured at 5 points along the radial direction of the test

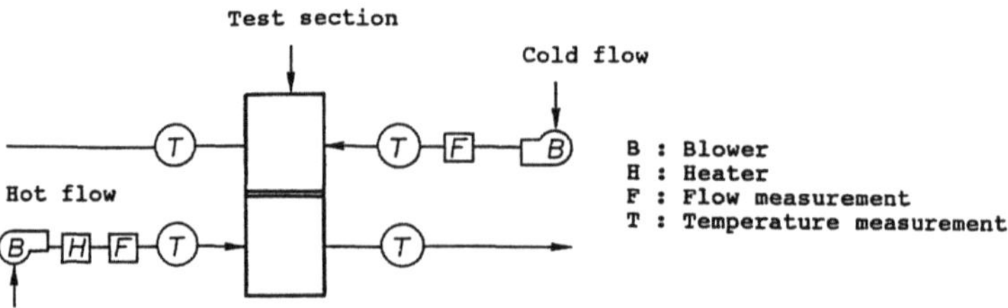

Fig.2 Test setup of the heat exchanger

cross-section by a Pitot tube and a precision pressure gauge. By integrating the velocity distribution, the air volumetric flowrate and average velocity in each channel was therefore determined. All the steady-state measurements were recorded by a hybrid recorder for 4 hours.

Results and Discussion

Typical temperature variations of the two air flows and the operating temperatures of the thermosyphon in each row are shown in Fig.3. The solid and the dashed lines were calculated by the program, and the symbols were experimental data.In Fig. 3, the measured operating temperatures of the thermosyphons were near the calculations. The experimental effectiveness was 0.62, which is 15% lower than the prediction.These discrepancy are partly due to the idealization of the model, the irregular fin geometry and the poor fin contactness of some thermosyphon tubes which were damaged during transportation. Figure 4 shows the variations of the effectiveness vs capacity rate ratio C_c/C_h.The experimental data shows that the effectiveness has a minimum when C_c/C_h approaches 1. The difference between

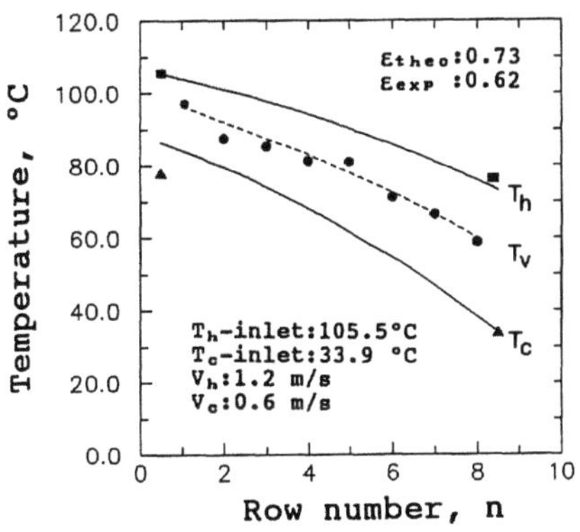

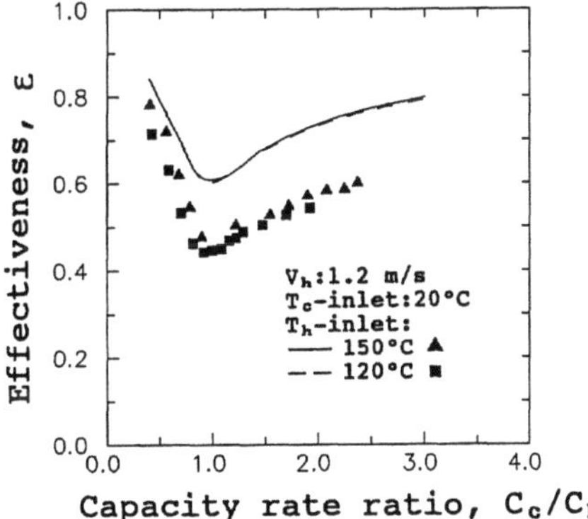

Fig.3 Temperature variations of flows and thermosyphons

Fig.4 Effectiveness vs capacity rate ratio

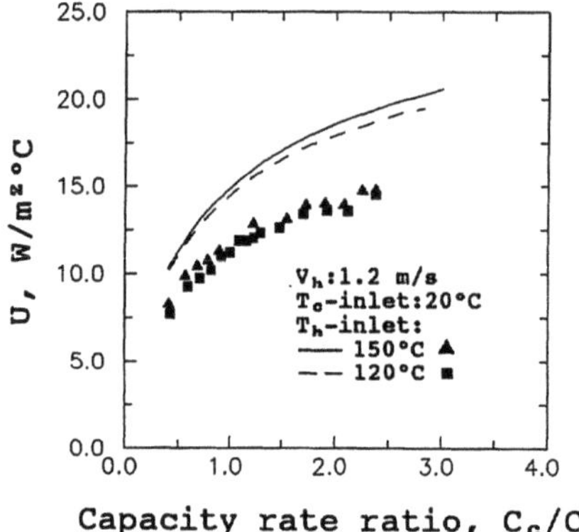

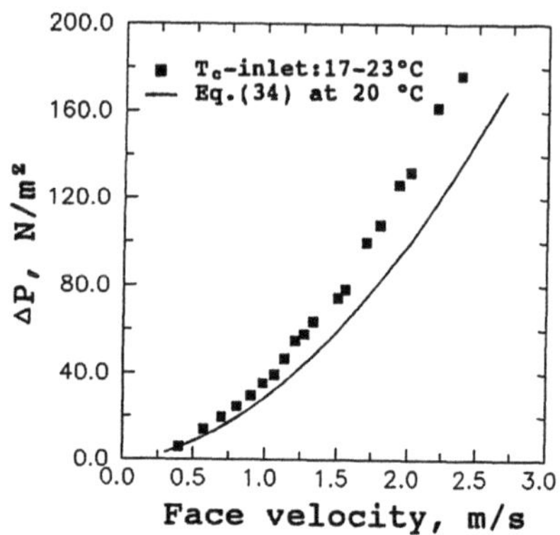

Fig.5 Overall heat transfer coefficient

Fig.6 Pressure drop of the cold flow vs face velocity

experimental data and theoretical predictions (dashed and solid lines) seems to be larger for lower hot flow inlet temperature and higher C_c/C_h. Under these operating conditions ,parts of the thermosyphons did not work normally as the corresponding vapor temperature in the tube was less than the generally recommended lowest limit (40°C for water [7]), and the start-up of these thermosyphons became difficult. Figure 5 shows the overall heat transfer coefficient of the heat exchanger. The experimental value ranged from 7.7 to 14.7 W/m²°C. Figure 6 represents the pressure drop of the cold air across the heat exchanger vs its face velocity. The pressure drop was underestimated by 32% maximum at a face velocity 2.2 m/s.

Conclusions

A computer program based on the thermal conductance model was developed to predict the performance of a two-phase closed thermosyphon heat exchanger. The experimental effectiveness and pressure drop of the heat exchanger showed good agreement with the theoretical prediction from the engineering point of view.

Acknowledgement

The present work is supported by The Energy Commission of Ministry of Economic Affairs in Taiwan R.O.C. under a contract No. 798W4.

Nomenclature

A	Area	m^2
A_t	Total area of evaporator outer surface	m^2
C	Flow capacity rate	$W/°C$
c_p	Specific heat	J/kgK
D	Diameter	m
f	Friction factor	-
F	Fill ratio	-
g	Gravitational acceleration	m/s^2
G_{max}	Mass velocity based on minimum free-flow area	kg/sm^2
h	Heat transfer coefficient	W/m^2K
H	Fin height	m
k	Thermal conductivity	W/mK
l	Length	m
L	Latent heat	J/kgK
$\dot{M}$	Mass flow rate	kg/s
n	Number of rows	-
P	Pressure	N/m^2
ΔP	Pressure drop	N/m^2
Pr	Prandtl number	-
q	Heat transfer rate of individual thermosyphon	W
Q	Heat transfer rate	W
R	Thermal resistance	$°C/W$
Re	Reynolds number	-
S_l	Longitudinal tube pitch	m
S_t	Transverse tube pitch	m
T	Temperature	$°C$
U	Overall heat transfer coefficient of a heat exchanger	$W/m^2°C$
V	Face velocity	m/s
y	Mean fin gap	m
δ	Fin thickness	m
ϵ	Heat exchanger temperature effectiveness	-
ρ	Density	kg/m^3
$\emptyset_2$	Condensation figure-of-merit $(Lk_l^3\rho_l^2/\mu_l)^{0.25}$	$kg/s^{2.5}K^{0.75}$
$\emptyset_3$	Boiling figure-of-merit $(\rho_l^{0.65}k_l^{0.3}c_{Pl}^{0.7}P_v^{0.23}/\rho_v^{0.25}L^{0.4}\mu_l^{0.1}P_{at}^{0.23})$	$kg^{0.6}/m^{0.2}s^{1.4}K$

Subscripts

c	Condenser side, cold flow
e	Evaporator side
f	Fin
h	Hot flow
i	Inner
j	Row number
l	liquid
o	outer
t	total
v	vapor

References

1. Lee Y. and Bedrossian A., The Characteristics of Heat Exchangers Using Heat Pipes or Thermosyphons, Int. J. Heat Mass Transfer, 21 (1978) 221-229.
2. Hung B. J. and Tsuei J. T., A Method of Analysis for Heat Pipe Heat Exchangers, Int. J. Heat Mass Transfer, 28 (1985) 553-562.
3. Stulc P.,Vasiliev L. L.,Kiseljev V. G. and Matvejev Ju. N., Heat Pipe Heat Exchangers in Heat Recovery Systems, Heat Recovery Systems, 5 (1985) 415-418.
4. Azad E. and Geoola F.,A Design Procedure for Gravity-Assisted Heat Pipe Heat Exchanger,Heat Recovery Systems,4 (1984) 101-111.
5. Briggs D. E. and Yang E. H., Convection Heat Transfer and Pressure Drop of Air Flowing Across Triangular Pitch Banks of Finned Tubes, CEP Symposium, 41 (1963) 1-10.
6. Rabas T. J. and Taborek J., Survey of Turbulent Forced-Convection Heat Transfer and Pressure Drop Characteristics of Low-Finned Tube Banks in Cross Flow, Heat Transfer Engineering, Vol.8, 2 (1987) 49-62.
7. Thompson N.,Heat Pipes-Performance of Two-Phase Closed Thermosyphons, Engineering Science Data Unit, Item No. 81038 (1981).

An Application of Semi-Empirical Turbulence Theory to the Hydrodynamics and Heat Exchange in Gas-Liquid Foam

A. ALABOVSKY, E. GALPERIN, V. SALO, N. SULGIK

KIEV POLYTECHNICAL INSTITUTE, KIEV, USSR

ABSTRACT

The following work presents an application of isotropic and
homogenous turbulence theory to the investigation of hydrodynamics and heat exchange inside gas-liquid foam. Fixed expressions for interfacial area between phases at the foam strata
and heat transfer coefficient (from warmed surface, which
was placed inside foam) were obtained.

One of the ways to intensify heat transfer between a warmed
surface and gas, is to use gas bubbling throug intermediate
liquid heat carrier. During this process heat exchange surface
should be placed inside the bubbling stratum. Experimantal data
/ 8 / shows that the temperature of the heat exchange surface is
descreased. This means that heat exchange intensity during foam
conditions are hight than during large volume boiling.

We therefore need certain physical models of heat and mass
transfer for foam transfer equipment.

The following work presents an application of Kolmogorov-Obuhov
isotropic and homogenius turbulence theory. Expressions for
interfacial area inside the foam stratum, a , gas content φ
were obtained together with basic equtions for the heat transfer coefficient between the warmed surface and the foam stratum.

We must emphasise that the foam regime is involves variations
in bubble shape, bubble volume, gas velocity, interfacial area,
gas content etc. Hence we have considered the following turbulence exchange scheme for the gas and liquid foam : from the

408

pulsations with large scale energy are transformed to the pulsations with small energy without dissipation. According to this principle the large pulsations are forming the interfacial area with an empirical dependence on the resistance to the gas movement throughout the liquid stratum. The small pulsation are dissipated according to thin liquid film viscosity. Generally, we consider the turbulence of this gas-liquid foam as isotropic and homogenous. Assuming that the final energy is dissipated at the liquid films we can write on the following expression for dissipation energy:

$$\varepsilon \sim \nu' \cdot \left(\frac{W_D}{\ell_D}\right)^2 \tag{1}$$

where: ν' = kinematic viscosity of the liquid ;

W_D = pulsation velocity, where dissipation takes place;

ℓ_D = geometrical scale of these pulsation.

To determine the expresions for W_D , we will assume that it is proportional to the dynamic velocity of the liquid on the film interface / 3 /

$$W_D \sim \left(\frac{\tau_B}{\rho'}\right)^{0,5} \tag{2}$$

where: τ_B = tangential stress ;

ρ' = density of the liquid.

Now $\tau_B = C_f'' \cdot \rho'' \cdot W_{oT}^2$, so

$$W_D \sim \left(\frac{\rho''}{\rho'}\right)^{0,5} \cdot W_{oT} \tag{3}$$

where: W_{oT} = gas phase velosity relative to the liquid ;

ρ'' = density of the gas phase.

As equation (3) were recieved according to the principle that the process of gas interaction does not depend on the way of energy supply (gas supply, acoustic pulsation, mechanical mixing etc.) it follows that the dissipation length should be defined only by properties of the liquid film and the ineet acceleration of the system.

$$\ell_D = f(g, \sigma, \nu', \rho') \qquad (\ 4\ )$$

Using this method of dimensional analysis / 3 / one can show that the dissipation length is given by

$$\ell_D = \ell_K \cdot Mo^{n} \qquad (\ 5\ )$$

where : n = exponent index (defined practicaly) ;

$\ell_K = (\sigma / \rho' \cdot g)^{0,5}$ = capillarity constant ;

$Mo = \dfrac{g \cdot \nu'^{4} \cdot \rho'^{3}}{\sigma^{3}}$ = Morton's criterion which shows out interrelation capillarity, viscosity and gravitation forces in the film of gas and liquid foam.

The following data defenitly shows that using n = 0.07 in equation (5) give us a chance to recieve, from the unit position, theoretical dependence in accordance with numercus experimental data of gas content, interfacial area and heat transfer in foaming conditions. Substituting expressions (3) and (5) in (1) gives the following equation for dissipation energy:

$$\mathcal{E} \sim \nu' \cdot \frac{\rho''}{\rho'} \cdot \frac{W_{oT}^{2}}{\ell_K^{2}} \cdot \frac{1}{Mo^{2n}} \qquad (\ 6\ )$$

This equation can be applied only for foam-turbulence conditions.

We can use equation (6) to determinate gas content , φ , the interfacial area, a and the heat transfer coefficient, α in the foam stratum.

Gas content

We need to use on stationary equations system of two-phase mixture motion with the common pressure / 3 / to find out gas content of the foam stratum:

$$- \varphi \cdot \frac{dP}{dz} + \varphi \cdot f_{21} + \rho'' \varphi \cdot g = 0 \qquad (7 \cdot)$$

$$-(1-\varphi) \cdot \frac{dP}{dz} + \rho' (1-\varphi) \cdot g - \varphi \cdot f_{21} = 0 \qquad (8)$$

where : f_{21} = interface force for unit volume;
P = system presure.

Eliminating the pressure gradient from the equations (7) and (8) gives

$$g (\rho' - \rho'') \cdot (1-\varphi) = f_{21} \qquad (9)$$

This relationship shows equality between gravity and resistance forces during stationary movement of two-phase system.

There are many suggestions at the literature about determination the ratio between different conditions of drops streaming or bubbles streaming. However, for the foam-turbulence regime, which have irratic bubble movement, there are very few equa-

tions. Authors / 4 / summarizing the results of more than 100
experimantal results, suggested that this resistance force, acting on the gas phase, showed be linked with the velocity of the
mixture rather than the velocity of the gas phase relative to
the liquid. With this assumption the resistance force should be:

$$f_{21} = \frac{C' \cdot \rho' \cdot W_\varphi \cdot |W_\varphi|}{d_e}$$
(10)

where : W_φ = drift velocity ; $W_\varphi = (1 - \varphi) \cdot W_{oT}$

C' = effective resistance coefficient ;

d_e = volume-surface average of the gas
bubble diameter.

To determine the average bubble diametr we used the Kolmogorov
result / 2 /. This shows that during the process of gas disper-
giration within turbulent liquid flow, the average gas-bubble
volume is determinated by the ratio of surface stretch force to
the turbulence pulsation force, within two-phase streams. The
theoretical equation for the volume-surface average diameter is:

$$d_e = k_e \cdot \frac{\sigma^{0,6}}{\varepsilon^{q4} \cdot \rho'^{0,6}}$$
(11)

where : σ = coefficient of surface tension.

Using the ratio for relative velocity of phases:

$$W_{oT} = \frac{W_o''}{\varphi} - \frac{W_o'}{1-\varphi}$$
(12)

where : W_o'' = adopted gas phase velocity ;

W_o' = adopted liquid phase velocity.

Combining equation (12), (11), (6) and (10), and using the results for f_{21} to determinate (9) we will recieve obtain the following equation for the gas content within two-phase foam flow.

$$\frac{g}{1-\varphi}\cdot\left(1-\frac{\rho''}{\rho'}\right)=C_\varphi'\left(\frac{\rho''}{\rho'}\right)^{0,4}\cdot\left(\frac{Wo''}{\varphi}-\frac{Wo'}{1-\varphi}\right)^{2,8}\cdot Mo^{0,1-0,8\cdot n} \qquad (13)$$

As for a non-stream bubbler, this means when $Wo' = 0.$, we obtain the following equation

$$\frac{\varphi^{2,8}}{1-\varphi}=C_\varphi\cdot\left(\frac{\rho''}{\rho'}\right)^{0,4}\cdot\left(Wo''\cdot\sqrt[4]{\frac{\rho'-\rho''}{g\cdot\sigma}}\right)^{2,8}\cdot Mo^{0,1-0,8\cdot n} \qquad (14)$$

Approximate solution of equation (14) within foam condition ($0.2 < \varphi < 0.75$) helps us to obtain the following expression ($\Pi = 0.07$) :

$$\varphi = k_\varphi\cdot\left(\frac{\rho''}{\rho'}\right)^{0,11}\cdot\left(Wo''\cdot\sqrt[4]{\frac{\rho'-\rho''}{g\cdot\sigma}}\right)^{0,73} \qquad (15)$$

This equation is in accordance whis the following correlations, by Kutateladze and Sterman / 3 , 5 /, which are finally confirmed by numerous experimental studies

$$\varphi = 0,26\cdot\left(\frac{\rho''}{\rho'}\right)^{0,12}\cdot\left(Wo''\cdot\sqrt[4]{\frac{\rho'-\rho''}{g\cdot\sigma}}\right)^{0,72} \qquad (16)$$

$$\varphi = 0,4\cdot\left(\frac{\rho''}{\rho'}\right)^{0,15}\cdot\left(Wo''\cdot\sqrt[4]{\frac{\rho'-\rho''}{g\cdot\sigma}}\right)^{0,7} \qquad (17)$$

It is clear from equations (16), (17) that the suggested

method to determinate gas content withing a foam stratum give
is compatible with experimental correlations.

Interfacial area

The specific interfacial area is one of the most important
quantities which directly influences the process of heat and
mass transfer inside direct contact gas-liquid equipment. At
the present time there is no complete turbulence theory for two-
phase flow, but there are several important results can which
can be used to calculate the specific interfacial area. Kolmogorov / 2 / obtained theoretical equation (11) to define the
surface-volume average diameter of gas bubbles. We used this
equation to determine the expression for gas content in the
foam stratum. We will use it once more to obtain an equation of
interfacial area.
It is obvious that the interfacial area within a two-phase
stratum is developed inversely proportional to the sagnificance
of average gas bubble diameter :

$$a \sim \frac{1}{d_e} \qquad\qquad (\ 18\ )$$

Taking into account the equation (11) we obtain :

$$a = k_a \cdot \frac{\varepsilon^{0,4}\,\rho'^{0,6}}{\sigma^{0,6}} \qquad\qquad (\ 19\ )$$

For a non-stream bubbler ($W_{oT} = W_o''/\varphi$) using expression
(17) to obtain the following equation for the interfacial area

414

$$a \cdot \ell_{\kappa} = k \cdot \left(\frac{W_0'' \cdot M'}{\sigma} \right)^{0,24} \cdot \left(\frac{\rho''}{\rho'} \right)^{0,28} \cdot Mo^{0,04-0,8 \cdot \Pi} \qquad (\ 20 \ )$$

where : M' = coefficient of liquid viscosity ;

$K = \dfrac{W_0'' \cdot M'}{\sigma}$ = capillary criterion.

Assuming that Π = 0.07 we find that the influence of Mo on a can be expressed by an exponent index of -0.017. Therefore it is reasonable to neglect effect of Mo on a . Finally we obtain the following theoretical expresion:

$$a \cdot \ell_{\kappa} = k \cdot \left(\frac{W_0'' \cdot M'}{\sigma} \right)^{0,24} \cdot \left(\frac{\rho''}{\rho'} \right)^{0,28} \qquad (\ 21 \ )$$

Figure 1 and 2 show the agreement between the theoretical results and experimental data (according to Korolevich and Rodionov / 6, 7 /). Figure 1 shows the agreement of $a \cdot \ell_{\kappa} / K^{0,24}$ with varying gas-liquid density ratio / 6 /. It is clear from the figure that the agreement is both qualitative and quantitative. Figure 2 presents the dependence of $a \cdot \ell_{\kappa} / \left(\frac{\rho''}{\rho'} \right)^{0,28}$ on K . The figure also shows the experimental results of specific interfacial area from Rodionov / 6, 7 /.

These results also confirm the suggested dependence, although the common curves are a little lower. This deffence can be explained by the chemical method used in the research work. This method gives only the "active" part of a value, but not the full meaning of a quantity.

Heat transfer coefficient

In order to define the heat transfer coefficient from warmed surface to foam stratum, we propose divididing the two-phase

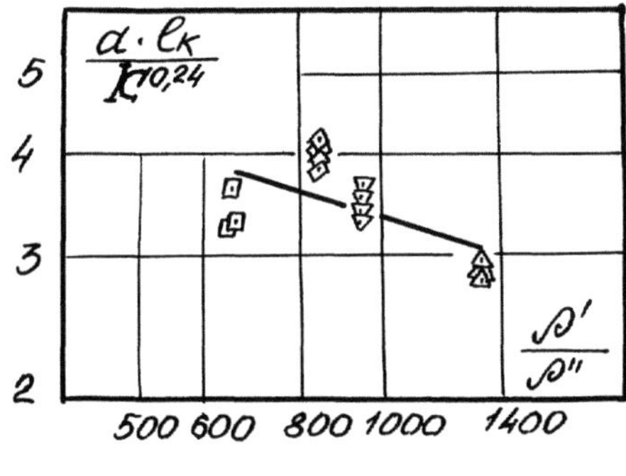

Fig 1.

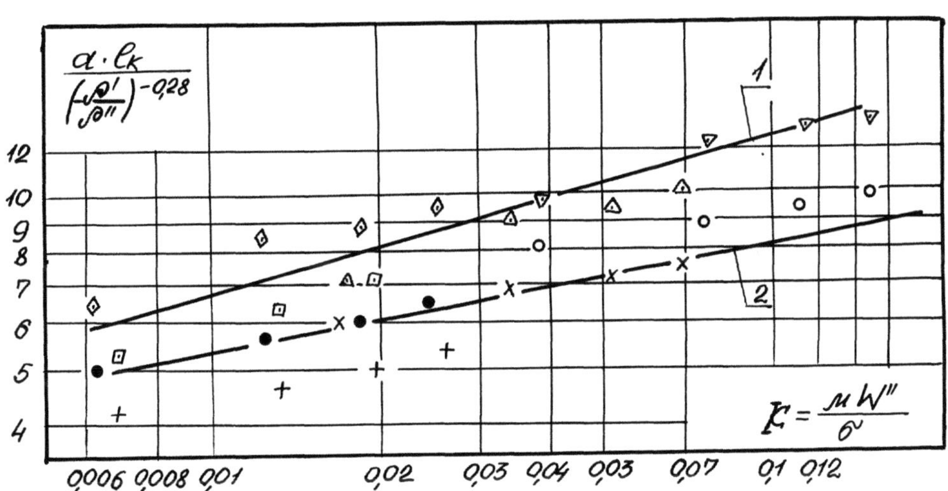

Fig. 2.

stream into two regions: clear liquid stratum "near the wall surface", and two-phase kernel, which has a large gas content and turbulence which is not influenced by the heat exchange process.

The main thermal resistance is close to the wall in the quasi-laminar liquid layer. In a case of constant heat exchange, the heat transfer coefficient is defined as:

$$\alpha = \frac{\lambda'}{\delta_T}$$

(22)

The thermal layer thickness can be related to the hydrodynamic layer, by the single phase analogy:

$$\tilde{\delta_T} = \frac{\tilde{\delta_g}}{P_{r_2}^{0,33}}$$

(23)

The quasi-laminar liquid film thickness is made proportional to $\tilde{\delta_0}$ to reduce Kholmogorov's scale / 1 /:

$$\tilde{\delta_g} = k_g \cdot \left(\frac{\nu'^3}{\varepsilon}\right)^{0,25}$$

(24)

Using an expression for gas content and dissipation energy (17) and (6) we obtain, after simplification, the equation for α :

$$\frac{\alpha}{\lambda'} \cdot \left(\frac{\nu'^2}{g}\right)^{1/3} = k_\alpha \cdot \left(\frac{W_0''}{(g \cdot \nu')^{1/3}}\right)^{0,15} \cdot \left(\frac{\rho''}{\rho'}\right)^{0,175} \cdot P_{r_2}^{0,33} \cdot Mo^{0,054-0,5 \cdot \Pi}$$

(25)

If $\Pi = 0.07$ the index for Mo in above equation is 0.018 and can therefore be neglected, given finally:

Nu
(W₀″ / (g·V′)^⅓)^0,15 Pr^0,33
0,23
0,21
0,17
0,15
ϑ″ / ϑ′
0,1   0,5   0,9   1,3   1,7
o — |10|

fig. 3.

0,65
Nu / Pr^0,33
0,45
0,25
o — |10|
◑ — |8|
△ — |9|
□ — |12|
● — |11|
W₀″ / (g·V′)^⅓
10   110   210   310

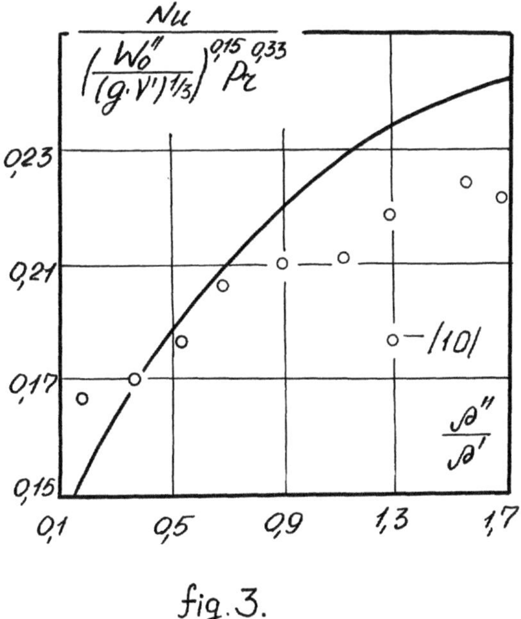

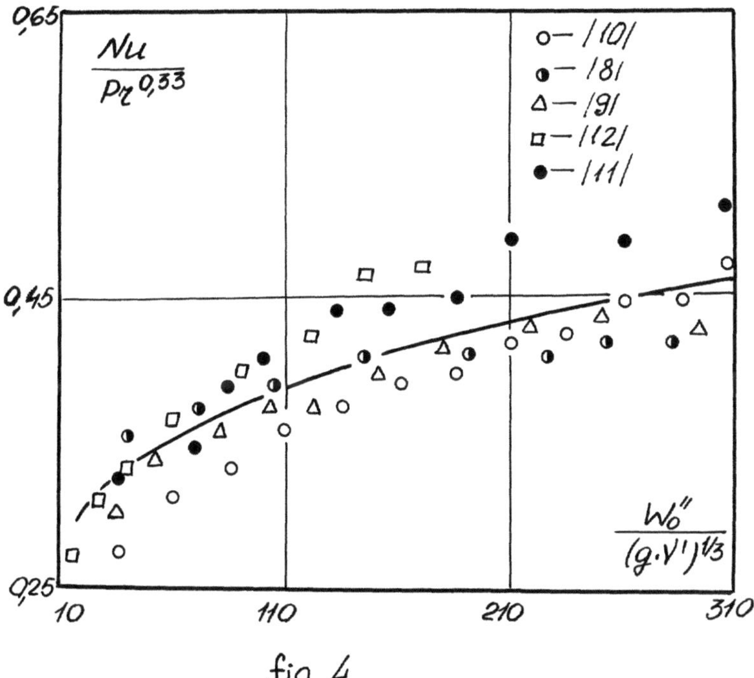

fig. 4.

418

$$\frac{\alpha}{\lambda'} \cdot \left(\frac{\nu'^2}{g}\right)^{1/3} = k_\alpha \cdot \left(\frac{W_0''}{(g \cdot \nu')^{1/3}}\right)^{0,15} \cdot \left(\frac{\rho''}{\rho'}\right)^{0,175} \cdot Pr^{0,33} \qquad (26)$$

This equation show that the heat transfer coefficient is indepedent of equipment diameter, the geometrical of the bubbling
grids and the heat exchange height and length above the grid.
It is also indepedent from foam stratum height H_f and the temperature difference between a warmed surface and foam. This is
in accordance with numerous experimental studies of different
authors / 11 , 12 /.

The length scale for the Nusselt number in equation (26) is
based upon the: $\ell_y = \left(\nu'^2/g\right)^{1/3}$. Reynolds-number in equation
(26) was also defined with the help of linear scale . The
physical properties in equation (26) are calculated at the
mean temperature: $(t_f + t_w)/2.$.

Figure 3 shows us an comparsion of experimental data / 8 / with
equation (26). The maximum error is 20 %. Figure 4 shows us
an comparsion between theoretical equation (26) and experimental data of papers / 8 - 12 /. It is clear that theoretical
equation agrees both qualitatively and quantitatively with the
experimental data.

So, this way of theoretical description of heat transfer process and hydrodynamics inside gas-liquid heat exchangers permits us from the unit position, to describe that physical
process, which early was two-phase one.

Literature:

1. Kolmogorov A.N., DAN SSSR , 1941, t.30, N 4, s. 299-303.
2. Kolmogorov A.N., DAN SSSR , 1949, t.66, N 5, s.1326-1334.
3. Kutatelaze S.S, Styrykovich M.A. - M., Energiya, 1976.
    - 296 s.
4. Ishii M., Zuber N.// AiCheJ, 1979, V.25, N 5, p.843-855

5. Kutepov A.M., Sterman A.S., Stushin N.G. - M. ,Vish. shkola, 1986. - 448c.
6. Alabovsky A.N., Korolevich A.J., Salo V.P., Izvestya vuzov, Energetyka, 1990, N 10, s. 71-76
7. Rodionov A.I., Vinter A.A., Trudy MHTI, 1966, vip. 51, s. 423-431.
8. Hoze A.N., Sharov J.I., JPM i TF, 1969, N 1, s.122-124.
9. Nakoryakov V.E., Muhin B.A., Kim I.G., Sbornyk nauchnih trudov, 1983, Novosibirsk, s. 104-115.
10. Hoze A.N., JPM i TF, 1971, N 5, s. 173-176.
11. Sokolov V.N., Domansky I.V. - L.: Mashinostroenie, 1976. - 216 s.
12. Budtov B.P., Konsetov V.V. - L.: Mashinostroenie, 1982. - 324 s.

Prediction of Heat Transfer Rates in a Liquid-Liquid Direct-Contact Heat Exchanger

J. HUTCHINS*, L. MORESCO**, K. PICKENS, and E. MARSCHALL

Department of Mechanical & Environmental Engineering
University of California, Santa Barbara
Santa Barbara, CA 93106

*Presently with Delco Electronics, Goleta
**Presently with Hewlett & Packard, Palo Alto

Summary

A heat transfer analysis of a liquid-liquid spray column operating as a direct-contact heat exchanger is presented. The analysis is based upon an iterative procedure originally proposed by Bühler. A major modification of the previously used analytical procedures is based on the assumption of free convection heat transfer inside the drops rather than of forced convection. Predicted heat transfer coefficients and heat transfer rates are compared with experimentally obtained data. A sensitivity analysis is used to demonstrate the degree of dependence of the analytical results on input parameters.

I.   Introduction

Direct-contact heat transfer is defined within the context of this paper as heat transfer between two liquids which are immiscible and which are in direct contact. The notion of direct contact implies that these liquids are not separated by a wall. Current interest in direct-contact heat transfer originates mainly from the fact that this heat transfer mode permits extraction of thermal energy from geothermal brines containing large amounts of dissolved solids. Due to the high concentration of dissolved solids conventional heat exchangers are usually subject to severe scale formation, whereas the absence of a wall between the two phases in a liquid-liquid direct-contact heat exchanger eliminates or greatly reduces these problems.

Since the prevention of problems associated with scale formation is a major incentive for utilizing direct-contact heat transfer, direct-contact heat exchangers usually consists of a vertical spray column containing as few devices as possible which could scale up.

The following discussion is restricted to liquid-liquid direct-contact heat transfer in spray columns. A schematic diagram of a vertical direct-contact heat exchange column is shown in Figure 1. The continuous phase enters the column at the top of the direct-contact heat exchanger, exiting through the bottom. The dispersed phase enters at the bottom of the column through nozzles or a bubble plate, since it is chosen such that it has a lower density than the continuous phase. It rises in droplet form due to buoyancy and exits at the top of the column. While, in principle, either phase can be the "hotter" one, in the following it has been assumed that the continuous phase is the hotter fluid. Due to the temperature difference between the two phases, the continuous phase loses energy to the dispersed phase over the entire length of the column.

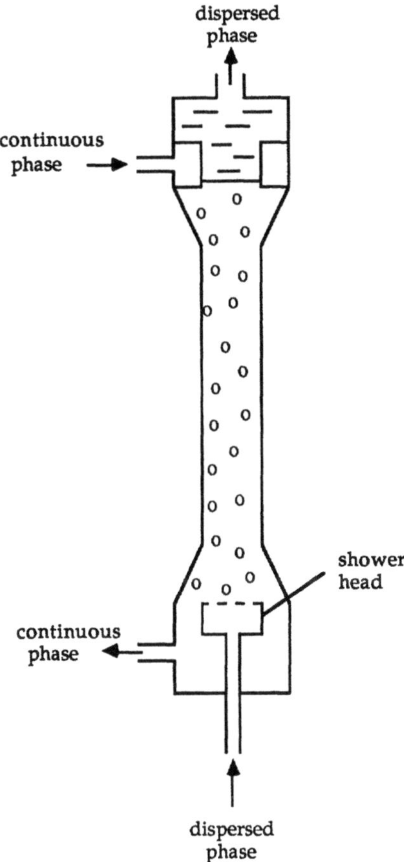

Fig. 1. Schematic of a spray column.

Major reviews of the state of the art of direct-contact heat transfer were provided by E. Kehat and S. Sideman [1] and by E. Marschall, G. Johnson and W. Culbreth [2]. Valuable information on direct-contact heat transfer and related subjects are also contained in Ref. [3-5].

Two approaches exist for the calculations of heat transfer in a direct-contact heat exchanger. The first approach relies on volumetric heat transfer coefficients, while the second approach is based on area related heat transfer coefficients. Volumetric heat transfer coefficients are simply calculated from

$$U_v = \dot{q} / V_c \Delta T_m \tag{1}$$

where $\dot{q}$ is the rate of heat transferred in the heat exchanger, V_c is the heat exchanger volume, and ΔT_m is the log mean temperature difference given by the temperatures of dispersed and continuous phase at the heat exchanger inlets and outlets. Volumetric heat transfer coefficients are readily determined, although incorrect measurements of the inlet and outlet temperatures frequently introduce appreciable errors. Unfortunately, volumetric heat transfer coefficients are only useful if they are applied to situations identical to the one in which they were found. A discussion of problems involved with use of volumetric heat transfer coefficients was provided in Ref. [1].

The second approach relies on the area related heat transfer coefficients h_i and h_0, which in turn depend on local flow fields and physical properties of continuous and dispersed phase. Since area related heat transfer coefficients in the past could not be as readily determined as volumetric heat transfer coefficients this second method has only recently become more feasible.

The analysis of direct-contact heat exchangers presented here relies on area related heat transfer coefficients and is based upon an iterative procedure originally proposed by Bühler [6]. Specifically, the analysis considers a vertical spray-column with known cross-sectional area in which a hot fluid is to be cooled from a high temperature T_{CI} to a low temperature T_{CO}. Thermal energy is to be transferred to a second fluid which is available at a temperature T_{DI}. The hot fluid is assumed to constitute the continuous phase. A first step in the analysis is to determine fluid flow characteristics.

II. Flow Conditions

To calculate the heat transfer, it is necessary to know the terminal velocity of the drops. The terminal velocity of the droplet is related to the density of the two phases, the drag coefficient of the droplets, and the volumetric flowrates of the two phases. Once this information is available, a force balance done on a droplet yields the terminal velocity of the droplet. A further result of this calculation is the hold-up which depends on the terminal drop velocity.

Hold-up is defined as the volume fraction of the dispersed phase to the total volume in a section of the column. Hold-up can vary as a function of height in the column if the densities of the phases change as a function of temperature. It is a key parameter in the design of any direct-contact heat exchanger.

If a drop rises in a column, the bouyant forces on the drop are sufficient to overcome the drag force on the drop imposed by the counter-current flow past the drop. If the buoyancy force is not sufficiently large to overcome the drag force, the droplet is carried downward in the column and swept out of the column. This phenomenon is known as "flooding". Other forms of flooding exist. For example, the dispersed phase flowrates may be so high that drops coalesce as they rise in the column. This may lead to an inversion of the phases, that is, the dispersed phase may turn into the continuous phase. The conditions which lead to flooding must be avoided in any operation of a direct-contact heat exchanger.

Flowrates, velocities, hold-up, and drop diameter are related to each other in the following manner:

Conservation of mass yields for the relative velocity between the two phases (slip velocity) the following relationship:

$$V_s = \dot{V}_C / (1 - \varepsilon) / A + \dot{V}_D / \varepsilon / A \tag{2}$$

where

$\dot{V}_C$ = volumetric flowrate of continuous phase
$\dot{V}_D$ = volumetric flowrate of dispersed phase
ε = hold-up
A = cross-sectional area of column

The relative velocity V_s can also be obtained from a force balance. For a spherical drop rising through the continuous phase a force balance yields:

$$\Sigma F_y = ma_y = F_{BOY} - F_D - mg \tag{3}$$

where:

m = mass of drop
a_y = acceleration of drop
F_{BOY} = buoyant force on the drop
F_D = drag force
g = gravitational acceleration

The buoyant force on the drop is:

$$F_{BOY} = \rho_{AVE} V_T g \tag{4}$$

where: V_T = volume of the drop.

The average density ρ_{AVE} is obtained from

$$\rho_{AVE} = \varepsilon \rho_D + (1 - \varepsilon) \rho_C,$$

where ρ_D and ρ_C are the densities of the dispersed and continuous phase, respectively.

Since spherical drops are assumed:

$$V_T = \frac{\pi d_T^3}{6}$$

where: d_T is the drop diameter.

The drag force on the drop can be expressed as:

$$F_D = \frac{1}{2} \rho_C C_D V_s^2 \, \pi \frac{d_T^2}{4} \tag{5}$$

where: C_D = drag coefficient of the drop

The mass of the drop is:

$$m = \rho_D \pi \frac{d_T^3}{6} \tag{6}$$

Substituting equations (4), (5), and (6) into (3) and assuming the acceleration of the drop (for terminal velocity) is zero, yields:

$$V_s = \left[\frac{4}{3} \frac{d_T g (\rho_C - \rho_D)(1-\varepsilon)}{\rho_C C_D} \right]^{1/2} \tag{7}$$

Equations (2) and (7) represent two independent equations for the terminal drop velocity. Equation (2) was derived on the basis of conservation of mass. Equation (7) was derived by a force balance. From equations (2) and (7) one obtains that

$$\dot{V}_C / (1-\varepsilon) / A + \dot{V}_D / \varepsilon / A = \left[\frac{4}{3} \frac{d_T \, g ((\rho_C - \rho_D)(1-\varepsilon)}{\rho_C C_D} \right]^{1/2} \tag{8}$$

Equation (8) relates volumetric flowrates to the hold up.

Two difficulties arise with the procedure presented so far. First, the drag coefficient for a drop is a function of Reynold's number of the continuous phase. Two commonly used empirical correlations for the drag coefficient were presented by Holland [7] and by Soo [8]. They are, respectively,

$$C_D = 24 \cdot \frac{\left(1 + 0.15 \cdot Re_C^{0.687}\right)}{Re_C} \tag{9}$$

and

$$C_D = \frac{32}{Re_C}\left[1 + 2\frac{\mu_D}{\mu_C} - 0.314\left(\frac{1 + 4\mu_D/\mu_C}{Re_C^{1/2}}\right)\right] \tag{10}$$

where: Re_C = Reynold's number of the continuous phase

$$= \frac{\rho_C V_s d_T}{\mu_C}$$

Other equations have been presented in the published literature. Thus, the question arises as to which drag coefficient equation properly describes the drag phenomena for two given fluids. To illustrate this problem, the drag coefficient depending on the Reynolds number of the continuous phase was calculated from the experimental data obtained by Bühler [6]. Bühler carried out experiments using a spray column which had been used before by Ferrarini [9] and by Hupfauf [10]. The dispersed phase was injected into the column through up to 162 nozzles with diameter ranging from 2×10^{-3} to 2×10^{-3} m. The continuous phase was water while the mineral oil Esso Somentor 33 served as the continuous phase. Bühler reported flow rates, temperatures, hold-up, and drop diameters in addition to information on the size of the nozzles and the column. A least squares fit of the calculated drag coefficient provided the following relationship:

$$C_D = 16.61 \, Re_C^{-0.401} \tag{11}$$

Drag coefficients obtained from experimental data presented by Bühler [6] and drag coefficients determined with help of equations 9, 10, and 11 are shown in Fig. 2. Drag coefficients obtained from the experimental data presented by Bühler are about twice as large as the coefficients predicted with help of equation (9). Equation (10) provides drag coefficients which are not even of the same order of magnitude as the experimental coefficients.

Predicted flowrates are very sensitive to the choice of the drag coefficient values. This will be demonstrated next. Using again data for hold-up and continuous flow rate which were obtained experimentally by Bühler [6], equation (8) was solved for the volumetric flow rate of the dispersed phase. Equations (9) and (11) were used to provide the necessary drag coefficients. Fig. 3 shows predicted flow rates versus experimental flow rates when equation (11) is employed. Fig. 4 presents a comparison of predicted and measured flow rates when equation (9) was used to calculate the drag coefficient. While Fig. 3 shows a good agreement between predicted and measured flow rates, as one might have expected, flow rates predicted on the basis of equation (9) are up to 100% higher than measured flow rates. This result underscores the importance of using the correct representation of the drag coefficient.

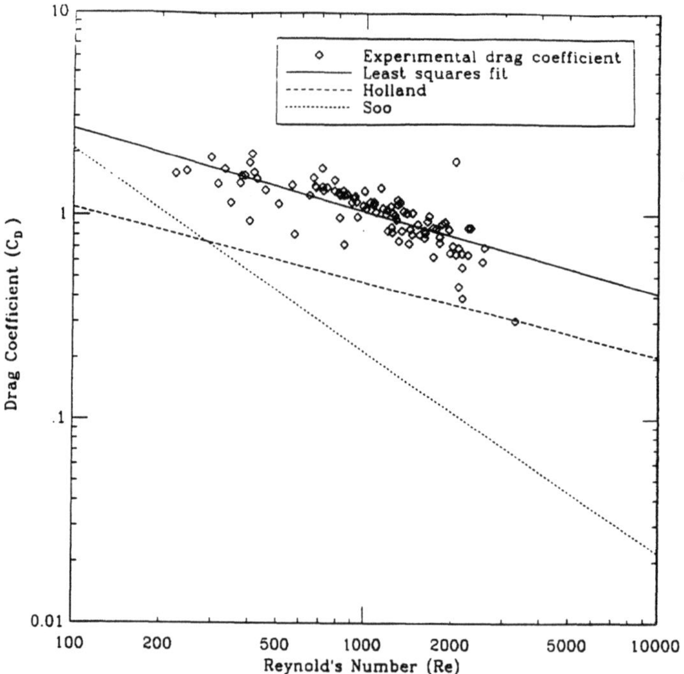

Fig. 2. Measured and predicted drag coefficients versus Reynold's number.

The second difficulty arises in the prediction of drop diameter. Even though numerous equations have been proposed for the prediction of drop volume in liquid-liquid two-phase flow, the agreement between predicted and experimentally obtained drop volumes ranges only from good to marginal. No single equation provides consistantly accurate predictions of drop volumes.

Walters [11] carried out drop formation experiments in the range of periodic drop formation. He compared his results with values predicted with equations available in the literature. While none of the equations predicted the true variation of drop diameter with nozzle flow rates, the best agreement was achieved when drop diameters were calculated with the de Chazal and Ryan [12] equation. Bühler [6] used a rather complicated correlation equation to represent measured drop diameters. his equation fits his own data reasonably well over a wide range of nozzle flow rates.

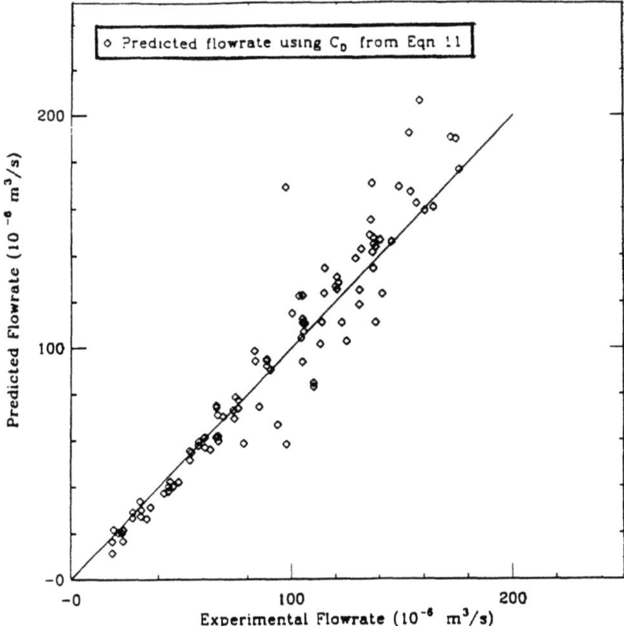

Fig. 3. Predicted versus experimental dispersed phase flow rates.

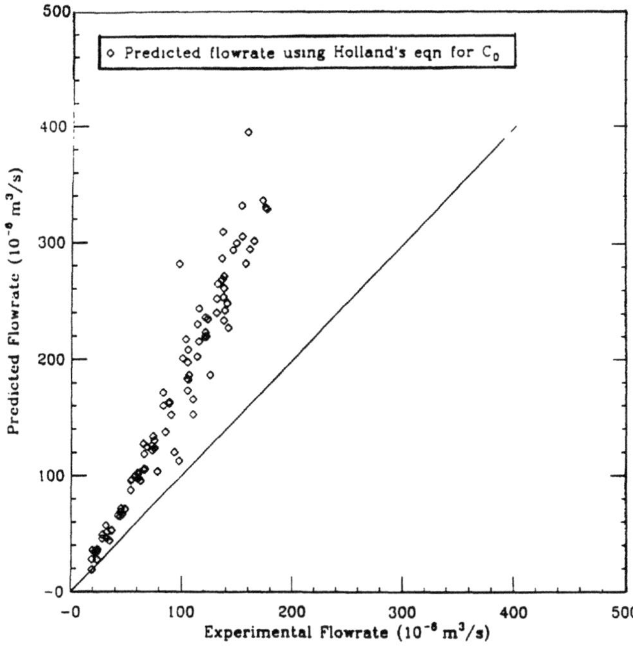

Fig. 4. Predicted versus experimental dispersed phase flow rates.

For drop formation from vertical jets, the following equations have been used with reasonable success [6]:

$$\frac{d_T}{d_N} = \frac{2.07}{0.485 \times E_o + 1} \text{ for Eö} < 0.615 \tag{11a}$$

$$\frac{d_T}{d_N} = \frac{2.07}{1.51 \times E_o^{1/2} + 0.12} \text{ for Eö} > 0.615 \tag{11b}$$

where

$$\text{Eö} = \frac{g(\rho_c - \rho_d) \cdot d_N^2}{\sigma} = \text{Eötvös} - \text{number}$$

g = gravitational acceleration
ρ_d = density of dispersed phase
ρ_c = density of continuous phase
d_N = nozzle diameter
d_T = drop diameter
σ = liquid-liquid interface tension

Virtually all equations were found for non-wetting, vertically oriented, circular nozzles with developed exit flow. Obviously, the drop diameter depends on the interface tension which is frequently not well known. In addition, while empirical equations predict a dropsize, this dropsize is only an average value. If the distribution about this average value is large, the prediction of heat transfer based on this average value may introduce appreciable errors.

Finally, it should be noted that the densities of the two phases may change significantly over the length of the column due to temperature variations. If density variations are large, the terminal velocities, the volumetric flow rates, the column hold-up and the drop diameter will be affected. In that case, the principle of conservation of mass can be used to account for the influence of density variation on flow parameters.

III. <u>Heat Transfer and Temperature Profile</u>

The calculation of the heat transfer between the two phases is done on a stepwise basis. Consider the column divided into m vertical sections of height Δz as shown in Figure 5. Consider the heat transfer from the continuous phase to the dispersed phase:

$$\Delta Q = hA_d (T_C - T_D) \tag{12}$$

where: h = overall heat transfer coefficient

A_d = surface area of the drops

T_C = average temperature of continuous phase for section n

T_D = average temperature of dispersed phase for section n

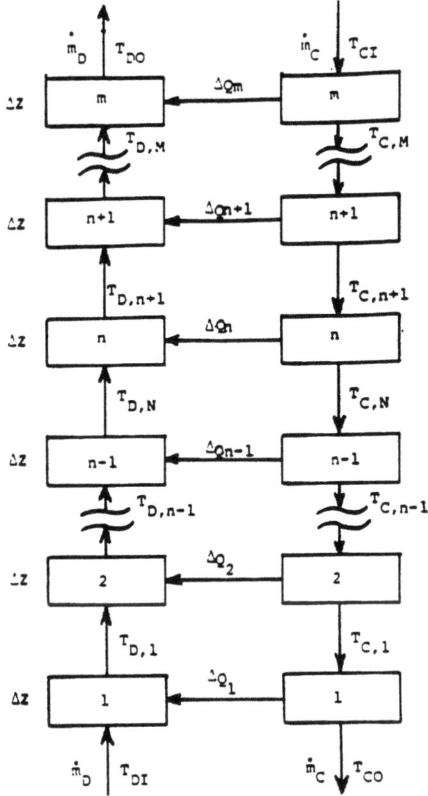

Fig. 5. Schematic of column sections.

This represents the heat transfer in section n. Performing an energy balance on section n yields:

$$\Delta Q = \dot{m}_C C_{PC} \Delta T_C \tag{13}$$

$$\Delta Q = \dot{m}_D C_{PD} \Delta T_D \tag{14}$$

where: ΔT_C = temperature change of the continuous phase over section n

ΔT_D = temperature change of the dispersed phase over section n

Substituting equation (12) into equations (13) and (14), one can solve for ΔT_C and ΔT_D:

$$\Delta T_C = \frac{hA_d(T_C - T_D)}{\dot{m}_C C_{PC}} \tag{15}$$

$$\Delta T_D = \frac{hA_d(T_C - T_D)}{\dot{m}_D C_{PD}} \tag{16}$$

Calculation of the temperature profiles becomes a stepwise procedure. Starting at the base of the column where the temperature of both phases is known, one can use equations (15) and (16) to calculate the temperature change over each vertical section. The overall heat transfer coefficient h is related to the individual thermal resistances of the continuous phase and the dispersed phase by

$$h = \frac{1}{\frac{1}{h_C} + \frac{1}{h_D}} \tag{17}$$

where h_C and h_D are the heat transfer coefficients of the continuous phase and the disperse phase respectively. Various correlations for the prediction of the heat transfer coefficients exist. Reviews of such correlations were presented by Ferrarini [9] and by Clift [4]. However, selecting the appropriate correlation poses a problem. This problem is caused by some uncertainties regarding the liquid-liquid interface characteristics. In the past, it has been commonly assumed that the liquid-liquid interface is fully mobile and, therefore the tangential velocity and shear stress are equal at the interface in both phases. As a consequence, correlations for external and internal Nusselt numbers were usually of the form

$$Nu_C = \frac{h_C}{k_C} D = f(Re_C, Pr_C, \mu_C / \mu_D)$$

and

$$Nu_D = \frac{h_D}{k_D} D = f(Re_D, Pr_D, \mu_D / \mu_C)$$

respectively. A typical example for this is the popular Handlos and Baron equation [13], which reads

$$Nu_D = \frac{0.00375 \, Re_D \, Pr_D}{(1 + \mu_D / \mu_C)} \tag{18}$$

This equation is not only based on velocity fields which were obtained for isothermal conditions but it also assumes velocity fields which were obtained for a "fully mobile" liquid-liquid interface.

where: h = overall heat transfer coefficient

A_d = surface area of the drops

T_C = average temperature of continuous phase for section n

T_D = average temperature of dispersed phase for section n

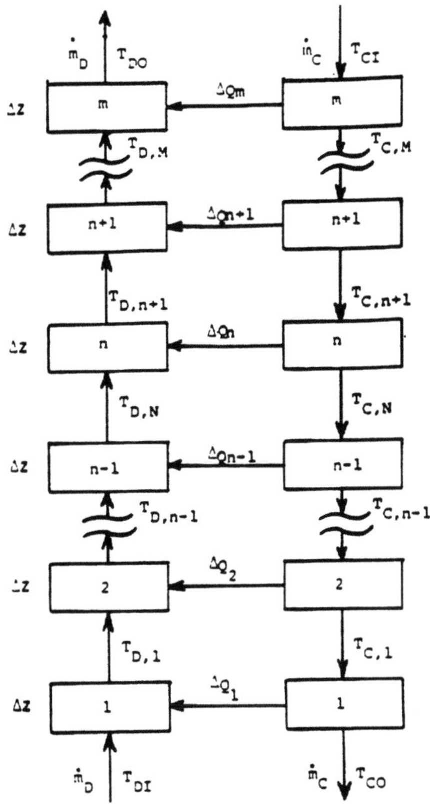

Fig. 5. Schematic of column sections.

This represents the heat transfer in section n. Performing an energy balance on section n yields:

$$\Delta Q = \dot{m}_C C_{PC} \Delta T_C \tag{13}$$

$$\Delta Q = \dot{m}_D C_{PD} \Delta T_D \tag{14}$$

where: ΔT_C = temperature change of the continuous phase over section n

ΔT_D = temperature change of the dispersed phase over section n

432

was flowing downwards as the continuous phase. Excellent agreement between experimentally obtained Nusselt numbers and Nusselt numbers predicted with equation (20) was found, as is apparent from Fig. 8.

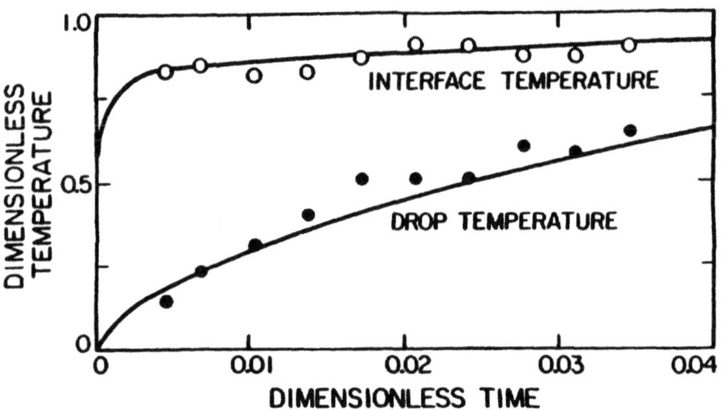

Fig. 6. Predicted and measured temperature variations in a liquid-liquid spray-column.

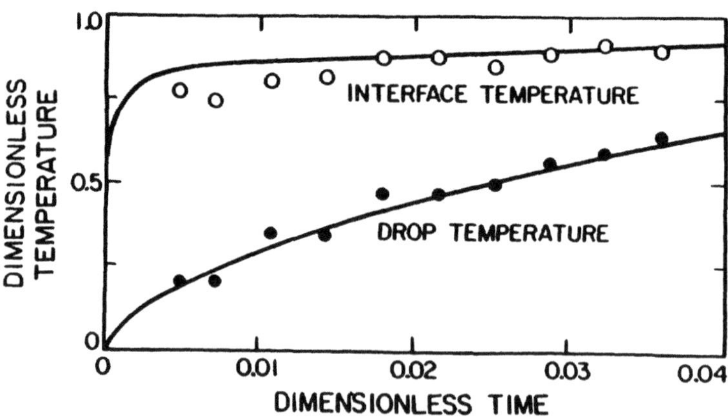

Fig. 7. Predicted and measured temperature variations in a liquid-liquid spray-column.

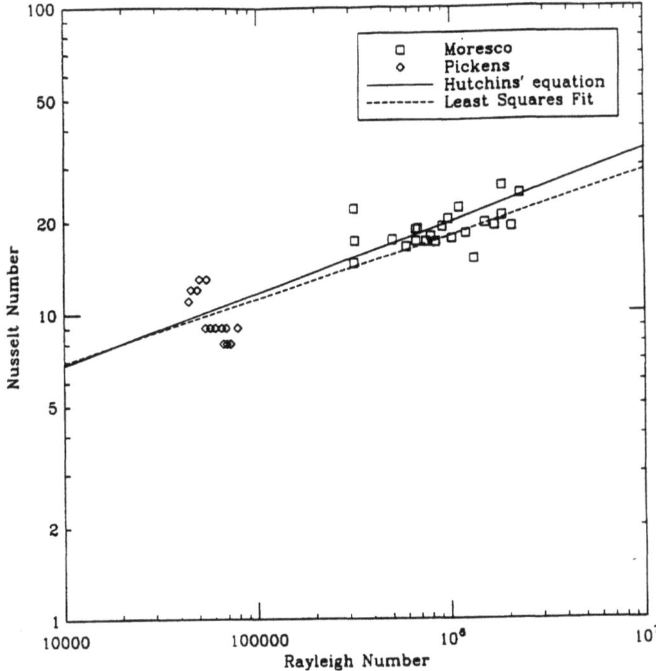

Fig. 8. Experimentally derived Nusselt Numbers from Moresco and Pickens.

The following equations are recommended for the prediction of heat transfer coefficients in liquid-liquid direct-contact heat transfer calculations:

$$Nu_D = 0.806 \ Ra^{.2215} \tag{20}$$

and

$$Nu_C = 1 + 0.752 \ Re_C^{0.472} \left(1 + \frac{1}{Re_C Pr_C}\right)^{1/3} Pr_C^{1/3} \tag{21}$$

Equations (20) and (21) assume a rigid or nearly rigid liquid-liquid interface. Equation (21) is cited in Ref. [4].

IV. Heat Transfer Sample Calculations

In order to support the choice of the recommended heat transfer equations, equations (20) and (21), heat transfer rates in a liquid-liquid direct-contact heat exchanger were determined. The calculations were based on experimental data obtained by Bühler. Thirty-two of his experiments were selected. The main criteria for the selection was that in these experiments the

Rayleigh number of the drops after drop formation was greater than 50,000. In addition, experiments which appeared to show large heat losses to the outside were discarded. This resulted in a selection of experiments with a wide variety of flow rates and hold-ups at various column conditions.

The dispersed phase outlet temperature was calculated using the measured values for the dispersed phase inlet temperature, the continuous phase outlet temperature, column hold-up, and flow rates. The column was divided in twenty vertical sections, and an iterative technique used by Hutchins [21] before, was employed for the step-wise calculations. To characterize the deviation between the calculated and measured dispersed phase outlet temperatures, the following expression was used:

$$E_1 = \frac{T_{DO,C} - T_{DO,E}}{T_{DO,E} - T_{Di,E}}$$ (22)

where

$$\begin{aligned}
T_{DO,C} &= \text{dispersed phase outlet temperature, calculated} \\
T_{DO,E} &= \text{dispersed phase outlet temperature, experimental} \\
T_{DI,E} &= \text{dispersed phase inlet temperature, experimental}
\end{aligned}$$

E_1 is simply the ratio of the difference between measured and calculated dispersed phase outlet temperatures divided by the measured temperature change of the dispersed phase. The standard deviation of E_1 for the thirty-two experiments was 15.8%.

In a variation of these calculations, the dispersed phase heat transfer coefficient was determined which would give the measured dispersed phase outlet temperature given the measured flow rates, hold-up dispersed phase inlet temperature and continuous phase temperatures. Fig. (9) shows, that the calculated dispersed phase Nusselt numbers agree very well with the Nusselt number predicted with equation (22).

V. Sensitivity Analysis

A sensitivity analysis was performed to determine the effect of variations of the drag coefficient, the drop diameter and the overall heat transfer coefficient on the heat transfer. The dispersed phase outlet temperature was calculated for the 32 cases previously discussed. Each of the parameters were varied individually by 10% and 20% from the nominal conditions and the dispersed phase outlet temperature was again calculated. The parameter E2 was used to compare the dispersed phase temperature of the nominal case to the perturbed cases where:

$$E_2 = \frac{T_{DO,C} - T_{DO,B}}{T_{DO,B} - T_{DI,E}}$$ (25)

where:

$T_{DO,C}$ = dispersed phase outlet temperature, calculated

$T_{DO,B}$ = dispersed phase outlet temperature, calculated, nominal conditions

$T_{DI,E}$ = dispersed phase inlet temperature, experimental

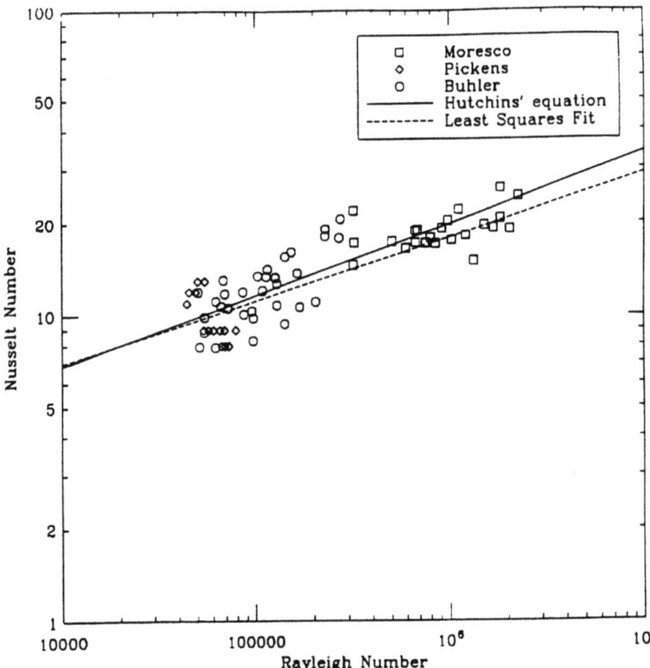

Fig. 9. Experimentally derived Nusselt Numbers from Moresco, Pickens, and Buhler.

The standard deviation of E2 for using the data from all 32 cases was calculated and is reported in Table 1. Examination of Table 1 shows that the variation in the drop diameter causes the greatest variations in outlet dispersed phase temperature. The variation of the overall heat transfer coefficient is second in importance. The greater influence of variations in drop diameter is understood by realizing that a drop diameter variation affects not only the calculation of the flow conditions but also of liquid-liquid interface area and of the heat transfer coefficients.

V. Conclusion

The prediction of liquid-liquid direct-contact heat transfer in a vertical spray column has been discussed. The importance of a careful selection of equations predicting drop diameter, drag coefficient, and heat transfer coefficients has been pointed out. A sensitivity analysis demonstrated that heat transfer calculations are especially sensitive to an incorrect choice of the drop diameter.

Table 1
Results of Sensitivity Analysis

% Variation	Standard deviation of E2
+10% drop diameter	.06
+20% drop diameter	.10
-10% drop diameter	.09
-20% drop diameter	.26
+10% drag coefficient	.01
+20% drag coefficient	.02
-10% drag coefficient	.02
-20% drag coefficient	.03
+10% heat trans. coef.	.04
+20% heat trans. coef.	.09
-10% heat trans. coef.	.04
-20% heat trans. coef.	.08

References

1. Kehat, E., and Sideman, S., "Heat Transfer by Direct Liquid-Liquid Contact," Recent Advances in Liquid-Liquid Extraction, Pergamon Press, Oxford, 1971, pp. 455-494.

2. Marschall, E., Johnson, G., and Culbreth, W., "Direct-Contact Heat Transfer," Progress in Chemical Engineering, Vol. 20, 1982, pp. 41-57.

3. Letan, R., "Liquid-Liquid Processes," Direct-Contact Heat Transfer, Hemisphere, Washington, 1988, pp. 83-118.

4. Clift, R., Grace, J.R., and Weber, M.E., Bubbles, Drops, and Particles, Academic Press, New York, 1978.

5. Goodfrey, J.E. and Hanson, C., "Liquid-Liquid Systems," Handbood of Multiphase Systems, Hemisphere, Washington, 1982, pp. 4-3 - 4-41.

6. Bühler, B.S., "Hydrodynamik and Wärmeaustausch in einem Flüssig-Flüssig Sprühturm," Doctoral Thesis, Eidgenössische Technische Hochschule Zürich, 1977.

7. Holland, F.A., "Fluid Flow for Chemical Engineers," Edward Arnold, London, 1973, p. 160.

8. Soo, S.L., "Fluid Dynamics of Multiphase Systems," Blaisdell Publishing Company, London, 1967, p. 87.

9. Ferrarini, R., "The Calculation of Flow and Heat Exchange in Liquid-Liquid Spray Columns," VDI-Forschungsheft 551, Düsseldorf, 1972.

10. Hupfauf, A., "Lokaler Wärmeübergang und Rückvermischung in Flüssig-Flüssig Sprühkolonnen," Doctoral Thesis, Eidgenössische Technische Hochshule Zürich, 1973.

11. Walters, T.W. and Marschall, E., "Drop Formation in Liquid-Liquid Systems," Experiments in Fluids, Vol. 7, 1989, pp. 210-214.

12. Chazal, L. de, and Ryan, J., "Formation of Organic Drops in Water," AIChE J., Vol. 17, 1971, pp. 1226-129.

13. Handlos, A.E., and Baron, "Mass and Heat Transfer from Drops in Liquid-Liquid Extraction," AIChE J., Vol. 3, 1957, pp. 127-135.

14. Savic, P., "Circulation and Distortion of Liquid Drops Falling Through a Viscous Medium," Ref. No. MT-22, Natl. Res. Council Can. Div. Mech. Engng., 1953.

15. Gal-Or, B., "On Motion of Bubbles and Drops," Can. J. Chem. Engr., 1970, pp. 526-531.

16. Scriven, L.E., "Dynamics of a Fluid Interface," Chemical Engineering Science, Vol. 12, 1960, pp. 98-108.

17. McWaid, T., and Marschall, E., "Improved Photochromic Flow Visualization Technique," Experimental Thermal and Fluid Science, Vol. 3, 1990, pp. 232-241.

18. Hutchins, J.F., "Transient Natural Convection Inside Drops in a Liquid-Liquid Direct-Contact Heat Exchanger," Doctoral Thesis, University of California, Santa Barbara, 1988.

19. Moresco, L., "The Effect of Dissolved Solids on Direct-Contact Heat Transfer," Doctoral Thesis, University of California, Santa Barbara, 1979.

20. Pickens, M.K., "Experimental Determination of Nusselt Numbers in a Liquid-Liquid Direct-Contact Heat Exchanger," M.S.-Thesis, University of California, Santa Barbara, 1990.

21. Hutchins, J.F., "Computer Simulation for the Design of a Direct-Contact Heat Exchanger," MS-Thesis, University of California, Santa Barbara, 1981.

Nomenclature

a_y	drop acceleration
A	cross-sectional area of column
A_d	surface area of drops
C_D	drag coefficient
d_N	nozzle diameter
d_T	drop diameter
Eö	Eötvös number
F_D	drag force
g	gravitational acceleration
h	overall heat transfer coefficient
h_C	continuous phase heat transfer coefficient
h_D	dispersed phase heat transfer coefficient
m	mass of drop
$\dot{m}_C$	mass flow rate, continuous phase
$\dot{m}_D$	mass flow rate, dispersed phase
Nu_C	Nusselt number, continuous phase
Nu_D	Nusselt number, dispersed phase
Pr_C	Prandtl number, continuous phase
Pr_D	Prandtl number, dispersed phase
$\dot{q}$	rate of heat transfer
Ra	Rayleigh number
Re_C	Reynolds number, continuous phase
Re_D	Reynolds number, dispersed phase
T_C	temperature, continuous phase
T_D	temperature, dispersed phase
U_v	volumetric heat transfer coefficient
V_C	heat exchanger volume
V_s	slip velocity, relative velocity
V_T	drop volume
$\dot{V}_C$	volumetric flow rate, continuous phase
$\dot{V}_D$	volumetric flow rate, dispersed phase
z	height
ε	hold-up
μ_C	kinematic viscosity, continuous phase
μ_D	kinematic viscosity, dispersed phase
σ	liquid-liquid interface tension
ρ	density
ρ_C	density, continuous phase
ρ_D	density, dispersed phase

Indices

i	inlet
o	outlet